U.S.NRC
United States Nuclear Regulatory Commission
Protecting People and the Environment

A Proposed Risk Management Regulatory Framework

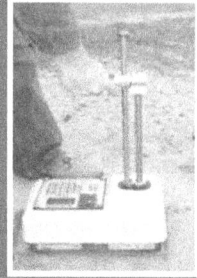

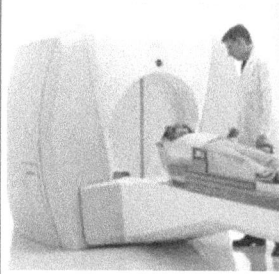

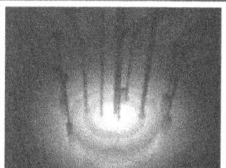

A Proposed Risk Management Regulatory Framework

A report to NRC Chairman Gregory B. Jaczko
from the Risk Management Task Force

Commissioner George Apostolakis, Head

Members

Christiana Lui, RMTF Executive Director
Mark Cunningham
George Pangburn
William Reckley

With contributions from

John Adams
Michel Call
Dennis Damon
Don Dube
Earl Easton
Timothy McCartin
Geary Mizuno
Joel Piper

April 2012

ABSTRACT

At the request of Chairman Gregory B. Jaczko, a task force headed by Commissioner George Apostolakis prepared this report. The task force's charter was to develop a strategic vision and options for adopting a more comprehensive, holistic, risk-informed, performance-based regulatory approach for reactors, materials, waste, fuel cycle, and transportation that would continue to ensure the safe and secure use of nuclear material. The proposed risk management regulatory framework builds upon well established practices, such as the NRC's defense-in-depth philosophy and its policies to incorporate risk-informed and performance-based approaches into the agency's regulation and oversight of byproduct, source, and special nuclear materials. Risk management is being adopted by many different organizations, including Federal agencies, and would seem to be the logical next step in the evolution of the NRC's regulatory programs. The report describes a proposed risk management regulatory approach that could be used to improve consistency among the NRC's various programs and discusses implementing such a framework for specific program areas.

We Athenians, in our own persons, take our decisions on policy and submit them to proper discussions; for we do not think that there is an incompatibility between words and deeds; the worst thing is to rush into action before the consequences have been properly debated. And this is another point where we differ from other people. We are capable at the same time of taking risks and of estimating them beforehand. Others are brave out of ignorance; and, when they stop to think, they begin to fear. But the man who can most truly be accounted brave is he who best knows the meaning of what is sweet in life and what is terrible, and then goes out undeterred to meet what is to come.

Funeral Oration Delivered by Pericles circa 430 B.C.
Thucydides, *History of the Peloponnesian War,*
Book B, Paragraph 40

CONTENTS

LIST OF FIGURES

LIST OF TABLES

EXECUTIVE SUMMARY

Introduction

The Atomic Energy Act (the "Act") and other applicable laws establish the fundamental basis by which the U.S. Nuclear Regulatory Commission (NRC) regulates the civilian uses of nuclear materials. The implementation of these laws has evolved considerably since the original Act, with concepts such as "defense in depth" and methods such as risk assessment emerging as important aspects of this evolution.

In early 2011, at the request of Chairman Gregory B. Jaczko, Commissioner George Apostolakis agreed to lead a Task Force to evaluate how the agency should be regulating 10 to 15 years in the future. More specifically, the Task Force was chartered "to develop a strategic vision and options for adopting a more comprehensive and holistic risk-informed, performance-based regulatory approach for reactors, materials, waste, fuel cycle, and transportation that would continue to ensure the safe and secure use of nuclear material."

This report describes the findings and recommendations of this evaluation. The underlying analysis was performed by a team of NRC staff (named the Risk Management Task Force (RMTF)) under the direction of Commissioner Apostolakis, with contributions from additional NRC staff. The RMTF also benefitted from comments and suggestions provided by members of the public and other NRC staff.

The accident at the Fukushima Dai-ichi nuclear power plants in Japan occurred shortly after the RMTF was established. The team's analysis has been influenced by the events at Fukushima and the subsequent studies, including the NRC Near-Term Task Force, and the continuing discussions on the accident's implications for U.S. nuclear power plants.

The findings and recommendations are compiled into two groups, with the most important described below.[1] The first group addresses agencywide, more strategic issues, describing a structure of goals and objectives that could be the framework for NRC regulatory activities 10 to 15 years in the future. The second group addresses what changes would be needed in specific program areas (e.g., power reactors, materials) in the next several years to ensure that the framework is implemented. It is acknowledged that these recommendations, if adopted, would require actions and resources to fully address their implementation that go well beyond the charter of the RMTF.

Agencywide Findings and Recommendations

The RMTF found many positive attributes of the individual regulatory activities now performed by the NRC. As a result, the RMTF determined that its proposed strategic vision could and should be more evolutionary than revolutionary. The findings of the Task Force are listed below.

- Finding: Whether used explicitly, as for power reactors, or implicitly, as for materials programs, the concept of defense in depth has served the NRC and the regulated industries well and continues to be valuable today. However, it is not used consistently, and there is no guidance on how much defense-in-depth is sufficient.

---

1   The complete set of findings and recommendations is provided in Appendix J.

- Finding: Risk assessments provide valuable and realistic insights into potential exposure scenarios. In combination with other technical analyses, risk assessments can inform decisions about appropriate defense-in-depth measures.

Considering these findings, the RMTF proposes a framework shown in Figure ES-1 and discussed below.

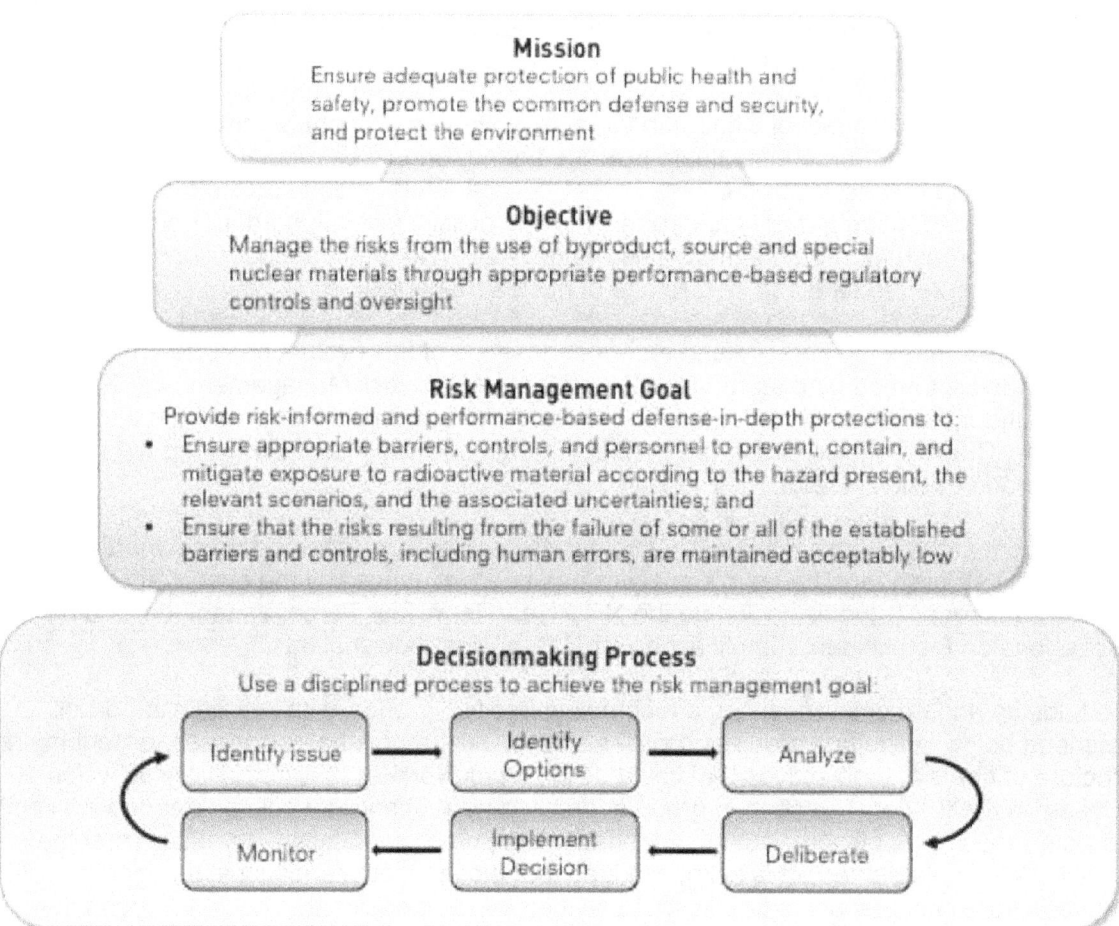

Figure ES-1 A Proposed Risk Management Regulatory Framework

The RMTF proposes that risk management should be stated as the NRC's objective ("Objective" in Figure ES-1). Declaring that this is the agency's objective is a natural next step in the evolution of the NRC's regulatory practices. This term is widely encountered in the management literature and is gaining greater use in other Federal agencies. In addition, it explicitly recognizes that adequate protection of public health and safety is not synonymous with absolute safety and that the NRC's role is to ensure that risks from the use of nuclear materials are well managed. Finally, establishing a common language of risk management across all NRC activities is consistent with the principles of good regulation.

As used in the proposed framework, the concept of risk consists of answers to the three standard questions: What can go wrong? How likely is it? What are the consequences? It is acknowledged that some program areas, such as power reactors and high-level waste repositories, take a more explicit and quantitative approach to answering all three questions, whereas other program areas take more qualitative approaches. However, all NRC programs practice risk management.

The RMTF recognizes the importance of translating the concept of risk management into more operational terms. The task force proposes doing this by integrating the traditional concept of defense in depth and the methods of risk assessment ("Risk Management Goal" in Figure ES-1). Traditional analysis techniques are combined with risk assessment methods to define appropriate personnel training and qualifications, barriers and controls, including ensuring that the risks from the failure of these protections are acceptably low. This combination brings forth the systems analysis approach and provides a way to decide how much defense in depth is sufficient. The decision of what are acceptably low risks is not necessarily based on quantitative probabilistic metrics. For nuclear power reactors and waste repositories, the existing quantitative goals and requirements help define acceptably low levels of risk. Risk management for those program areas dealing with lesser amounts of radioactive materials is often achieved largely by standard radiation protection practices, which are established by all licensees in accordance with the NRC regulations defined in Title 10 of the Code of Federal Regulations (10 CFR) Part 20, "Standards for Protection Against Radiation."

There is value in describing the basic steps to be performed in risk management decisionmaking ("Decisionmaking Process" in Figure ES-1). Although these steps have many similarities to current NRC decisionmaking processes (e.g., Regulatory Guide 1.174, "An Approach for Using Probabilistic Risk Assessment in Risk-Informed Decisions on Plant-specific Changes to the Licensing-basis"), there are important benefits to standardizing these steps across all NRC program areas.

The proposed framework should be implemented for both safety and security-related issues. Although estimating the risk of security-initiated events is difficult, and methods are not well established, the NRC nonetheless should have as a goal managing the appropriate amount of security defense in depth and better integrating security vulnerability assessments and risk assessments for other safety issues.

As a first step in implementing the proposed Risk Management Regulatory Framework, the RMTF proposes the following recommendation:

> ### Recommendation (Risk Management Regulatory Framework):
>
> The NRC should formally adopt the proposed Risk Management Regulatory Framework through a Commission Policy Statement.

Consistent with normal agency practice, obtaining stakeholder input will be valuable in the development of the proposed policy statement. Recognizing the regulatory authority of Agreement States in the materials program area, they should have an early role in this process.

This proposed framework includes several important benefits:

- Updated knowledge from contemporary studies, such as risk assessments, would be incorporated into the regulations and guidance, thereby improving their realism and technical basis.

- Implementation of a systematic approach would foster a consistent regulatory decisionmaking process throughout the agency and improve resource allocation.

- Consistency in language and communication would be improved across the agency and externally.

- Support of issue resolution would be achieved in a systematic, consistent, and efficient manner.

Implementation of the proposed framework would also pose challenges:

- A change would be required within the agency and externally to increase understanding of the value and use of risk concepts and risk management language.

- The proposed risk-informed and performance-based concept of defense in depth may require the development of additional decision metrics and numerical guidelines.

- The approach would likely require developing new or revised risk-assessment consensus codes and standards.

- A long-term commitment from the Commission and senior agency management would be required for implementation.

Program Area Findings and Recommendations

As noted above, the RMTF has developed findings and recommendations on what changes would be needed to ensure that the proposed risk management framework would be implemented in 10 to 15 years. The RMTF did not assess some NRC activities in detail (e.g., environmental reviews, decommissioning), but these activities could be addressed in a manner similar to those presented below. The program area findings reflect what the RMTF considers to be important gaps between how specific types of licensees are regulated today and how they would be regulated in the future using the proposed risk management framework. The recommendations suggest ways in which the gaps could be closed.

Power Reactors

Power reactors in the United States have been licensed using 10 CFR Part 50, "Domestic Licensing of Production and Utilization Facilities." Implementation of 10 CFR Part 50 has been achieved, for the most part, using deterministic methods and acceptance criteria. From a safety perspective, a set of licensing-basis events was established that was intended to ensure conservatism in design and protection from a wide spectrum of postulated events, up to and including design-basis accidents (DBAs). These postulated accidents are highly stylized and generally do not consider multiple failures of safety systems. Qualitative approaches for ensuring reliable safety systems, such as the single failure criterion, were implemented.

Testing plans and operational limits were established in technical specifications to ensure that safety systems would perform as intended if called upon. In addition to the failure or malfunction of plant equipment, the challenges to and protections of power reactor designs include consideration of external hazards, such as earthquakes, which may initiate a variety of plant transients while also challenging one or more of the defense-in-depth barriers provided to mitigate or contain potential releases of radioactive materials.

The NRC's current regulatory approach for power reactors includes a set of events and accidents necessary for "adequate protection" and additional events and accidents licensees are required to address to provide an additional amount of safety. The former set consists mostly of the set of "design-basis" events and accidents (described above) established more than 30 years ago. The latter set has emerged in more recent times to address specific issues, such as station blackout accident risks.

The NRC's power reactor regulatory program has been the subject of considerable work to increase the use of risk assessment methods and results. Requirements have been added, modified (including the development of risk-informed alternative rules), and deleted. Risk information is used in some licensing activities. The power reactor oversight program has important risk considerations included in resource allocation and the evaluation of inspection findings.

The RMTF assessed this regulatory structure in the context of the proposed risk management framework and developed a number of findings. The most important findings are listed below.

- Finding: The concept of design-basis events and accidents continues to be a sound licensing approach, but the set of design-basis events and accidents has not been updated to reflect insights from power reactor operating history and more modern methods, such as probabilistic risk assessment (PRA).

- Finding: Requirements for beyond-design-basis accident scenarios (e.g., station blackout) were established at different times and in different ways. Differences in implementation approaches have reduced the efficiency and consistency of the NRC's regulatory and oversight activities.

- Finding: The extent to which licensee activities undertaken as part of voluntary industry initiatives can be credited has been a source of contention in the Reactor Oversight Process and has reduced the efficiency of that process.

- Finding: The process for establishing the external hazard design basis does not use consistent event frequency or magnitude methods.

- Finding: Differences in regulatory language and approaches between power reactor security and safety regulation may have reduced the efficiency and effectiveness of the NRC's work.

In light of these findings, the RMTF offers the following recommendations for power reactors.

Recommendations (Power Reactors):

- The set of design-basis events and accidents should be reviewed and revised, as appropriate, to integrate insights from the power reactor operating history and more modern methods, such as probabilistic risk assessment (PRA).

- The NRC should establish through rulemaking a *design-enhancement category* of regulatory treatment for beyond-design-basis accidents. This category should use risk as a safety measure, be performance-based (including the provision for periodic updates), include consideration of costs, and be implemented on a site-specific basis.

- The NRC should reassess methods used to estimate the frequency and magnitude of external hazards and implement a consistent process that includes both deterministic and PRA methods. Consideration of the risks from beyond-design-basis external hazards should be included in the proposed design-enhancement category.

- The NRC should develop and implement guidance for use in its security regulatory activities that uses a common language with safety activities and harmonizes methods with risk assessment and the proposed risk-informed and performance-based defense-in-depth framework.

A significant change from current practice probably will be required for determining the design basis for Generation IV reactor designs. However, the NRC should be amenable to, and promote where practical, the use of a revised set of design-basis events and accidents for operating reactors and certified reactor designs.

In addition to the framework benefits and challenges noted above, the following observations are also relevant to the power reactor regulatory program:

- The proposed design-enhancement category would clarify the attributes of all requirements established as substantial safety (beyond-design-basis) improvements. This approach may contribute to the resolution of the "patchwork" issue identified by the Fukushima Near-Term Task Force.

- Consideration of cost in the proposed design-enhancement category would necessitate a reconsideration of the agency's tools for performing cost-benefit analysis.

Nonpower Reactors

Nonpower reactors (NPRs), generally known as research and test reactors, are nuclear reactors primarily used for research, training, development, and isotope production. They contribute to almost every field of science, including physics, chemistry, biology, medicine, geology, archeology, and environmental sciences.

NPRs have been licensed using a process that has many similarities to the process used to license power reactors. The concepts of defense in depth and design-basis events and accidents are used, as is the concept of a "maximum hypothetical accident." However, NPRs have a substantially smaller radiological hazard than power reactors, because of their much smaller power rating and intermittent use.

The most important findings are:

- Finding: The analysis of design basis and the maximum hypothetical accidents based on conservative design limits, acceptance criteria, safety margins, and assumptions in conjunction with the application of a defense-in-depth philosophy continues to be a sound but highly conservative licensing approach to ensuring adequate safety of NPRs.

- Finding: While PRAs have been performed for NPRs by other organizations, modern risk assessment methods have not been used in NRC NPR licensing decisions.

In light of these findings, the RMTF's recommendations for nonpower reactors are provided below.

Recommendations (Nonpower Reactors):

- The proposed defense-in-depth framework should be applied to the NPR licensing process to ensure that the current amount of defense in depth is appropriate given the relatively small radioactive hazard. This application should include safety and security licensing matters.

- The NRC should evaluate the utility of performing a pilot risk assessment, including consideration of external hazards, using modern risk assessment methods at an NPR. This evaluation would assess the value of the risk insights gained from the risk assessment on the basis of possible safety enhancements and possible contributions to a more efficient and effective risk-informed and performance-based regulatory framework for NPRs.

Materials

Reactor and accelerator-produced nuclear materials are used extensively throughout the United States for industrial applications, basic and applied research, manufacture of consumer products, academic studies, and medical diagnosis, treatment, and research. In addition, source materials are used in the production of processed uranium for nuclear fuel fabrication and a wide variety of other uses. The regulatory framework for use of these materials is contained in the Commission's radiation protection standards in 10 CFR Part 20, in application-specific regulations in 10 CFR Part 30 through Part 39, and in related guidance and policies. The NRC's materials regulatory programs are designed to ensure that licensees use these materials safely and securely so that they present no undue risk to public health and safety and the environment.

Under the provisions of Section 274 of the Atomic Energy Act, the NRC may enter into agreements with States under which the NRC discontinues its authority over certain radioactive materials and a State assumes that authority. The Agreement States, as they are known, now regulate more than 85 percent of the materials licensees in the United States. Presently, there are 37 Agreement States and, in concert with the NRC, they play a major role in the regulation of most materials uses.

The risks presented by the use of nuclear materials differ significantly from the risks presented by power reactors. The risk environment for reactors focuses predominantly on the prevention of low-frequency, high-consequence scenarios, whereas the risk environment for materials focuses primarily on higher-frequency, lower-consequence scenarios. Furthermore, in the materials area, risk assessments are largely qualitative; in the reactor program, such assessments are generally quantitative. Traditionally, the basis for the materials program largely has been a deterministic one, with rules and guidance developed over time and as a result of operational experience. Beginning in the 1990s, the NRC undertook a number of initiatives to better risk-inform and performance-base its nuclear materials (and other) regulatory programs. These initiatives led to fundamental changes in inspection frequency and approach, as well as licensing policies and practices, and regulations and guidance. Today, risk insights and performance considerations continue to be significant factors in materials program development and implementation.

The RMTF assessed this regulatory program in the context of the proposed risk management framework and developed a number of findings. The most important of these findings are listed below.

- Finding: The terminology of defense in depth is not used consistently across the NRC's materials regulatory programs.

- Finding: The materials program could benefit from a more structured application of the risk management process in resource allocation. This process would allow program managers to more systematically apply resources to those areas where the safety or security risk warrants it.

- Finding: Buy-in of the 37 Agreement States is essential to the success of the risk management process implementation.

In light of these findings, the RMTF's recommendations for materials are provided below.

> ### Recommendations (Materials):
>
> - The NRC materials program should continue to apply risk insights and performance-based considerations, as appropriate, in rulemaking, guidance and policy development, and implementation in accordance with the proposed risk management framework. This consideration should include both safety and security licensing processes.
>
> - The development and rollout of the recommended Risk Management Policy Statement should be closely coordinated with the leadership of the Agreement States.

Low-Level Waste

The NRC regulates the management and disposal of LLW through regulations, licensing, inspection, enforcement, guidance, and policy development. The primary regulations in this area are 10 CFR Part 20, 10 CFR Part 35, "Medical Use of Byproduct Material" (decay-in-storage provisions), and 10 CFR Part 61, "Licensing Requirements for Land Disposal of Radioactive Waste." The Agreement States also play a major role in LLW regulation since they regulate all the operating commercial LLW disposal sites, as well as the major LLW processors. The Low-Level Radioactive Waste Policy Act of 1980, and amendments to that act in 1985, established that LLW disposal was a State responsibility. The LLW Act encouraged States to enter into regional compacts that would develop common disposal facilities for use by member States of a compact. The Commission has directed the staff to expand its effort to bring a clearer risk-informed approach to 10 CFR Part 61 in a staff requirements memorandum issued in January 2012.

The RMTF assessed this regulatory program in the context of the proposed risk management framework and developed a number of findings. The most important of these findings are listed below.

- Finding: The regulatory framework for LLW disposal has, for the most part, implicitly followed a risk-informed and performance-based approach. However, changes in the LLW environment and the maturing of the performance assessment method over the past 30 years have underscored the need to provide a stronger risk basis to the program.

- Finding: Certain aspects of the LLW regulatory framework readily lend themselves better to the risk management approach, such as waste classification or concentration averaging. Applying the proposed risk management approach to comprehensive LLW licensing decisions, however, may be more challenging because it involves estimating facility performance due to events potentially far into the future.

• <u>Finding</u>: The interlocking and reinforcing systems approach in 10 CFR Part 61 (site suitability, waste form and classification, intruder barrier, and institutional controls) represent an implicit consideration of defense-in-depth features, based on the risk posed by various classes of waste.

In light of these findings, the RMTF recommendations for low-level waste are provided below.

Recommendations (Low-Level Waste):

• The NRC should adopt the concept of risk management to the LLW program, as well as any revisions proposed to 10 CFR Part 61 (including performance assessment requirements) and related guidance documents.

• The NRC should develop an explicit characterization of how defense in depth, within the proposed risk management framework, applies to the LLW program and build this into current and future staff guidance documents and into training and development activities for the staff.

• The NRC should include environmental reviews within the scope of its risk management framework.

High-Level Waste

U.S. policies governing the permanent disposal of high-level waste (HLW) are defined by the Nuclear Waste Policy Act of 1982, as amended (NWPA). This Act specifies that HLW will be disposed of underground, in a deep geologic repository, and that Yucca Mountain, NV, will be the single candidate site for characterization as a potential geologic repository. In light of the Federal Government's decision to discontinue activities directed at ultimate disposal of HLW at the Yucca Mountain site, the future direction and options for the NRC's regulation of HLW disposal are not certain at this time. The following paragraphs discuss how risk-informed and performance-based factors were considered in the development of the current regulations and guidance for HLW disposal and how the proposed risk management framework might interface with future regulatory development.

Under the NWPA Act, the NRC is one of three Federal agencies that have a role in the disposal of spent nuclear fuel and HLW from the Nation's nuclear weapons production activities:

• The U.S. Department of Energy (DOE) is responsible for designing, constructing, operating, and decommissioning a permanent disposal facility for HLW, under NRC licensing and regulation.

• The U.S. Environmental Protection Agency (EPA) is responsible for developing site-specific environmental standards for use in evaluating the safety of a geologic repository.

• The NRC is responsible for developing regulations to implement EPA's safety standards and for licensing and overseeing the construction and operation of the repository.

In the late 1970s, the NRC began developing regulations for HLW disposal. Since that time, risk information and risk concepts have been used increasingly to assist and guide the development of the HLW program. NRC regulations for geologic disposal are found in 10 CFR Part 60, "Disposal of High-Level Radioactive Wastes in Geologic Repositories," (generic regulations for all sites other than Yucca Mountain, issued in 1983), and 10 CFR Part 63, "Disposal of High-Level Radioactive Wastes in a Geologic Repository at Yucca Mountain, Nevada" (specifically for Yucca Mountain, issued in 2001). Revisions to the standards and regulations for geological disposal to a potential repository at Yucca Mountain made significant use of risk information to support more effective and efficient standards and regulations. In particular, the NRC's regulations in 10 CFR Part 63 reflected a risk-informed and performance-based approach that focused on regulatory compliance for the post-closure period on the dose to a reasonably maximally exposed individual (dose limit provided in the EPA standard for a potential repository at Yucca Mountain under Title 40, "Protection of Environment," of the *Code of Federal Regulations* (40 CFR), Part 197, "Public Health and Environmental Radiation Protection Standards for Yucca Mountain, Nevada"). As experience with performance assessment methods and results (e.g., risk information) increased over time, so did the use of risk information to resolve technical issues and assist in the development of regulatory guidance.

The RMTF assessed this regulatory program in the context of the proposed risk management framework and developed the following findings:

- Finding: The development of regulations for geologic disposal of HLW at Yucca Mountain was based significantly on risk information developed from performance assessments and closely followed the proposed Risk Management Regulatory Framework.

- Finding: The NRC's regulatory philosophy of defense in depth is reflected in the multiple-barrier requirement for post-closure in 10 CFR Part 63. Compliance with the multiple barrier requirements is demonstrated through the performance assessment.

- Finding: As performance assessment capabilities and experience increased at the NRC during the past 30 years, so did the use of risk insights to help guide the HLW program. Risk insights and performance-assessment capabilities have been used to improve the efficiency and effectiveness of guidance documents, inform pre-licensing interactions with DOE, and help identify and direct data needs and experimental activities.

In light of these findings, the RMTF recommendation for high-level waste is provided below.

Recommendation (High-Level Waste):

- Any future revisions to the regulatory framework for geologic disposal of HLW should be done in accordance with the proposed risk management framework to ensure that risk information continues to be appropriately considered in the development of requirements and appropriately reflect any future HLW disposal paradigm.

Uranium Recovery

Uranium recovery is the first step in the nuclear fuel cycle. It involves extracting uranium from its parent ore and processing it into a physical and chemical form (yellowcake) that will allow additional processing and fabrication to become nuclear fuel. The NRC's regulatory authority under the Atomic Energy Act (the Uranium Mill Tailings Radiation Control Act of 1978 (UMTRCA)) and 10 CFR Part 20 and 10 CFR Part 40, "Domestic Licensing of Source Material," does not extend to the mining of uranium-bearing ore by conventional methods. That activity is regulated primarily by States and other Federal agencies. Rather, the NRC regulates the processing of uranium ore to concentrate uranium and the disposal of the large amount of tailings resulting from that processing.

Both NRC and EPA regulations in this area are largely deterministic and do not reflect risk-informed, performance-based considerations. The Commission considered a revision to the regulatory framework for uranium recovery in 2000 that would have updated and risk-informed the regulations, but the agency chose not to pursue it. A staff risk assessment in 2001 concluded that in situ recovery (ISR) facilities are of inherently low risk. The NRC licensing guidance for ISR is currently being revised to reflect risk insights and licensing experience. ISR facility licenses also include a performance-based license condition that allows licensees to make certain changes without requesting NRC approval in the form of a license amendment.

The RMTF assessed this regulatory program in the context of the proposed risk management framework and developed the following findings:

- Finding: Uranium recovery facilities are of low radiological risk to workers and members of the public under normal operational conditions and most accident scenarios.

- Finding: Although the NRC staff has made inroads to risk-informed, performance-based licensing of uranium recovery facilities, the regulatory framework is largely a deterministic one. NRC regulations in 10 CFR Part 40 and Appendix A to 10 CFR Part 40 reflect the requirements of UMTRCA and EPA's rules in 40 CFR Part 192, which pre-date the Commission's move to a risk-informed, performance-based framework in the mid-1990s. Similarly, program guidance, especially for conventional mills, could benefit from a greater risk basis.

- Finding: The exact nature of the ISR rule under development by the EPA is not clear at this time and, therefore, presents an uncertainty to adoption of the proposed risk management regulatory framework.

- Finding: Consideration of environmental risks is a central part of the uranium recovery regulatory program.

In light of these findings, the RMTF recommendations for uranium recovery are provided below.

> ## Recommendations (Uranium Recovery):
>
> • Notwithstanding the current uncertainty associated with the EPA rulemaking, the NRC should adopt the proposed risk management regulatory framework to the uranium recovery program to provide greater efficiency, effectiveness, and predictability in policy development and regulatory decisionmaking.
>
> • The NRC should work closely with the Agreement States and the regulated community to guide implementation of risk management in the uranium recovery program.
>
> • The NRC should include environmental reviews within the scope of its risk management framework.

Fuel Cycle

The NRC regulates major fuel cycle facilities, including those involved in conversion of uranium ore to UF_6, gaseous centrifuge and diffusion enrichment, reactor fuel fabrication, plutonium processing, and UF_6 deconversion. Reactor fuel fabrication facilities include those that produce low-enriched uranium, high-enriched uranium, and mixed-oxide (Pu + U) fuels. The NRC also regulates possession of small amounts of special nuclear material (SNM), usually for research purposes. This regulatory program is primarily governed by regulations contained in 10 CFR Part 70, "Domestic Licensing of Special Nuclear Material," and 10 CFR Part 76, "Certification of Gaseous Diffusion Plants," as well as by 10 CFR Part 40.

The regulations applicable to licensing of fuel cycle facilities vary because of the differences in the nature and amounts of the materials licensed and the historical origins of the facilities. Regulation of risk to workers is a major focus for these facilities, since workers typically are in close proximity to the hazards. The regulations in 10 CFR Part 40 apply to conversion and deconversion facilities as licensees authorized to possess source material. The safety requirements for 10 CFR Part 40 uranium conversion licenses are general, not prescriptive, and do not require performance of any risk assessment. 10 CFR Part 70 applies to major low-enriched uranium fuel fabrication facilities, high-enriched uranium processing facilities, and new enrichment facilities (gaseous centrifuge). For these major fuel cycle facilities, 10 CFR Part 70, Subpart H, "Additional Requirements for Certain Licensees Authorized to Possess a Critical Mass of Special Nuclear Material," requires licensees to perform an integrated safety analysis (ISA). ISAs share some elements in common with PRAs in that all significant accident scenarios must be evaluated, and consequences must be estimated.

The RMTF assessed this regulatory program in the context of the proposed risk management framework and developed the following findings:

- Finding: The current fuel cycle regulatory approach incorporates several elements of the proposed risk management regulatory framework, such as the use of ISAs to identify safety significant items, and the implementation of a revised fuel cycle oversight program as directed by the Commission.

- Finding: The concept of defense in depth, as embedded in fuel cycle regulatory requirements and practices, is consistent with Commission guidance. Its implementation changes as the processes change at the fuel cycle facilities.

In light of these findings, the RMTF recommendation for fuel cycle is provided below.

Recommendations (Fuel Cycle):

The fuel cycle regulatory program should continue to evaluate the risk and the associated defense-in-depth protection by using insights gained from ISAs. ISAs should continue to evolve to support regulatory decisionmaking.

Spent Nuclear Fuel Storage

Spent nuclear fuel (SNF) dry storage systems are designed to be robust, passive systems. They are designed to withstand the effects of "worst-case" events or design-basis events and phenomena while still maintaining the capabilities to provide adequate shielding and confinement of radioactive contents and prevent nuclear criticality. The systems are designed to perform these functions while requiring minimal maintenance or repair. The regulations in 10 CFR Part 72, "Licensing Requirements for the Independent Storage of Spent Nuclear Fuel, High-Level Radioactive Waste, and Reactor-Related Greater Than Class C Waste," include requirements based on risk to some degree. Both the NRC and the Electric Power Research Institute have conducted PRAs of dry cask storage systems and concluded that the risk associated with them is very low. The NRC has employed a conservative approach in its regulations, guidance, and licensing practices for independent spent fuel storage installations to minimize the likelihood of adverse consequences to public health and safety. The Commission has directed the NRC staff to revisit the paradigm for SNF storage and transportation to include evaluating the dry storage of SNF for periods significantly in excess of those previously envisioned.

The RMTF assessed this regulatory program in the context of the proposed risk management framework and developed the following findings:

- Finding: The regulatory approach for SNF storage is largely based on meeting applicable industry consensus standards and conservative guidance to ensure adequate safety margins in the facility and cask designs and operations. More recently, insights from a limited number of risk studies have been gradually factored into this regulatory approach. Furthermore, though qualitative, a systematic approach that parallels answering the risk triplet was used in the latest revision of the Standard Review Plan.

- <u>Finding</u>: The concept of defense in depth is not explicitly or consistently applied in the SNF storage regulatory program.

In light of these findings, the RMTF recommendations for spent fuel storage are provided below.

<u>Recommendation (Spent Fuel Storage):</u>

- While elements of the proposed risk management approach have been used in the SNF storage regulatory approach to evaluate the acceptable level of risk and the sufficiency of defense in depth (physical barriers, controls or margins) more consistently, the NRC should develop the necessary risk information, the corresponding decision metrics, and numerical guidelines. This is important in guiding further changes to the existing SNF storage regulatory approach and the evaluation of strategies for extended SNF storage activities.

- As part of the implementation of the proposed risk management regulatory framework, the NRC should more consistently consider the concept of defense in depth explicitly and evaluate its proper use in the SNF storage regulatory program. The NRC should also improve appropriate parts of staff training to make this concept a central part of such training.

Transportation

The transportation of radioactive materials within the United States is regulated jointly by the U.S. Department of Transportation (DOT), the NRC, DOE, and State and local governments. The approval or certification of shipping package designs for radioactive materials is shared jointly by the NRC, DOT, and DOE. NRC and DOT responsibilities for the certification of shipping package designs are delineated in a 1979 Memorandum of Understanding between DOT and the NRC. The NRC's primary "licensing" role in transportation safety is the review and certification of Type B and fissile material shipping package designs. The NRC inspects 10 CFR Part 71, "Packaging and Transportation of Radioactive Material," certificate holders and package component fabricators to ensure that transportation casks are fabricated and tested in accordance with the package specifications in the NRC certificate. The United States, which is a participating International Atomic Energy Agency (IAEA) member state, has also endorsed the concept that its domestic transportation regulations in Title 49, "Transportation," of the Code of Federal Regulations (49 CFR) Part 173,"Shippers—General Requirements for Shipments and Packaging," should be compatible with IAEA's transportation regulations to the greatest extent practicable. The practical result is that the NRC and DOT periodically undertake a joint rulemaking effort to revise their respective regulations to be compatible with the latest revision to IAEA transportation regulations.

The basic physical tests implemented in the 1964 IAEA regulations remain as the primary tests used today for approving Type B and fissile material shipping packages. The continued use of these physical tests has been supported by numerous risk studies done in the United States, as well as in other IAEA member states. All of the studies have shown that the risk of shipping spent fuel is very low.

The RMTF assessed this regulatory program in the context of the proposed risk management framework and developed the following findings:

• Finding: While the U.S. transportation regulatory approach is governed by the IAEA transportation regulations, the current NRC transportation regulatory approach uses several elements of the proposed risk management framework.

• Finding: Risk assessments have been conducted on the safety of transportation of spent fuel. However, there is a lack of risk information on the transportation of other radioactive materials.

In light of these findings, the RMTF recommendations for transportation are provided below.

> ### Recommendations (Transportation):
>
> • Considering the strong international regulatory basis for transportation and the need to conform U.S. standards to those of the IAEA and other member states, application of the proposed risk management framework should focus on implementation guidance.
>
> • The risk management process should be used to influence the future outcome of IAEA deliberations on proposed changes in international transportation regulations.
>
> • The NRC should explore the value of using risk insights to justify regulations different from the IAEA's for domestic use only, such as regulations dealing with domestic storage and transportation of high burnup fuel. Risk information could be used to develop a more flexible approach toward implementing and making gradual changes to current transportation regulations.

Concluding Remarks

The NRC has made progress in its efforts to implement risk-informed and performance-based approaches into its regulation of the various uses of byproduct, source, and special nuclear materials. Nevertheless, it is necessary to reassess that progress and the underlying strategic vision from time to time.

The RMTF has found that the NRC's programs do not require radical or revolutionary changes, but they could benefit from continuing the evolution that has been embraced throughout the agency's history. To that end, a Risk Management Regulatory Framework is recommended as the next logical step for the NRC. This proposed framework uses a disciplined risk management process to identify and evaluate issues and make decisions on appropriate defense-in-depth protections for various radiological hazards. The risk-informed and performance-based defense-in-depth protections provide sufficient barriers, controls, and personnel to prevent, contain, and mitigate the exposure of workers or the public to radioactive materials. The appropriate barriers, controls, and personnel are based on the hazards present, the relevant scenarios leading to possible exposures, and the associated uncertainties to ensure that the risks resulting from the failure of some or all of the established barriers are maintained acceptably low.

To a certain degree, some resistance to change is natural and perhaps even desirable for the NRC to maintain a clear and stable regulatory environment. In addition, the ongoing work activities within the agency and limited resources can impede the development and implementation of the proposed regulatory framework. However, a patchwork of regulatory requirements has been created as a result of addressing problems on a case-by-case basis for many years. The RMTF has concluded that cost-effective changes are possible and recommends that they be undertaken in a holistic manner across the NRC's programs. With this in mind, implementation of the proposed Risk Management Regulatory Framework can be pursued in a planned and deliberate manner so that it does not disrupt the NRC's mission, but ensures that the NRC continues to improve on how it protects public health and safety, promotes the common defense and security, and protects the environment.

ACKNOWLEDGMENT

I thank Chairman Jaczko for proposing this work and the agency for providing resources to carry it out. The Risk Management Task Force (RMTF) consisted of Christiana Lui, RMTF Executive Director, Mark Cunningham, George Pangburn, and William Reckley. Their knowledge, innovative thinking, and dedication confirmed, once again, the high regard that I have for the NRC staff. Our sessions debating risk-informed and performance-based concepts will be among my fond memories of my tenure as a commissioner.

The RMTF benefited from contributions by the following staff members: John Adams, Michel Call, Dennis Damon, Don Dube, Earl Easton, Timothy McCartin, Geary Mizuno, and Joel Piper. Their contributions helped us cover the large range of regulatory activities of the agency, a nearly impossible task for any small group.

I also thank the many other individuals and organizations who provided comments and insights. In that regard, our meetings with NRC regional offices were particularly valuable.

George Apostolakis

1. INTRODUCTION

1.0   Background

The wide-scale use of radioactive materials has evolved over only the last several decades of human history. During that time, many beneficial applications of radioactive materials and nuclear technology have been developed and put into use. Experiences during this period also have demonstrated the need to protect workers and the public from inadvertent exposures to radioactive materials. The importance of regulatory controls to govern the use or generation of radioactive materials is reflected in the congressional findings of the Atomic Energy Act ("the Act"), which states:

> The processing and utilization of source, byproduct, and special nuclear material must be regulated in the national interest and in order to provide for the common defense and security and to protect the health and safety of the public.

The increasing use of radioactive materials during the decades that followed implementation of the Act is reflected in the various regulatory programs of the U.S. Nuclear Regulatory Commission (NRC). This includes a wide range of uses of radioactive materials in industrial, medical, and research applications; the use of nuclear reactors for electricity production and research; the transportation and storage of radioactive materials; and the disposal of radioactive wastes. The challenge undertaken by the NRC and its licensees is to enable the beneficial uses of radioactive materials while ensuring those activities pose no undue risk to the public health and safety.

The NRC has established agencywide regulations and policies to help ensure that civilian uses of radioactive materials pose no undue risk. The Commission's basic radiation protection regulations are contained in Title 10 of the Code of Federal Regulations (10 CFR) Part 20, "Standards for Protection Against Radiation." Regulations in 10 CFR Part 20 establish exposure limits and require licensees to make every reasonable effort to maintain exposures to radiation as low as is reasonably achievable. Other regulations, such as those contained in 10 CFR Parts 30 through 70, describe additional basic safety requirements for NRC-licensed materials and facilities. Specific regulatory controls reflect the differences in the radiological hazards associated with different types of NRC-licensed materials, devices, and facilities.

In the 1970s, the NRC completed its first probabilistic risk assessment of two nuclear power reactors, which introduced a new way to measure nuclear safety and the effectiveness of the NRC's regulations. The Commission subsequently established a policy on how risk assessment methods should be used to complement the NRC's established regulations in all its regulatory programs. This policy, coupled with additional Commission guidance[1] issued in 1999, has resulted in a variety of program-specific improvements.

While progress has been made, the NRC's Strategic Plan and Principles of Good Regulation make it clear that improvements in efficiency, effectiveness, and reliability continue to be agency

---

1   The NRC's "White Paper on Risk-Informed and Performance-Based Regulation" (NRC,1999) contained descriptions of many terms, including defense-in-depth.

goals. The NRC Strategic Plan (NRC, 2012) notes that the expanded use of risk-informed and performance-based insights and the use of state-of-the-art technologies are the means by which the agency enhances the effectiveness and realism of NRC actions. The Principles of Good Regulation reinforce these points, noting that regulatory activities should be consistent with the degree of risk reduction they achieve. Furthermore, regulations should be based on the best knowledge available from research and operational experience.

In a memorandum dated February 11, 2011, NRC Chairman Gregory Jaczko created a task force headed by Commissioner George Apostolakis to develop a strategic vision and options for adopting a more comprehensive and holistic risk informed, performance-based regulatory approach for reactors, materials, waste, fuel cycle, and transportation that would continue to ensure the safe and secure use of nuclear material (NRC, 2011). The task force was afforded the flexibility to provide options ranging from a complement to or alternative to the existing regulatory framework. The objectives of the task force were defined as follows:

> Task Force Objectives:
>
> The task force should identify the options and specific actions that the NRC could pursue to achieve a more comprehensive and holistic risk-informed, performance-based regulatory structure.

About 1 month after the creation of this task force, hereafter referred to as the Risk Management Task Force (RMTF), a significant accident occurred at the Fukushima Dai-ichi nuclear power facility in Japan. The Commission established a Near-Term Task Force (NTTF) "to conduct a systematic and methodical review of U.S. Nuclear Regulatory Commission processes and regulations to determine whether the agency should make additional improvements to its regulatory system and to make recommendations to the Commission for its policy direction, in light of the accident at the Fukushima Dai-ichi Nuclear Power Plant." Recommendation 1 of the NTTF (NRC, 2011a) is listed below.

> NTTF Recommendation 1:
>
> The Task Force recommends establishing a logical, systematic, and coherent regulatory framework for adequate protection that appropriately balances defense-in-depth and risk considerations."

The RMTF has benefited from the discussion accompanying Recommendation 1 in the NTTF report. The RMTF report could contribute to the implementation of the NTTF's recommendations.

1.1   The Risk Management Task Force Report

The report that follows is organized into five chapters. In Chapter 2, the RMTF describes the current regulatory approach and proposes a regulatory framework based on its findings. The proposed framework centers on the concept of risk management.

Chapter 3 provides a general discussion of risk management concepts, similar to those commonly used by other organizations, including Federal agencies. These concepts are then translated into the NRC-specific framework, should the Commission decide to pursue a risk management focus for the NRC's regulatory programs. The chapter and related appendices discuss a common NRC goal related to risk-informed and performance-based defense-in-depth principles and techniques that can be used to evaluate risks, qualitatively or quantitatively, for different radiological hazards and applications. In addition, the chapter discusses possible factors used to determine the tolerability or acceptability of the risk estimations or characterizations resulting from the staff's evaluations. The chapter details several approaches and possible actions that might be taken to develop and implement a risk management framework for the NRC's regulatory programs. The RMTF offers its findings and recommendations in this chapter on the adoption of a risk management framework at the NRC.

Chapter 4 discusses how the NRC would implement a risk management framework in the regulation and oversight of licensees within various regulatory program areas. Licensees would ultimately manage risks to workers and the public based on the specific uses of byproduct, source, and special nuclear materials. Chapter 4 also discusses how the NRC would integrate a risk management framework into its processes used to regulate and oversee its licensees. The RMTF offers its findings and recommendations on specific regulatory programs and general or crosscutting topics or activities.

Chapter 5 summarizes the RMTF findings and discusses possible implementation strategies.

To maintain the report's focus on the chartered tasks, many of the more detailed descriptions supporting the discussions in the chapters and related specific topics are presented in the following appendices:

*   Appendix A: Risk Management Systems

*   Appendix B: Risk Management—Analyses and Deliberation

*   Appendix C: Defense-in-depth

*   Appendix D: Performance-based Regulation

*   Appendix E: Nuclear Power Reactors—Risk-Informed Initiatives

*   Appendix F: Nuclear Power Reactors—Licensing-basis Events

*   Appendix G: Nuclear Power Reactors—Safety Classification

*   Appendix H: Nuclear Power Reactors— Chapter 4 Alternatives

- Appendix I: Comment Summary

- Appendix J: Findings and Recommendations

1.2 References

(NRC,1999) U.S. Nuclear Regulatory Commission, Staff Requirements Memorandum Regarding SECY-98-144, "White Paper on Risk-informed and Performance-Based Regulation," March 1, 1999, Agencywide Documents Access and Management System (ADAMS) Accession No. ML003753601.

(NRC, 2011) U.S. Nuclear Regulatory Commission, "Assessment of Options for More Holistic Risk-Informed, Performance-based Regulatory Approach," Memorandum dated February 11, 2011, to R.W. Borchardt and S.G. Burns from Chairman Gregory B. Jaczko, ADAMS Accession No. ML1104606111.

(NRC, 2011a) U.S. Nuclear Regulatory Commission, "Recommendations for Enhancing Reactor Safety in the 21st Century; The Near-Term Task Force Review of Insights from the Fukushima Dai-Ichi Accident," July 2011, ADAMS Accession No. ML112510271.

(NRC, 2012) U.S. Nuclear Regulatory Commission, "Strategic Plan: Fiscal Years 2008–2013," NUREG–1614, Vol. 5, February 2012, ADAMS Accession No. ML12038A003.

2. A PROPOSED RISK MANAGEMENT REGULATORY FRAMEWORK

2.0   Background

In its efforts to propose options for a more holistic approach to the development and implementation of risk-informed and performance-based regulation of NRC licensees, the Risk Management Task Force (RMTF) identified a general framework based on established risk management concepts. The NRC mission is focused on protecting the public from possible hazards introduced by the use of radioactive materials. While some regulatory provisions, such as 10 CFR Part 20, cut across all NRC programs, the regulatory systems for reactors, materials, and other NRC program areas have been developed somewhat independently from each other. The first challenge for the RMTF was to develop a framework that could be used throughout the NRC's regulatory programs.

The value of soliciting insights from the NRC staff and external stakeholders regarding the RMTF activities was recognized from its formation. The RMTF prepared an internal survey and distributed it to a cross-section of NRC staff and managers in several key program offices, and issued a notice in the *Federal Register* (published on November 22, 2011 (76 FR 72220)) to solicit public input. A summary of the responses to the internal survey and *Federal Register* notice is provided in Appendix I. The RMTF also benefited greatly from numerous informal discussions with NRC staff and managers, and external stakeholders. Furthermore, the RMTF visited all the NRC regional offices. These interactions with the regional staffs were especially useful because they provided the RMTF with an opportunity to hear firsthand the experiences and issues associated with the use of risk-informed and performance-based approaches in the NRC's oversight process.

2.1   Considerations and Findings

The RMTF started with the NRC Mission Statement, which is derived from the Atomic Energy Act and is defined as follows in the NRC Strategic Plan (NRC, 2012):

NRC Mission

License and regulate the Nation's civilian use of byproduct, source, and special nuclear materials to ensure adequate protection of public health and safety, promote the common defense and security, and protect the environment.

The Strategic Plan describes the NRC's Principles of Good Regulation as follows:

Principles of Good Regulation

The safe and secure use of radioactive materials and nuclear fuels for beneficial civilian purposes is made possible by the agency's adherence to the following principles of good regulation: independence, openness, efficiency, clarity, and reliability. In addition, regulatory actions are effective, realistic, and timely.

The mission and principles of good regulation naturally translate into a high-level objective for the NRC to establish regulatory controls based on risks to the public that are introduced by licensed activities. The consideration of risks and tailoring regulations and oversight to manage those risks is inherent in current NRC programs. However, they are sometimes expressed using different terminology. This approach is consistent with several proposals for effective government, including Executive Order 13563, "Improving Regulation and Regulatory Review" (Executive Order, 2011).

An example of the challenges related to terminology can be found in the often used terms "safety" and "risk" within the context of NRC programs and goals. The dictionary definitions of these two words highlight the relationship between and the difference in the terms. Such distinctions have resulted in occasional disagreements among knowledgeable practitioners about the terminology and related actions.

- • Safety: the condition of being free from danger, injury, or damage

- • Risk: the possibility of damage, injury, or loss

Recognizing that it is not possible to achieve absolute safety or zero risk while benefitting from the use of radioactive materials, terms such as "adequate protection" and "no undue risk" are used in conclusions or decisions reached by the NRC. No matter an individual's preference for "adequate protection" or "acceptable risk," assessments and regulatory controls will involve asking the same basic questions, which are commonly referred to as the risk triplet (Kaplan and Garrick, 1981). These questions are listed below.

- • What can go wrong?

- • How likely is it?

- • What are the consequences?

While the RMTF observes that the NRC can and has treated "adequate safety" and "acceptable risk" as synonymous, it also recognizes that these concepts present challenges to the development and implementation of any generic framework. Related issues, such as the selection of analysis techniques, are discussed in later sections of this report.

The NRC has developed regulations that limit radiation exposure to workers and the public by requiring its licensees to limit the time persons are exposed to radiation, limit access to areas containing radioactive materials, and provide physical barriers between radioactive materials and individuals. These controls are associated with (1) requirements for radioactive materials to be contained within devices or facilities to prevent their inadvertent release, and (2) measures to address the possible degradation of barriers and controls intended to protect workers and the public. NRC rules and programs also address, to varying levels of specificity, the need for personnel to be trained and qualified for the tasks they perform. The totality of these requirements constitutes the way in which risk from the use of radioactive materials is managed.

Requirements related to barriers, controls, and personnel to limit exposure to radioactive materials vary among NRC licensees, depending on the hazard present (e.g., the type and amount of radioactive materials), the relevant scenarios that might lead to exposures, and the associated uncertainties. The risks presented by most uses of byproduct materials differ significantly from the risks presented by nuclear power reactors. The risk environment for reactors focuses predominantly on the prevention and mitigation of low-frequency, high-consequence events whereas the risk environment for materials uses focuses primarily on high-frequency, low-consequence events.

The potential for releases to the environment is a major focus of the reactor program. This potential is not as significant for the materials program because of the much smaller amounts of radioactive material possessed by most materials users and the lower risk posed by many of those specific uses.

The RMTF considered a number of approaches to accomplishing its charter. It considered the agency's historic approach to consideration of risk, researched and reviewed literature across the field of risk, and received input from internal and external stakeholders to develop alternative approaches for a more comprehensive and holistic approach to its regulatory programs. These included: (1) no action, (2) a purely deterministic approach, (3) a single risk-based numerical criterion, and (4) an approach based on the concept of risk management. A more detailed discussion of these approaches is contained in Appendix A.

The U.S. Department of Homeland Security Risk Lexicon (DHS, 2010) defines risk management as follows:

> Risk management is the process for identifying, analyzing, and communicating risk and accepting, avoiding, transferring, or controlling it to an acceptable level considering associated costs and benefits of any actions taken.

Risk management allows for various approaches to consider risks to the public in the NRC's regulatory decisionmaking, including the use of both quantitative and qualitative tools. Such flexibility is essential given the broad range of NRC regulatory programs and related decisions. It represents a logical evolution from the risk-informed, performance-based philosophy that has governed NRC activities for many years. It may also provide program managers with a more systematic approach to resource allocation, whether in budget formulation, response to events, or licensing decisions. Risk management offers the potential for an improved regulatory framework. The sections that follow describe the general attributes of that framework.

Whether the use of radioactive materials is in complex facilities, such as nuclear power reactors, or in simple sealed sources and devices, such as moisture density gauges, the NRC's risk management approach has been (and still is) guided by the fundamental principle of defense-in-depth. The NRC white paper on risk-informed and performance-based regulation (NRC, 1999) described many terms, including defense-in-depth. This description was modified in the statements of consideration for NRC's final rule 10 CFR 50.69, "Risk-Informed Categorization and Treatment of Structures, Systems and Components for Nuclear Power Reactors," (NRC, 2004), as follows:

> Defense-in-depth is an element of the NRC's safety philosophy that employs successive measures to prevent accidents or mitigate damage if a malfunction, accident, or naturally caused event occurs at a nuclear facility. Defense-in-depth is a philosophy used by the NRC to provide redundancy as well as the philosophy of a multiple-barrier approach against fission product releases. The defense-in-depth philosophy ensures that safety will not be wholly dependent on any single element of the design, construction, maintenance, or operation of a nuclear facility. The net effect of incorporating defense-in-depth into design, construction, maintenance, and operation is that the facility tends to be more tolerant of failures and external challenges.

While this defense–in-depth definition comes from a power reactor document, it can be applied across all NRC regulatory areas by broadening "nuclear facility" to include other radioactive materials uses and by replacing "fission product releases" with "radioactive material releases" or "exposure to radioactive materials."

Although some NRC regulatory programs may not have used the defense-in-depth terminology, the general approach of identifying barriers to protect workers and the public from exposure to radioactive material is common to all NRC activities. For some radioactive materials, the radiological hazard may be minimal and the defense-in-depth determinations may result in simple containers and labeling systems to maintain the barrier. For complex reactor facilities, the defense-in-depth determinations may result in multiple layers, robust structures, and many support systems and procedures to maintain the basic barriers. Later sections of the report will discuss specific regulatory programs in more detail, but the basic approach being proposed by the RMTF is for the agency to adopt a common explanation of our risk management goal, which is defined in simple terms of which barriers, controls, and personnel are needed to protect individuals from exposure to radioactive materials.

Even for the nuclear power reactor program, in which the defense-in-depth terminology is most ingrained, the concept is implemented through an elaborate system of regulations, guidance documents, and general industry practices. After decades of use, there is no clear definition or criteria on how to define adequate defense-in-depth protections. The traditional approach used by the NRC and industry to provide confidence in a reactor design's defense-in-depth capabilities is based on analyzing stylized accident scenarios using approved conservative codes and criteria. The conservatisms added to design limits, acceptance criteria, and safety margins are intended to manage the uncertainties associated with accidents, including possible "unknown unknowns," at the time a plant was designed. Safety margins are included in the analyses such that specific barriers are designed and constructed to ensure actual failures are not expected until key parameters well exceed the values assumed in the supporting engineering evaluations. Important limitations of this traditional regulatory approach are that (1) significant accident scenarios may not be identified or addressed by the defined barriers and controls, and (2) the stylized analyses and related barriers and controls may misdirect resources to address low-risk scenarios.

The RMTF review of NRC programs leads to the following finding regarding defense-in-depth.

> Finding 2.1: Whether used explicitly, as for power reactors, or implicitly, as for materials programs, the concept of defense-in-depth has served the NRC and the regulated industries well and continues to be valuable today. However, it is not used consistently, and there is no guidance on how much defense-in-depth is sufficient.

Risk-informed approaches have been developed over the last several decades to supplement the traditional regulatory approach by doing a more methodical assessment of the risk triplet questions (what can go wrong, how likely is it, and what are the consequences). The NRC white paper mentioned previously provides the following definition:

> A "risk-informed" approach to regulatory Decisionmaking represents a philosophy whereby risk insights are considered together with other factors to establish requirements that better focus licensee and regulatory attention on design and operational issues commensurate with their importance to health and safety.

The term "risk insights" refers to the results and findings of risk assessments. A risk assessment is intended to be as realistic as possible, as opposed to the bounding, stylized approach of traditional methods. It explicitly and systematically assesses (quantitatively or qualitatively) the likelihood of a given scenario and considers failures of multiple barriers and controls.

The white paper also discusses the difference between prescriptive and performance-based regulatory requirements:

> A regulation can be either prescriptive or performance-based. A prescriptive requirement specifies particular features, actions, or programmatic elements to be included in the design or process, as the means for achieving a desired objective. A performance-based requirement relies upon measurable (or calculable) outcomes (i.e., performance results) to be met, but provides more flexibility to the licensee as to the means of meeting those outcomes.

The benefits of using risk-informed and performance-based approaches throughout the NRC's regulatory programs have been recognized and are reflected in various policy statements and initiatives.

Within the nuclear power reactor program, there has been increased use of formal risk assessment techniques, such as probabilistic risk assessment (PRA).[1] In contrast to the

---

1   Internationally, this approach is known as probabilistic safety assessment (PSA). For the purposes of this report, there is no difference between PRA and PSA.

traditional methods described above, a PRA models many credible accident sequences by considering the facility or operation as a "system of systems" consisting of structures, systems, components, and personnel. The risk assessments have resulted in the consideration of additional accident sequences during licensing (e.g., intersystem loss of coolant accident, station blackout), informed program areas, such as emergency planning, and have led to various initiatives to incorporate risk-informed, performance-based activities into the NRC's regulation and oversight of nuclear power reactors.

Risk assessments and related insights also have been used in several NRC nonreactor programs. NUREG/CR-6642, "Risk Analysis and Evaluation of Regulatory Options for Nuclear Byproduct Materials Systems" (NRC, 2000), documented a programmatic quantitative analysis of tasks, hazards, barriers, and doses (to workers and members of the public) associated with a variety of byproduct materials and devices. These risk insights were used to inform the NRC inspection programs and were incorporated into licensing guidance in the NUREG-1556 series, "Consolidated Guidance About Materials Licensees" (NRC, 1998). In another example, the NRC has required fuel cycle facilities to perform and document a type of risk assessment referred to as integrated safety assessments (ISAs). The high-level waste, spent fuel storage, and transportation programs also have all performed different types of risk assessments in their development of rules, guidance, or policy over the years.

In addition to the tools and methods used to perform risk assessments, the NRC has developed criteria and guidance to help judge the acceptability of the risks identified by those analyses. The publication of the Commission's quantitative safety goals (NRC, 1986) (and the adoption of the subsidiary objectives regarding core damage frequency and large, early release frequency) has allowed the use of PRA in helping to address a limitation of the traditional defense-in-depth approach for power reactors; namely, how much defense-in-depth is sufficient. The regulations in 10 CFR Part 63 established high-level waste repository performance objectives and a means to evaluate the adequacy of defense-in-depth in the context of those performance objectives. In the area of byproduct materials, NUREG-1556 provides risk-informed guidance on the barriers and controls that should be in place for meeting the dose criteria set forth in 10 CFR Part 20 for a wide variety of applications.

In summary, the introduction of risk assessments to supplement traditional approaches has provided systematic evaluation processes and more balanced considerations in several NRC regulatory programs. Some of these insights resulted in changes to NRC regulations and licensee programs that may have prevented serious challenges to or breaches of the barriers and controls put into place to prevent the release of radioactive materials. Other evaluations have shown that some existing regulatory requirements were not actually needed. These proactive changes are better than relying solely on events and operational experience to assess the effectiveness of current regulatory requirements and licensee practices. Therefore, the RMTF finds the following:

> Finding 2.2: Risk assessments provide valuable and realistic insights into potential exposure scenarios. In combination with other technical analyses, risk assessments can inform decisions about appropriate defense-in-depth measures.

2.2   Recommendations

In addition to the NRC's longstanding goal to move toward more risk-informed regulatory approaches, the benefits of performance-based approaches also have been recognized and encouraged. The NRC has recognized that purely deterministic and prescriptive approaches can limit the flexibility of both the regulated industries and the NRC to respond to lessons learned from operating experience and support the adoption of improved designs or processes. The RMTF recommends the following:

> Recommendation 2.1: The goal to adopt risk-informed and performance-based approaches, where practical, should continue and should be incorporated into the revised regulatory framework.

As discussed above, the RMTF evaluated several options and determined that a framework based on risk management principles provided the best approach. Therefore, the RMTF offers the following recommendation:

> Recommendation 2.2: The general regulatory approach of the NRC should be defined in terms of "managing the risks" posed to workers and the public from the various uses of byproduct, source, and special nuclear materials.

To build on the mission and the Principles of Good Regulation from the NRC Strategic Plan and past agency performance, the RMTF adopted an "objective" for the agency. The objective is intended to help promote improved management by the NRC and its licensees of radiological hazards and associated risks. Key concepts to be captured by the agency's objective include:

•   Risk management as the vehicle to accomplish the NRC mission

•   Applicability to all NRC programs (expressed in terms of byproduct, source, and special nuclear material)

•   Maintaining the longstanding agency goal to use, whenever practical, performance-based approaches to regulation

•   Applicability to NRC processes that establish requirements for or oversight of licensee programs

To support the above recommendation, the RMTF proposes the following statement as the objective for the NRC's regulatory framework:

> *Objective*
>
> *Manage the risks from the use of byproduct, source, and special nuclear materials through appropriate performance-based regulatory controls and oversight.*

To a large degree, this objective has been part of the NRC's existing practices. Therefore, it is not expected that refining the focus to risk management would require extensive changes in either the NRC's philosophy or the regulatory requirements placed on its licensees. However, establishing a common language of risk management across all NRC activities would be beneficial. Consistent with Recommendation 2.1, the goal to use performance-based regulatory approaches whenever practical has been incorporated into the objective for the revised framework.

As stated in Finding 2.1, the concept of defense-in-depth has served the NRC and the industry well. This concept is a logical way to implement a risk management vision for the NRC, and the RMTF recommends that it be used across NRC programs and the life cycle of the related materials, devices, and facilities. The life cycle of regulatory activities generally can be described in terms of (1) design, fabrication, and construction, (2) operations, and (3) decommissioning. Defense-in-depth is used in this context to mean the barriers, controls, and personnel used to prevent, contain, and mitigate possible inadvertent exposure from radioactive materials. Although the principle of defense-in-depth can be applied to the NRC's various regulatory programs, the nature of the hazards—in terms of the amount and form of radioactive materials—means that the actual barriers, controls, and requirements for personnel will vary considerably among the different types of NRC-licensed activities.

Another important difference between NRC regulatory programs involves the amount of uncertainty associated with possible inadvertent exposure to radioactive materials. Some byproduct materials are used in small amounts, in stable forms, and within relatively simple devices. The radiological hazard in these cases is limited. There is a good understanding of release scenarios and possible pathways to the public for both normal operation and the loss of barriers and controls established for such devices. In contrast, power reactors are complex facilities with a large inventory of many different radioactive isotopes and possible scenarios that could interrupt safety functions. These functions include the control of reactivity, removal of heat from the reactor core, and confinement of fission products within the facility. These complexities also result in uncertainties associated with the response of plant equipment and personnel that comprise the barriers and controls to prevent the release of radioactive materials as a result of possible accidents. Risk assessment techniques have proven to be a valuable tool in the identification and resolution of uncertainties and in the evaluation of what are reasonable and sufficient levels of defense-in-depth for particular hazards. The RMTF, therefore, recommends the following:

> Recommendation 2.3: In defining requirements for the protection of workers and the public, the NRC should recognize and address uncertainties associated with the hazards and the events, including human errors, which could challenge or degrade barriers and controls. A balanced approach that considers traditional and risk assessment techniques should be used to identify barriers and controls so that appropriate requirements are defined to prevent, contain, and mitigate exposures to radioactive materials.

To further define a revised regulatory framework, the RMTF defines a goal for the risk management objective that addresses Recommendation 2.3. Building on existing NRC practices, a risk-informed and performance-based defense-in-depth principle was identified as the best approach. Key concepts to be captured by the risk management goal include:

- Support implementation of the risk management framework

- Provide a logical way to implement the agency's objective to manage the risks from the use of byproduct, source, and special nuclear materials

- Apply to all NRC programs (expressed in terms of hazard present)

- Incorporate performance-based objectives

- Address uncertainties associated with radiological hazards

- Address possible failure of some or all established barriers and controls (including "cliff-edge" effects)

The RMTF proposes the following risk management goal to address Recommendation 2.3 and the attributes for a risk management goal:

Risk Management Goal

Provide risk-informed and performance-based defense-in-depth protections to:

- Ensure appropriate barriers, controls, and personnel to prevent, contain, and mitigate exposure to radioactive material according to the hazard present, the relevant scenarios, and the associated uncertainties.

- Ensure that the risks resulting from the failure of some or all of the established barriers and controls, including human errors, are maintained acceptably low.

In program areas such as nuclear power reactors and waste repositories, the NRC has established quantitative goals to help define acceptably low levels of risk. Risk management for those program areas dealing with lesser amounts of radioactive materials often is achieved largely by standard radiation protection practices, which are established by all licensees in accordance with NRC regulations defined in 10 CFR Part 20. The use of the above risk management goal is addressed in Chapter 4 discussions of each regulatory program. However, if the agency decides to pursue the Risk Management Regulatory Framework, the NRC program offices would need to support its implementation by translating the above risk management goal into more practical criteria and guidance for specific devices and facilities.

To ensure a consistent approach across the NRC's regulatory programs, a common process should be used to identify and evaluate issues and then to make decisions and implement actions to manage the risks associated with the use of radioactive materials. Although not always presented in this form, the decisionmaking process shown in Figure 2-1, and described in more detail in Chapter 3, generally reflects the NRC's way of regulating all of its licensees. This decisionmaking process is readily apparent in existing programs, such as regulatory analyses for rulemakings and the Reactor Oversight Process.

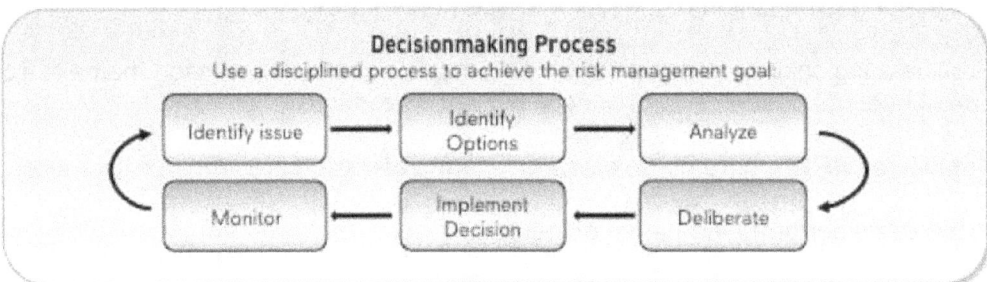

Figure 2-1 The Regulatory Decisionmaking Process

There is a logical flow from the NRC's mission to the proposed agency objective to manage risks, to the risk management goal provided by risk-informed and performance-based defense-in-depth, and, finally, to a standard risk management decisionmaking cycle as shown above. This flow forms the proposed Risk Management Regulatory Framework shown in Figure 2-2.

> Recommendation 2.4: The NRC should formally adopt the proposed Risk Management Regulatory Framework through a Commission Policy Statement.

A concerted effort will be needed to adopt the proposed Risk Management Regulatory Framework and improve its chances of success. This process would likely take years to develop and incorporate into NRC regulatory programs and procedures. An important first step in this process would be the formal adoption of the framework by the Commission using a vehicle such as a Policy Statement. This could then be followed by changes to guidance documents, such as management directives and office-level procedures, recognizing that many aspects of the proposed Risk Management Regulatory Framework are already in place, albeit using different terminology. For some regulatory programs, the implementation of the proposed framework could also involve changes to regulations, regulatory guides, and other processes that directly affect NRC licensees and other external stakeholders. Chapter 4 provides discussions of options for implementing the proposed framework for various NRC regulatory programs. This effort would also be an opportunity to update and consolidate previous policy statements, initiatives, and activities related to the adoption of risk-informed, performance-based approaches to improve the NRC's regulation of source, byproduct, and special nuclear material.

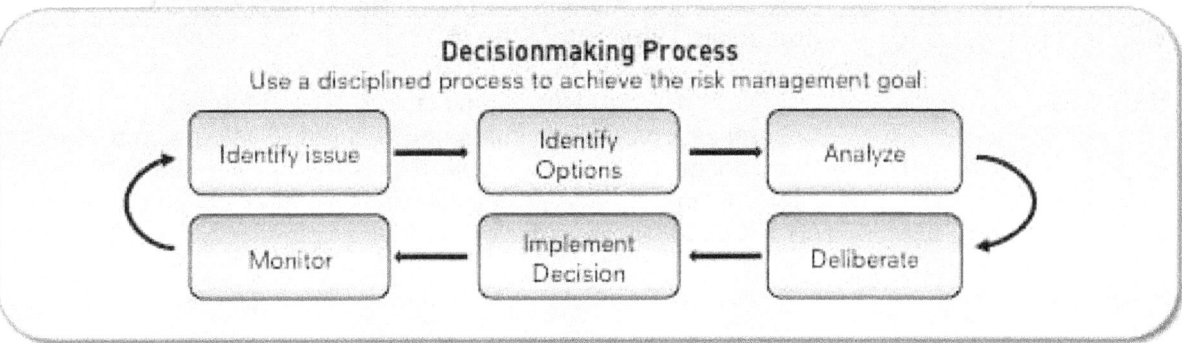

Mission
Ensure adequate protection of public health and safety, promote the common defense and security, and protect the environment

Objective
Manage the risks from the use of byproduct, source and special nuclear materials through appropriate performance-based regulatory controls and oversight

Risk Management Goal
Provide risk-informed and performance-based defense-in-depth protections to:
- Ensure appropriate barriers, controls, and personnel to prevent, contain, and mitigate exposure to radioactive material according to the hazard present, the relevant scenarios, and the associated uncertainties; and
- Ensure that the risks resulting from the failure of some or all of the established barriers and controls, including human errors, are maintained acceptably low

Decisionmaking Process
Use a disciplined process to achieve the risk management goal:

Identify issue → Identify Options → Analyze → Deliberate → Implement Decision → Monitor →

Figure 2-2 A Proposed Risk Management Regulatory Framework

2.3   References

(DHS, 2010)   U.S. Department of Homeland Security, "DHS Risk Lexicon,"
                                     September 2010.

(Executive Order, 2011)   Executive Order 13563, "Improving Regulation and Regulatory
                                     Review," January 18, 2011, published in the *Federal Register* on
                                     January 21, 2011 (76 FR 821).

(Kaplan and Garrick, 1981)   S. Kaplan and B.J. Garrick, "On the Quantitative Definition of
                                     Risk," *Risk Analysis*, 1:11–27, 1981.

(NRC, 1986)   U.S. Nuclear Regulatory Commission, "Safety Goals for the
                                     Operations of Nuclear Power Plants; Policy Statement," August
                                     1986, Agencywide Documents Access and Management System
                                     (ADAMS) Accession No. ML011210381.

(NRC, 1998)   U.S. Nuclear Regulatory Commission, "Consolidated
                                     Guidance About Materials Licensees," NUREG-1556, 1998
                                     and later, multiple volumes.

(NRC, 1999)   U.S. Nuclear Regulatory Commission, "White Paper on
                                     Risk-Informed and Performance-Based Regulation," Staff
                                     Requirements Memorandum Regarding SECY-98-144,
                                     March 1, 1999, ADAMS Accession No. ML003753601.

(NRC, 2000)   U.S. Nuclear Regulatory Commission, "Risk Analysis
                                     and Evaluation of Regulatory Options for Nuclear
                                     Byproduct Materials Systems," NUREG/CR-6642,
                                     February 2000, ADAMS Accession No. ML003693052.
                                     (Not publicly available.).

(NRC, 2004)   U.S. Nuclear Regulatory Commission, "Risk-Informed
                                     Categorization and Treatment of Structures, Systems, and
                                     Components for Nuclear Power Reactors, 10 CFR 50.69,
                                     published in the *Federal Register* on November 22, 2004,
                                     (67 FR 68008)

(NRC, 2012)   U.S. Nuclear Regulatory Commission, "Strategic Plan: Fiscal
                                     Years 2008-2013," NUREG-1614, Vol. 5, February 2012,
                                     ADAMS Accession No. ML12038A003.

3. RISK MANAGEMENT DECISIONMAKING PROCESS

3.0   Background

As discussed in Chapter 2, risk management provides the most useful framework for a comprehensive and holistic risk-informed, performance-based regulatory approach. Risk management concepts, process descriptions, and decisionmaking models are being widely used in various sectors, including Government agencies, financial institutions, and technology companies. There are several general descriptions of risk management processes and methodologies that share key points. These are summarized in Appendix A, "Risk Management Systems." A basic structure with typical stages of a risk management process is included in the Risk Management Regulatory Framework (Figure 2-2) and is repeated below in Figure 3-1:

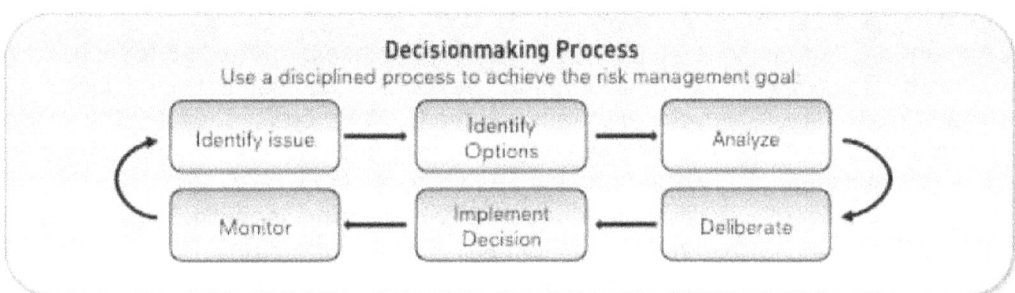

Figure 3-1 The Regulatory Decisionmaking Process

In describing the options available to the NRC for improving its regulatory programs by adopting a risk management framework, it is important to recognize that the protection of workers and the public from radiological hazards is provided by the owners or operators of the materials, devices, and facilities licensed by the NRC. However, the use of radioactive materials introduces particular risks and public concerns; therefore, the NRC's regulatory regime, adopted pursuant to the Atomic Energy Act and other statutes, provides the legal basis for ensuring that NRC licensees provide reasonable assurance of adequate protection of the health and safety of their workers and the public.

The NRC has developed regulatory requirements for licensees to provide radiation protection programs, monitoring programs, and other controls to limit the radiation exposure to workers and the public to acceptably low levels. Managing the risks associated with a totally controlled radioactive material and population would be relatively straightforward; exposures could be limited using standard radiation protection programs (controlling time, distance, and shielding). However, many of the uses of radioactive materials introduce possible conditions or events, including human errors, which might lead to a breakdown of established barriers and controls. A key role of a Risk Management Regulatory Framework, as described in Chapter 2, is to ensure that no undue risk to the public from such unplanned events occurs.

The Risk Management Task Force (RMTF) is recommending that the NRC meet its risk management goal by requiring its licensees to provide risk-informed defense-in-depth protections. Although the terminology may vary, defense-in-depth concepts have been used throughout the NRC's history and across its regulatory programs. As mentioned in previous findings and recommendations (Chapter 2), defense-in-depth concepts have been instrumental in establishing effective regulatory controls and oversight for the civilian use of byproduct, source, and special nuclear materials. The NRC also has recognized and encouraged using performance-based approaches. Additional discussions of defense-in-depth are provided in Appendix C, "Defense-in-depth," and other sections of this report.

As discussed in the remainder of this report, appropriate evaluations and actions will vary depending on the type and amount of radioactive material and the complexity of the device or facility. For some materials and devices, concerns about an inadvertent exposure are small, and few controls beyond routine radiation protection programs are necessary. Facilities such as commercial nuclear power reactors, however, contain large amounts of radioactive materials that warrant multiple levels of protection to prevent, contain, and mitigate possible releases. The NRC uses processes and programs, such as rulemaking and guidance, licensing, environmental reviews, and oversight, to define and maintain confidence in the appropriate defense-in-depth measures to be provided by licensees.

The risk management goal and related decisionmaking process are intended to be used universally for NRC actions on the regulation and oversight of licensed activities, including reactors, materials, waste, fuel cycle, and transportation. Therefore, if adopted, the proposed risk management framework would be used by major program offices and supporting offices whenever they make technical or other regulatory decisions. In many cases, the current NRC decisionmaking processes (e.g., integrated decisionmaking for nuclear reactors and risk-informed decisionmaking for materials licensees) are similar to the proposed risk management framework described in this report. Major actions, such as licensing of reactor or fuel cycle facilities, agency-level initiatives, and rulemakings, include project plans and processes that are often explained in terms of the six steps of the decisionmaking process (or in equivalent terms and steps). At first glance, the handling of some routine activities may appear simpler than the decisionmaking process, but this is probably because some steps—such as selecting technical analysis techniques and decisionmaking criteria—are already incorporated into procedures or guidance documents. Provided the existing guidance and processes ensure that appropriate barriers and controls are established and risks are maintained at an acceptably low level, the move to a risk management framework is unlikely to require significant or immediate changes to existing practices.

Although it was beyond the immediate scope of the RMTF, general concepts of risk management and methodical decisionmaking are applicable to other agency activities. For example, a consistent risk management approach across NRC programs could support changes in management practices so that program managers would have increased flexibility to allocate resources to address risk or safety concerns. Insights from inspections, operating experience, scientific studies, or risk assessments could inform timely changes to programs and focus areas as part of the implementation of budgeted resources. The NRC also uses risk management processes in providing security to the agency's information systems.

Additional discussions on the six stages of regulatory decisionmaking are provided below and in Appendix A and Appendix B, "Risk Management—Analyses and Deliberations." Insights on the implementation of the proposed framework for specific regulatory programs are provided in Chapter 4.

3.1   Identify Issues

The regulatory decisionmaking process begins with an identified issue, a proposal from a licensee or other stakeholder, or some other problem that requires a regulatory decision by the NRC staff or management. It is important to clearly define the problem to be analyzed and resolved by the subsequent steps in the process. Framing or characterizing the issue or problem refers to defining the scope of the problem (e.g., applicable licensees or facilities) and its potential implications in terms of the risk management goal identified in Figure 2-2 and restated below:

> *Risk Management Goal*
>
> *Provide risk-informed and performance-based defense-in-depth protections to:*
>
> - Ensure appropriate barriers, controls, and personnel to prevent, contain, and mitigate exposure to radioactive material, according to the hazard present, the relevant scenarios, and the associated uncertainties.
>
> - Ensure that the risks resulting from the failure of some or all of the established barriers and controls, including human errors, are maintained acceptably low.

Some problems, such as the determination of the appropriate response to the attacks of September 11, 2001, initially involved all NRC activities. Broadly applicable issues usually are divided into narrower problem statements for specific regulatory programs. Major licensing actions within NRC program areas often require an assessment of the overall risks associated with a device or facility, including how different measures or controls contribute to risk-informed and performance-based defense-in-depth. Most NRC decisions, however, relate to a problem or proposal that affects individual barriers or controls associated with specific devices or facilities already designed, licensed, and constructed. It is sometimes easy to forget that these decisions should also be made in the context of the overall risk-informed and performance-based defense-in-depth goal.

3.2   Identify Options

It is important to develop and consider choices in the resolution of issues or problems, or in the review of proposals from licensees. Some NRC processes, such as rulemaking and environmental reviews, include specific guidance for identifying and evaluating multiple options or alternatives, and they usually include a "no action" alternative. Other activities, such as the review of specific proposals from licensees, may not require the NRC staff to identify options beyond approving or denying an application, although discussion of alternatives with licensees is common even in these licensing-related deliberations.

The development or consideration of alternatives is an opportunity for the NRC staff and licensees to propose a performance-based approach to meet regulatory requirements. As discussed in Appendix D, "Performance-Based Regulation," performance-based approaches can improve the flexibility of both the regulated industries and the NRC to respond to lessons learned from operating experience and support the adoption of improved designs or processes. The use of performance-based approaches also can support more efficient regulation of licensees by reducing the number of licensing actions associated with the future operation of facilities (e.g., the technical specification improvement initiatives for nuclear power reactors). As such, the NRC staff should consider the use of performance-based approaches whenever practical and should identify and consider those approaches as alternatives as part of the risk management framework. This goal is reflected in the objective defined in Recommendation 2.1 and Figure 2-2, which states that, where possible, the management of risks from the use of byproduct, source, and special nuclear material should be evaluated using performance-based regulatory controls and oversight.

3.3   Analyze

A major element of a risk management program is the evaluation of risks associated with the subject system or activity. A more detailed discussion of technical analyses and current NRC approaches is provided in Appendix B. In the context of this report, technical analysis refers to the general task of analyzing a problem or issue to support deliberations. A key concept in the development of a risk management process is that technical analyses can be done in many ways, including traditional mechanistic analyses (e.g., thermal-hydraulic calculations), probabilistic risk assessments (PRAs), and other techniques selected to support specific decisions related to particular issues and hazards.

Within the proposed risk management framework, technical analyses will be used to support decisions on appropriate barriers and controls (including personnel) to prevent, contain, and mitigate exposures to radioactive materials. Technical analyses also will be used to ensure that the risks from events that degrade or challenge the barriers are maintained acceptably low. The appropriate barriers range from simple containers for some radioactive sources to complex structures for nuclear power plants. Likewise, the systems and actions taken to maintain barriers can range from labels and administrative controls to complex mitigation systems. Licensee organizations and individual workers are major contributors to the goal of preventing, mitigating, and containing the release of radioactive materials. Controls established for materials licensees can also address circumstances where the use of a device requires the temporary bypass of physical barriers (e.g., radiography and irradiation devices). The possible need for emergency preparedness requirements are based on the risk that barriers might be compromised in a way that the public could be exposed to radioactive materials and protective actions—such as sheltering or evacuation—might be warranted.

Just as the ultimate determination on appropriate barriers, controls, and personnel depends on the radiological hazard and possible challenges to those protections, the selection of technical analysis techniques for a specific regulatory decision will consider how the issue relates to the hazards and challenges. In general, decisions on simpler devices and frequent events (e.g., activities with many licensees) can be supported by analyses based on traditional engineering approaches, operating experience, and qualitative risk assessments. Decisions related to more complex facilities and infrequent events (e.g., nuclear power reactors) can benefit from analyses that include more robust quantitative risk assessments. More complex facilities also can introduce additional uncertainties about how the facility will respond to relevant scenarios and the ability of barriers and controls to contain and mitigate possible releases of radioactive material. Such uncertainties should be addressed, as practical, within the technical analyses and be identified and accounted for as part of the deliberative process. Sensitivity studies can be a useful tool to evaluate uncertainties and otherwise inform the decisionmaking process. The technical analysis step of the process is further discussed in Appendix B.

The key to selecting a risk assessment technique or combination of techniques is to ensure that the analysis will support the associated deliberation or decisionmaking process in an effective and efficient manner. Acceptance criteria or other rationale for decisionmaking should be established before using traditional methods, PRAs, or a combination of approaches. Many NRC decisions, in effect, are an evaluation of possible changes (e.g., license amendments, new understanding gained from operating experience) from an established baseline (e.g., initial licensing decision). In some cases, the evaluation (usually using traditional mechanistic techniques or engineering judgment) may conclude that the change does not have an adverse impact on a specific barrier or control and a broader assessment of the impact on overall risk may not be necessary. In other cases, a problem or proposal may involve a change to a barrier or control that requires consideration of the overall risk-informed and performance-based defense-in-depth protections. Previous NRC efforts to address appropriate balancing of traditional analysis methods and PRA methods include the integrated decisionmaking process in Regulatory Guide 1.174, "An Approach for Using Probabilistic Risk Assessment in Risk-Informed Decisions on Plant-Specific Changes to the Licensing-basis" (NRC, 2011), for nuclear reactors and risk-informed decisionmaking for materials programs. The topic of acceptance criteria is discussed further in the following section.

3.4   Deliberate

The risk management framework is primarily an approach to provide structure and logic to the decisionmaking process. The technical analysis and identification of options are key inputs to the process, but they can only be successful if there is an equally disciplined approach to the actual decisionmaking. As stated in Chapter 2, the NRC's decisions related to its core mission can be described in terms of the risk management goal, which is to provide risk-informed and performance-based defense-in-depth protections for the use of byproduct, source, and special nuclear materials.

The report, "Understanding Risk: Informing Decisions in a Democratic Society" (National Research Council, 1996), defines deliberation as "any formal or informal process for communication and collective consideration of issues." It further states that deliberations "formulate the decision problem, guide analysis to improve decision participants' understanding, seek the meaning of analytic findings and uncertainties, and improve the ability of interested and affected parties to participate effectively in the risk decision process." The concept of deliberation is called the "integrated decisionmaking process" in Regulatory Guide 1.174. A more detailed discussion of the deliberative processes to support a risk management framework is provided in Appendix B.

The technical analysis is an important input into the deliberation process, but it is not the only factor influencing the final decision. As Regulatory Guide 1.174 states:

> "The results of the different elements of the engineering analyses … must be considered in an integrated manner. None of the individual analyses is sufficient in and of itself."

Decisionmakers need to consider the assumptions, uncertainties, and sensitivities associated with the technical analysis, as well as the analytical results and how they compare to decision criteria established for mechanistic approaches, PRAs, or a combination thereof. The deliberative process may also include consideration of resources and schedules for the agency and its licensees and the input received from various internal and external stakeholders. Other factors include legal requirements and desired consistency with other guidance documents, treaties, or standards.

A representation of the deliberation process, including consideration of the different factors, is provided in Figure 3-2.

Figure 3-2 Deliberations

Guidance documents to help in the decisionmaking process include a variety of acceptance criteria for different regulatory programs, specific facilities, and for different types of events or operating conditions. Part of the long-term implementation of the proposed Risk Management Regulatory Framework would include updating guidance documents to reflect the goal of risk-informed and performance-based defense-in-depth. Given the variety of hazards, scenarios, and uncertainties associated with possible exposures to radioactive materials from different types of NRC-licensed devices and facilities, it would be necessary to update or develop such guidance within each regulatory program.

3.5   Implement Decision

Decisions resulting from deliberations on specific issues or proposals address whether the defense-in-depth protections associated with the radiological hazard are adequate, can be relaxed, or need to be strengthened. The NRC implements its decisions within regulatory processes that include the following:

- preparing regulations and guidance

- reviewing proposed licensing actions

- performing environmental reviews

- executing its oversight programs

The actual defense-in-depth protections are implemented and maintained by licensees who process and control the radioactive materials or related facilities. Findings and recommendations for specific regulatory programs are discussed in Chapter 4.

3.6 Monitor

An important part of any decisionmaking model is monitoring and feedback to identify issues and to gauge the effectiveness of decisions and implementation actions. Evaluations provide important information about the effectiveness of the framework, the regulatory programs, and specific decisions. Evaluation tools include existing programs, such as operating experience programs, inspection and oversight programs, and interactions with various stakeholders. In some cases, the implementation of specific actions to address risks may include feedback provisions, such as reporting requirements or periodic assessments. In either case, evaluation plans should be built into the overall implementation plan to specify when evaluations will be conducted, who will conduct them, and what will be evaluated. New information may emerge during an evaluation that is sufficiently important to require repeating some parts of the risk management process. In such cases, the six-stage process would not be sequential; instead, it would require flexibility and iteration as important new information comes to light.

An area of tension related to NRC processes and decisions is the need to balance goals for regulatory stability and predictability with the desire to identify and address new information, especially information related to risks associated with regulated activities. The RMTF recommendations on periodic risk assessments of the hazards and risks associated with specific uses of source, byproduct, and special nuclear materials are provided in Chapter 4. Such periodic assessments would then support the risk management framework to help determine if changes to regulatory requirements or oversight were in order.

3.7 Communications

Communications with internal and external stakeholders is vitally important during the decisionmaking process. It is notable that all of the risk management frameworks developed by various groups and outlined in Appendix A highlight the importance of communications, engagement of stakeholders, and collaboration as part of the deliberative process.

The development and implementation of the proposed Risk Management Regulatory Framework will present a significant communication challenge within the NRC and with external stakeholders. Experiences associated with previous initiatives related to risk-informed, performance-based regulation demonstrated the importance of effective communication in explaining and promoting the changes being introduced. Some of these challenges and possible implementation strategies, including key communications, are discussed further in Chapter 5.

3.8   References

(National Research Council, 1996)   "Understanding Risk: Informing Decisions in a Democratic Society," National Academy Press, 1996.

(NRC, 2011)   U.S. Nuclear Regulatory Commission, "An Approach for Using Probabilistic Risk Assessment in Risk-Informed Decisions on Plant-Specific Changes to the Licensing-basis," Regulatory Guide 1.174, Revision 2, May 2011, Agencywide Documents Access and Management System (ADAMS) Accession No. ML100910006.

4. IMPLEMENTATION OPTIONS

4.1 Introduction

This chapter discusses how the proposed Risk Management Regulatory Framework described in Chapters 2 and 3 could be implemented by the NRC in its licensing and oversight activities. Since the historical approach to licensing and oversight of particular facilities using nuclear materials has varied, the actions that would need to be taken to implement the framework would also vary.

For each major category of NRC-regulated activity, including reactors, materials, waste disposal, uranium recovery, fuel cycle, spent nuclear fuel storage, and transportation, the associated section provides the following:

• Background on regulatory context.

• Risk Management Task Force (RMTF) findings and associated recommendations on the potential implementation of the proposed framework (for both safety and security activities). The findings are related to what the RMTF considers to be important gaps between how NRC-licensed materials, devices, and facilities are regulated today and the recommendations are related to how they would be regulated in the future using the proposed framework.

• Recommendations on implementation approaches.

In making its recommendations, the RMTF considered factors, such as who would be affected (the NRC, licensees, or both), the administrative mechanism that would be used for implementation (e.g., rulemaking), and the resources (time and cost) that would be required for implementation. Using these factors, the RMTF defined three implementation options:

Option A: Continue the current approach ("patchwork")

This option would direct few, if any, additional resources to the development and implementation of a broad risk management approach across NRC programs. The efforts related to ongoing risk-informed and performance-based initiatives and activities related to the followup to the Fukushima accident would continue on their current courses.

Option B: Implement the proposed Risk Management Regulatory Framework through selected guidance and rule changes

This option would emphasize specific rule and guidance changes to implement the proposed Risk Management Regulatory Framework. For specific rule changes, this option would maintain the basic structure of current regulations and oversight programs, but it would develop changes to or additions of specific regulations applicable to NRC licensees. An historical example of this approach is the development and issuance of 10 CFR 50.65, "Requirements for Monitoring the Effectiveness of Maintenance at Nuclear Power Plants" (NRC, 1991), which added a significant new requirement within the existing licensing and oversight programs. For guidance, changes would be largely aimed at changing how the NRC operates, but new requirements or expectations on licensees would not necessarily be imposed. An example of a successful effort developed and implemented through guidance is Regulatory Guide 1.174, "An Approach for

Using Probabilistic Risk Assessment in Risk-informed Decisions on Plant-specific Changes to the Licensing-basis" (NRC, 2011). Guidance-oriented activities likely would involve revisions or additions to Commission Policy Statements and guidance documents, such as regulatory guides, review plans, management directives, and office-level procedures. This option would require development of a plan describing actions to be taken and allocation of staff resources for the plan's execution.

Option C: Implement the proposed Risk Management Regulatory Framework through broad-scale regulatory framework changes

This option would involve a significant revision of the basic framework used for the licensing and oversight of a particular type of NRC-regulated activity. NUREG-1860, "Feasibility Study for a Risk-informed and Performance-based Regulatory Structure for Future Plant Licensing," described such possible structural changes to the licensing approach for nuclear power reactors (NRC, 2007). Significant planning and resource allocations would be required to implement this option.

The following sections discuss the implementation of the proposed Risk Management Regulatory Framework for various NRC programs and are organized as follows:

- Power Reactors (Section 4.2.1)

    o   Currently operating reactors licensed under the provisions of 10 CFR Part 50, "Domestic Licensing of Production and Utilization Facilities,

    o   "New" reactors licensed under the provisions of 10 CFR Part 52, "Licenses, Certifications, and Approvals for Nuclear Power Plants,"

    o   Generation IV reactors.

- Nonpower Reactors (Section 4.2.2)

- Materials (Section 4.3)

- Waste Disposal (Section 4.4)

    o   Low-Level Waste

    o   High-Level Waste

- Uranium Recovery (Section 4.5)

- Fuel Cycle (Section 4.6)

- Spent Fuel Storage (Section 4.7)

- Transportation (Section 4.8)

4.1.1 References

(NRC, 1991)   U.S. Nuclear Regulatory Commission, "Monitoring the Effectiveness of Maintenance at Nuclear Power Plants," Final Rule, published in the *Federal Register* on July 10, 1991 (56 FR 31306).

(NRC, 2007)   U.S. Nuclear Regulatory Commission, "Feasibility Study for a Risk-informed and Performance-based Regulatory Structure for Future Plant Licensing," NUREG-1860, Vol. 1, December 2007, Agencywide Documents Access and Management System (ADAMS) Accession No. ML080440170.

(NRC, 2011)   U.S. Nuclear Regulatory Commission, "An Approach for Using Probabilistic Risk Assessment in Risk-informed Decisions on Plant-Specific Changes to the Licensing-basis," Regulatory Guide 1.174, Revision 2, May 2011, ADAMS Accession No. ML100910006.

4.2   Nuclear Reactors

4.2.1   Power Reactors

4.2.1a Background

The historical basis for power reactor licensing and oversight is described well elsewhere (Walker, 2010; Okrent, 1981) and will not be discussed extensively here. In brief, power reactors in the United States have been licensed using 10 CFR Part 50.[1] Implementation of 10 CFR Part 50 has been achieved, for the most part, using deterministic methods and acceptance criteria. From a safety perspective, a set of licensing-basis events was established that was intended to ensure conservatism in design and protection from a wide spectrum of postulated events, up to and including design-basis accidents (DBAs). Appendix F provides additional information on this spectrum of postulated events. These postulated accidents are highly stylized and generally do not consider multiple failures of safety systems. Qualitative approaches for ensuring reliable safety systems, such as the single failure criterion, were implemented. Testing plans and operational limits were established in technical specifications to ensure that, if called upon, safety systems would perform as intended. In addition to the failure or malfunction of plant equipment, the challenges to and protections of nuclear power plants from various external hazards are described in plant licensing documents (e.g., final safety analysis reports (FSARs), Chapter 2). External hazards are events that could initiate a variety of plant transients while also challenging one or more of the barriers provided to mitigate or contain potential releases of radioactive materials.

One key concept that emerged early in the licensing history of nuclear power reactors was "defense in depth." This concept was developed and applied to compensate for the recognized lack of complete knowledge of nuclear reactor operations and the consequences of potential accidents. Although both reactor operation experience and the knowledge base of potential (and actual) consequences have grown considerably since that time, the Risk Management Task Force (RMTF) concludes that the concept continues to be highly relevant today. That is, even with this increased experience, there continues to be knowledge gaps and uncertainties[2] that can be addressed by a defense-in-depth approach. Additional discussion of defense-in-depth is provided in Appendix C.

A question that arises when considering changes to nuclear reactor regulations and the categorization of events or equipment is if requirements are needed for adequate protection of the health and safety of the public. The NRC establishes regulatory requirements for protecting public health and safety and common defense and security using a two-tier structure. This two-tier structure is consistent with the statutory requirements in the Atomic Energy Act of 1954, as amended. The top tier consists of those requirements needed to ensure adequate protection of public health and safety, and to be in accordance with the common defense and security. The adequate protection standard is also described in terms of ensuring that licensed activities pose no undue risk to public health and safety or are not inimical to common defense and security. In a case brought before the Court of Appeals for the D.C. Circuit (Union of Concerned Scientists

---

1   All currently operating power reactors have been licensed under 10 CFR Part 50. The newest currently operating reactor began operations in 1996. The operating license for Watts Bar Unit 2 (and potentially Bellefonte, Unit 1) also are being pursued using the provisions of 10 CFR Part 50.

2   The conservatism embedded in reactor design by the DBAs, external hazard analyses, and defense-in-depth is also intended to protect the facility from accidents or conditions that the NRC has not thought of, the so called "unknown unknowns."

v. NRC, 824 F.2d 108, 120 (D.C. Cir. 1987)), the Court recognized that adequate protection (no undue risk) does not equate to "zero risk." The Court stated:

> Similarly, under the adequate-protection standard of section 182(a), the NRC need ensure only an acceptable or adequate level of protection to public health and safety; the NRC need not demand that nuclear power plants present no risk of harm.

Measures taken to prevent, contain, and mitigate events or concerns in this tier historically correspond to the traditional design-basis accidents described in Appendix F and several programmatic requirements for dealing with beyond-design-basis events (e.g., emergency planning and loss of large areas due to fires or explosions). NRC requirements to address concerns of adequate protection are developed and imposed without consideration of cost.

The second tier of the NRC safety structure was described by the Court as follows:

> If it so desires, however, the Commission may impose safety measures on licensees or applicants over and above those required by section 182(a)'s adequate-protection standard. As we have noted, section 161 of the Act empowers the Commission to issue rules, regulations, or orders to "protect health or to minimize danger to life or property." 42 U.S.C. Sec. 2201(b), (i). This section cannot be read simply to permit the Commission to provide adequate protection; another section of the Act requires the Commission to do that much. We therefore must view section 161 as a grant of authority to the Commission to provide a measure of safety above and beyond what is "adequate." The exercise of this authority is entirely discretionary. If the Commission wishes to do so, it may order power plants already satisfying the standard of adequate protection to take additional safety precautions. When the Commission determines whether and to what extent to exercise this power, it may consider economic costs or any other factor. The Commission, after all, need not exercise the authority granted by section 161 at all; given this fact, the Commission certainly may use cost-benefit analysis to decide whether exercising the authority conferred by section 161 makes economic or policy sense.

In the development of its findings and recommendations, the RMTF considered how some requirements would be associated with the first-tier mission of the NRC to ensure adequate protection of public health and safety and the common defense and security, and how other requirements would address the second tier by going beyond adequate protection in an attempt to further reduce risk. As described by the Court of Appeals and in longstanding NRC regulations and guidance documents, the requirements established to reduce risk (beyond measures needed for adequate protection) will not attempt to eliminate all risk but will instead pursue reasonable reductions. An evaluation of the costs and benefits of proposals falling within the second tier has been used as part of the determination of what is a reasonable requirement to reduce risks to public health and safety and the common defense and security.

The RMTF considered the two-tier safety structure discussed above, along with the risk management goal described in Chapter 2, when preparing its findings and recommendations. The risk management goal is to:

Provide risk-informed and performance-based defense-in-depth protections to:

- Ensure appropriate barriers, controls, and personnel to prevent, contain, and mitigate exposure to radioactive material according to the hazard present, the relevant scenarios, and the associated uncertainties; and

- Ensure that the risks resulting from the failure of some or all of the established barriers and controls, including human errors, are maintained acceptably low.

Taken together, the above criteria for risk-informed and performance-based defense-in-depth address both tiers of the safety structure. However, the first criterion does not align with the first tier (adequate protection) by itself and the second criterion does not align with the second tier (additional protections) by itself. Instead, a specific design feature or operating control can be seen as addressing one or more of the factors in the risk management goal. The nature and importance of the protection provided will determine whether it serves to ensure adequate protection or provides additional protections to further reduce risk to life or property.

The above observation on defense-in-depth is fully consistent with the general findings and recommendations in Chapter 2. However, as mentioned in earlier parts of this report, experience has also identified limitations with the traditional approaches, particularly in supporting specific decisions on what are reasonable and sufficient levels of defense-in-depth for particular hazards. As described in Recommendation 2.3, the RMTF recommends that a balanced approach, considering traditional and risk assessment techniques, is needed to identify appropriate barriers and controls.

Probabilistic risk assessment (PRA) was first used in the early 1970s for nuclear power reactors (NRC, 1975), after the designs of essentially all operating plants were fixed. The NRC uses risk assessment in a way that recognizes the particular strengths of this approach and complements the traditional, more deterministic, approach. That is, risk assessment provides:

- A systematic approach for assessing the wide variety of hazards that can challenge the safety of an NRC-licensed facility.

- A logical method for characterizing the capability of a facility's design and operation to respond to the identified hazards.

- A method for estimating the consequences of combinations of hazards and unsuccessful responses to these hazards (thousands of realistic accident sequences are investigated, in contrast to the limited number of stylized accidents considered in the traditional approach).

- A model for assessing the occurrence frequencies of these hazards, the probabilities of unsuccessful response (with the associated design or operational failures), and the consequences of the unsuccessful responses. These frequencies and probabilities can be estimated quantitatively.

- Rankings of accident sequences and of systems, structures, and components (SSCs) according to their contributions to risk.

- Valuable input to risk management through the rankings of accident sequences and SSCs that allow the allocation of resources to be focused on what really matters to risk.

Proposed new power reactor designs are light-water reactors similar in many respects to operating power reactors. New reactor designs are being licensed using the various subparts of 10 CFR Part 52, "Licenses, Certifications, and Approvals for Nuclear Power Plants," which in all technical areas essentially references 10 CFR Part 50. That is, new power reactors are licensed using the same set of licensing-basis events, including design-basis events that include assumptions such as the single failure criterion. Recognizing the value of PRA in providing a complementary view of safety, each new design is required to have a PRA, and any plant referencing that design is required to update the PRA by the time of fuel loading to reflect the final design and operational characteristics. As envisioned in the NRC's Severe Accident Policy Statement (NRC, 1985), new reactors must satisfy a regulatory requirement in 10 CFR Part 52 to include severe accident features in their designs. In addition, an analysis is required for new reactor designs to ensure that critical safety functions would be maintained following the impact of a large commercial aircraft. In general, the designs of these new power reactors are fixed and approved through an NRC-issued design certification, although none is in actual operation.

A subset of new reactors being pursued by the nuclear industry involves small modular designs that introduce some new features, compared to traditional large reactor designs. The small modular reactors (SMRs) currently being developed are integral pressurized-water reactors (iPWRs) with major components, such as steam generators, control rod drives, and a pressurizer contained within the reactor vessel, thereby eliminating the need for most reactor coolant piping. The power levels for SMRs generally are less than 300 megawatt-electric (MWe), and passive safety features are incorporated into the designs. Potential applicants and the NRC staff are planning on a licensing approach for iPWRs that is generally consistent with that used for larger new reactors with passive safety features, such as the Westinghouse AP1000. Some applicants are considering the two-phase 10 CFR Part 50 licensing process (i.e., a construction permit and operating license), but site-specific licensing would be linked to 10 CFR Part 52 design certifications for subsequent deployments. Therefore, CFR 10 Part 52 provisions mentioned above would be addressed for SMRs. Some iPWR vendors have expressed an interest in performing a Level 3 PRA. However, it is not clear how these efforts would be incorporated into the overall licensing approach.

Generation IV power reactor designs, including, for example, the Next Generation Nuclear Plant (NGNP), are being considered by designers, although none has yet been submitted officially for NRC review. These designs vary considerably from the current set of operating reactors and the proposed new reactor designs, including the consideration of coolants such as helium, liquid sodium, and molten salt. Reflecting these design differences, discussions have been held with the NRC on using alternative regulatory approaches, such as using event frequency information, to define the set of design-basis accidents to be analyzed. The designs of these advanced reactors are still conceptual.

The RMTF has considered the history and process for licensing and oversight of power reactors in the context of the risk management framework and has developed a set of findings and recommendations. The findings relate to what the RMTF has identified to be important gaps between how power reactors are regulated today and how they would be regulated in the future using the proposed Risk Management Regulatory Framework. The RMTF assessed each power reactor class (operating, new and SMR, and Generation IV) individually and found many similarities with respect to the findings. Implementation recommendations for each class are provided in the following sections of this report.

Design-basis Accident Licensing Approach

Methods used to certify new designs include the set of DBAs and qualitative reliability approaches developed more than 30 years ago, although more modern approaches exist and are used extensively in, for example, the oversight of operating reactors. Some licensees have used PRA to improve operational flexibility, even with the current set of DBAs and designs established long ago. Further, it appears that both licensees and the NRC benefit from the clarity that the DBA approach provides, making interactions between licensees and the NRC more efficient.

The RMTF examined power reactor regulatory requirements from a broad and strategic perspective. From this perspective, the RMTF observes a "within" design basis part of the power reactor regulatory framework that uses, among other things: qualitative approaches for ensuring system reliability (e.g., the single failure criterion) when more modern quantitative approaches exist; stylized considerations of human performance (e.g., operators are assumed to take no actions within, for example, 30 minutes of an accident's initiation); and a set of accidents not reflective of operating experience and modern understanding (e.g., the now-understood downside risks of automatic containment spray actuation in pressurized-water reactors with large dry containments). The continued use of the DBAs established over 30 years ago runs counter to the NRC Principles of Good Regulation that regulations (and, by extension, DBAs) should be based on the best available knowledge from research and operational experience. The longstanding use of the same design-basis events has, however, helped maintain stability in regulatory processes, which is another aspect of the NRC's Principles of Good Regulation. The RMTF, therefore, offers the following finding and recommendation for power reactors:

> Finding PR-F-1: The concept of design-basis events and accidents continues to be a sound licensing approach, but the set of design-basis events and accidents has not been updated to reflect insights from power reactor operating history and more modern methods, such as PRA.

> Recommendation PR-R-1: The set of design-basis events and accidents should be reviewed and revised, as appropriate, to integrate insights from power reactor operating history and more modern methods, such as PRA.

Beyond-design-basis Accidents

Several rules have been issued to address accidents identified in PRAs as having significant risks not recognized in or enveloped by the set of DBAs. An important example is 10 CFR 50.63, the station blackout rule (NRC, 1988). However, these events were not added to the set of DBAs, and a consistent framework was not developed to address them. As a result, a "patchwork" of agency guidance was established to address beyond-design-basis accidents. The Fukushima Near-Term Task Force (NTTF) report (NRC, 2011a) included a number of observations on the results of this ad hoc approach and recommended that "a logical, systematic, and coherent regulatory framework for adequate protection that appropriately balances defense-in-depth and risk considerations" be established.

One aspect of the NRC's consideration of beyond-design-basis accidents has been the use of voluntary industry initiatives. For example, in the 1990s, the NRC and power reactor licensees developed and implemented accident management guidelines, the purpose of which was "to enhance the capabilities of the licensee's emergency response organization to prevent and mitigate severe accidents and minimize any offsite releases" (NRC, 1996). As explained in that reference (SECY-96-088), the staff "accepted the industry commitment to implement A/M [accident management] at each NPP [nuclear power plant] pursuant to a formal industry position on A/M in lieu of pursuing other actions for obtaining improvements in industry A/M capabilities ...," such as incorporating requirements into a regulation or operating licenses. In 2011, the Fukushima NTTF noted that "NRC inspection programs give less attention to beyond-design-basis requirements and little attention to industry voluntary initiatives since there are no requirements to inspect against." During discussions between the RMTF and the NRC staff, the extent to which such voluntary initiatives could be credited in risk assessments (which are intended to be realistic) arose on several occasions. More specifically, the extent to which licensee activities undertaken in such initiatives can be credited has been a source of contention in the Significance Determination Process, which is part of the Reactor Oversight Process.

The Western European Nuclear Regulators' Association (WENRA) prepared a document that included discussion of "design extension" analysis (WENRA, 2008).[3] This analysis was intended to achieve the following:

> Examine the performance of the plant in specified accidents beyond the design basis, including selected severe accidents, in order to minimize as far as reasonably practicable radioactive releases harmful to the public and the environment in cases of events with very low probability of occurrence.

The document indicated that:

> [B]eyond design basis events shall be selected (based on a combination of deterministic and probabilistic assessments as well as engineering judgment) and considered in the safety analysis to determine those sequences for which reasonable practicable preventive or mitigative measures can be identified and implemented, and that realistic assumptions and modified acceptance criteria may be used for the analysis of the beyond design basis events.

---

3   WENRA is a network of Chief Regulators of European Union countries with nuclear power plants, and Switzerland, as well as of other interested European countries, that have been granted observer status. The main objectives of WENRA are to develop a common approach to nuclear safety, to provide an independent capability to examine nuclear safety in applicant countries, and to be a network of chief nuclear safety regulators in Europe exchanging experience and discussing significant safety issues.

The document also listed a set of events that, at a minimum, needed to be considered, including station blackouts. WENRA and other organizations continue to promote the adoption of standardized approaches, including the "design extension" category, throughout the international nuclear community (IAEA, 2010).

As discussed in several appendices to this report, the inclusion of a design-enhancement[4] category in the United States would result in the following framework for design-basis and beyond-design-basis events:

- Design-basis Events

  o   normal operation

  o   anticipated operational occurrences

  o   design-basis accidents

  o   design-basis external hazards

- Beyond-design-basis Events

  o   design-enhancement events

    –   internal events

    –   external hazards

  o   residual risk scenarios

    –   internal events

    –   external hazards

Design-basis events traditionally have been associated with mechanistic analyses, reliance on and protection of safety-related equipment, and the establishment of technical specifications and other licensing-related operational controls that have been deemed necessary for adequate protection of public health and safety. Beyond-design-basis events have more often included the use of best-estimate type analyses, PRAs, and in establishing additional plant protections to further reduce risks to the public health and safety and common defense and security (e.g., additional protections for station blackout conditions and aircraft impacts).

Two key concepts related to the identification of relevant scenarios, categorization, and subsequent design features and operating limits are 1) the threshold to define when a scenario needs to be considered within a category and 2) the acceptance criteria to define when a design feature or operating limit provides the desired protection from the defined scenario(s). The RMTF considered the two-tier structure described by the U.S. Court of Appeals (i.e., adequate protection and additional protections), as well as the risk-informed and performance-based defense-in-depth concept used in the risk management goal and developed Figure 4.2-1, which represents a general regulatory framework for nuclear power reactors.

---

4   The RMTF has chosen to use the term "design-enhancement" rather than "design extension."

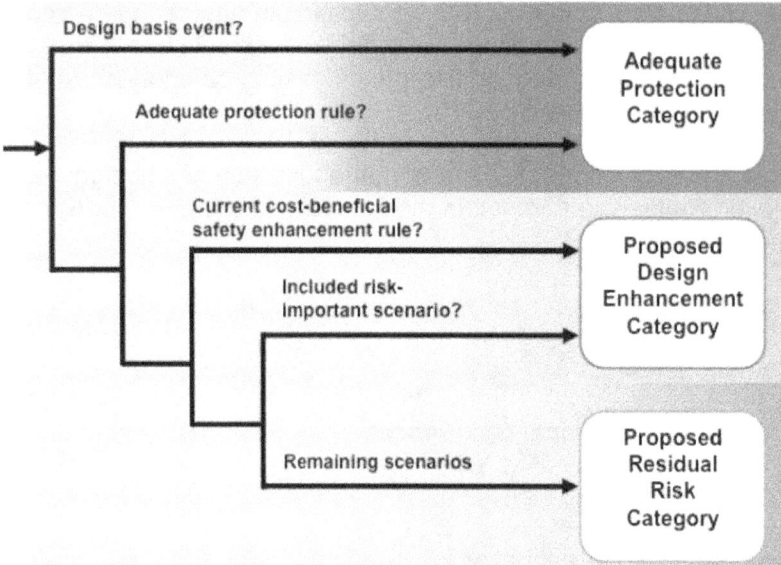

Figure 4-1 Regulatory Framework for Nuclear Power Reactors

As shown in Figure 4-2-1, the RMTF agrees with WENRA, the International Atomic Energy Agency (IAEA), and the Fukushima NTTF recommendation and proposes that a design-enhancement category be established for beyond-design-basis accidents. The RMTF has defined a number of desirable characteristics of the design-enhancement category. These characteristics reflect consideration of current agency practices, including those associated with cost-beneficial safety enhancements. The desirable characteristics are listed below.

- Consistency of regulatory approach. The RMTF recommends that the design-enhancement category be established by rulemaking to ensure appropriate regulatory controls, including, for example, change control and reporting requirements. Well-defined regulatory requirements would also clarify reliability and operational expectations and oversight and significance determination process considerations.

- Risk as safety measure. Where practical, PRA should be the tool used to measure the safety importance of accident sequences, SSCs, human actions, etc., in this category, and the potential safety improvements or relaxations associated with potential plant modifications not needed to ensure adequate protection. In the long term, these PRAs should be developed on a site-specific basis and be "full scope" and include all modes of operations and all initiating events (internal and external). They should be carried out to include offsite health consequences and should include uncertainty analyses. Toward this end, the approach defined in 10 CFR 50.71(h), which ties current PRA requirements to the availability of consensus standards, should be applied.

- Focus on performance. Regulations and other regulatory programs established in this category should be performance-based, using the guidance provided by the Commission in 1999 and discussed in Appendix D. Performance monitoring should include periodic reassessments of risk, considering new information from operating experience, research, and other sources. Toward this end, the approach defined in 10 CFR 50.71(h), which specifies that new reactors will perform a reassessment of risks (using PRAs meeting NRC-endorsed codes and standards) every 4 years, should be applied together with ad hoc reassessments with the development of new, potentially significant information (e.g., new seismic hazard information). Furthermore, a concept similar to the reliability assurance program (RAP) for new reactors should be developed to support the performance-based approach for SSCs identified as important in scenarios in this category. The SSCs included in RAP are identified as being risk-significant using a combination of deterministic, probabilistic, and other analysis techniques. Inclusion of SSCs in the RAP and the related controls and monitoring are intended to ensure that the subject SSCs are designed, constructed, and operated in a manner consistent with the supporting technical analyses (including reliability and availability).

* Consideration of costs. The Backfit Rule establishes the basis for the NRC's consideration of costs in reactor regulatory decisionmaking. Implementation of regulations and other regulatory programs included in the design-enhancement category could reflect an "as low as is reasonably achievable" assessment with cost–benefit considerations similar to the Backfit Rule and other NRC regulatory analyses included in the determination of what is "reasonable."

* Implementation on a site-specific basis. Regulations and other regulatory programs included in the design-enhancement category should be implemented, to the maximum extent practical, considering site-specific design and operations, as reflected in a site-specific PRA.

Finding PR-F-2: Requirements for beyond-design-basis accident scenarios (e.g., station blackout) were established at different times and in different ways. Differences in implementation approaches have reduced the efficiency and consistency of the NRC's regulatory and oversight activities.

Finding PR-F-3: The extent to which licensee activities undertaken as part of voluntary industry initiatives can be credited has been a source of contention in the Reactor Oversight Process and has reduced the efficiency of that process.

Recommendation PR-R-2: The NRC should establish through rulemaking a *design-enhancement category* of regulatory treatment for beyond-design-basis accidents. This category should use risk as a safety measure, be performance-based (including the provision for periodic updates), include consideration of costs, and be implemented on a site-specific basis.

The addition of a design-enhancement category and related requirements for identifying events and developing measures to keep risks as low as is reasonably achievable or below an established level may change the landscape currently defined by design-basis events and several selected conditions (e.g., station blackout, ATWS, and aircraft impact for new reactors). This change may, in turn, enable changes to the current handling of design-basis events by providing a defined regulatory program for items that may not warrant the full requirements currently associated with design-basis events but that warrant some regulatory measures to manage the risks associated with off-normal scenarios. A review of the existing design-basis events and the overall combination of events within all categories could be performed to determine whether some elements of the current design-basis events could be better addressed within the design-enhancement category. Such a review and subsequent changes to NRC guidance documents could serve as a way to implement Recommendation PR-R-1. Any relocations from the design-basis categories to the design-enhancement category would likely support a more performance-based way to ensure risk-informed and performance-based defense-in-depth.

External Hazards

NRC's licensed facilities are subject to a wide variety of safety challenges from naturally occurring hazards. The traditional licensing approach for power reactors establishes what were perceived as conservative bounds on the magnitude of particular hazards using concepts such as design-basis floods and safe shutdown earthquakes. The extent of conservatism varies. The design basis for earthquakes might be associated with lower frequencies than, for example, external floods. Although new information on the frequencies and magnitudes of these hazards has been obtained, this information has not been routinely evaluated and communicated.

As stated above, the NRC's approach to addressing the risk from naturally occurring hazards varies among the hazards. Methods to address earthquakes are the most advanced, perhaps because this hazard is judged to be of the highest risk significance. Information on earthquake frequencies is periodically updated. Even for this hazard, however, the risk assessment expertise within the NRC and the industry is very limited. In addition, studies have shown that there are serious limitations in attempting to address seismic hazards or other natural events (including related uncertainties) at extremely low frequencies (Johnson and Apostolakis, 2012).

> Finding PR-F-4: The processes for establishing the external hazard design bases do not use consistent event frequency and magnitude methods.
>
> Finding PR-F-5: New information that would provide the basis for external hazard frequency updates is not systematically collected, evaluated, and communicated.
>
> Finding PR-F-6: PRA methods for assessing external hazard risks are available, but expertise in performing such studies is very limited. Uncertainty analyses and the recognition of the limitations of available scientific knowledge are a key element of these methods.

Recommendation PR-R-3: The NRC should reassess methods used to estimate the frequency and magnitude of external hazards and implement a consistent process that includes both deterministic and PRA methods. Consideration of the risks from beyond-design-basis external hazards should be included in the design-enhancement category described in Recommendation PR-R-2.

Recommendation PR-R-4: The NRC should establish a program to systematically collect, evaluate, and communicate external hazard information.

Defense-in-depth

The term defense-in-depth traditionally has been used in the context of power reactor safety. As discussed in Appendix C, a number of different descriptions of this concept exist. For this and other reasons, the RMTF is recommending the risk-informed and performance-based defense-in-depth characterization discussed in Chapters 2 and 3.

Methods for quantitatively assessing risk have been developed for power reactors to a much greater degree than for other NRC-regulated activities. Recognizing this, the RMTF has expanded upon the Chapter 2 risk-informed and performance-based defense-in-depth philosophy to create possible defense-in-depth guidance specific to power reactors. In addition, the RMTF has extended this philosophy to considerations of power reactor security.

The more quantitative RMTF characterization of risk-informed and performance-based defense-in-depth for power reactors is shown in italics below:

1. Establish appropriate barriers, controls, and personnel to prevent, contain, and mitigate exposure to radioactive material according to the hazard present, the relevant scenarios, and the associated uncertainties.

    a. *Each barrier is designed with sufficient safety margins to maintain its functionality for relevant scenarios and account for uncertainties.*

    b. *Systems needed to ensure a barrier's functionality are designed to ensure appropriate reliability for relevant scenarios.*

    c. *Barriers and systems are subject to performance monitoring.*

2. Ensure that the risks resulting from the failure of some or all of the established barriers and controls, including human error, are maintained acceptably low.

Within the above construct of risk-informed and performance-based defense-in-depth for power reactors, safety margins refer to conservatisms added to ensure that plants and specific barriers are designed and constructed so that failures are not expected until key parameters well exceed the values assumed in the supporting engineering evaluations. Safety margins usually derive from the traditional approach to design-basis accidents, but they can be informed by risk assessment techniques. Measures to address the reliability of barriers and supporting systems have increasingly been introduced to the regulatory process for power reactors (e.g., maintenance rule and reliability assurance programs for new reactors), but additional improvements for establishing and monitoring reliability goals could be developed for some equipment considered important to safety (e.g., equipment used in response to the loss of large areas due to fires or explosions). The improvements related to reliability have resulted largely from risk assessments and their use in programs such as the Reactor Oversight Program.

An important component of the risk-informed and performance-based defense-in-depth approach for power reactors is the consideration of relevant scenarios in which some or all of the established barriers and controls are challenged or fail. PRAs can be used to evaluate the risk profile of a plant to ensure that the design and operating practices satisfy the NRC's safety goals for nuclear power reactors. In addition, PRAs and other evaluations can be used to identify potential cliff-edge effects in which failing a barrier or exceeding a design value (e.g., flood level) would lead directly to core damage and a release of radioactive material to the environment. Finally, if the risk from some sequences involving failures of defense-in-depth protections is significantly lower than what is "acceptably low," the possibility of relaxing some of these protections could be considered.

> Finding PR-F-7: The availability and broad-scale use of quantitative risk assessment methods (PRA) for power reactors provide an opportunity for a more quantitative characterization of defense-in-depth.

> Recommendation PR-R-5: The NRC should apply the risk-informed and performance-based defense-in-depth concept to power reactors in a more quantitative manner.

Security

The NRC's security requirements are established to protect the same radioactive hazards as safety requirements. However, the regulatory approach used is somewhat different.[5] In discussions with NRC staff, the RMTF found that differences in language and methods exist that reduce the efficiency of the NRC's interactions with licensees. The proposed RMTF characterization of risk-informed and performance-based defense-in-depth, applied to security regulatory activities, could help to remedy this inefficiency.

With respect to language, the NRC security staff endeavors to blend the language of the traditional security concepts with the agency's safety concepts, including defense-in-depth. While this approach has been used in recent licensing reviews, language differences may reduce efficiency.

With respect to methods, there are similarities between risk assessments and security vulnerability assessments. The latter must deal with a difficult challenge—estimating the

---

5   The traditional approach taken for security is somewhat analogous to the deterministic approach to external hazards such as flooding. In the case of security, the equivalent to a maximum water level to protect against is the "design-basis threat," which defines the number and capabilities of the attacking force.

frequency of the threat (the "initiating event," in safety terms) and the fact that the threat changes continuously and may actively seek to identify and exploit weaknesses in the established barriers.

Risk assessment methods may provide opportunities for addressing security-related vulnerabilities, as early risk studies identified important accident scenarios (e.g., interfacing systems loss-of-coolant accidents) not fully captured by consideration of design-basis events and accidents.

In the last decade, considerable research has been performed on estimating security risks, much of it sponsored by the U.S. Department of Homeland Security. Tools have been developed such as ITRA (Integrated Chemical, Biological, Radiological, and Nuclear Terrorism Risk Assessment Threat Estimation), a program aimed at providing an integrated quantitative assessment of the relative risk associated with chemical, biological, radiological, and nuclear terrorism to the homeland, and RAPID (Risk Assessment Process for Informed Decisionmaking), a program aimed at developing a strategic-level process to gauge future risks across the full range of DHS responsibilities" (DHS, 2010). Some of these methods focus on the display of information to decisionmakers, including "break-even analysis," a variant of cost–benefit analysis that estimates the threshold value at which a policy alternative's costs equal its benefits (DHS, 2010). While much remains to be accomplished in assessing security risks, the NRC should actively review and implement methods improvements for its own purposes and for ensuring that agency and licensee resources are focused on the most important issues.

Finding PR-F-8: Vulnerability assessments performed to assess security have important similarities in scope (e.g., facility equipment and radioactive hazards considered) and methods to risk assessments.

Finding PR-F-9: Differences in regulatory language and approaches between power reactor security and safety regulation may have reduced the efficiency and effectiveness of the NRC's work.

Finding PR-F-10: In the past decade, considerable research has been performed on estimating security risks, much of it sponsored by the U.S. Department of Homeland Security.

Recommendation PR-R-6: The NRC should develop and implement guidance for use in its security regulatory activities that uses a common language with safety activities and harmonizes methods with risk assessment and the proposed risk-informed and performance-based defense-in-depth framework.

4.2.1.1   Implementation Options—Operating Power Reactors

Background

The NRC's current policy on the use of risk information (NRC, 1995) has been implemented to a greater extent for operating power reactors than for other facilities or devices regulated by the agency. Guidance on backfitting requirements for these reactors (NRC, 2004), allowing changes to their licensing-basis (NRC, 2011), and defining acceptable levels of risk from their operation (NRC, 1986), combined with focused attention on establishing sophisticated, standardized, quantitative risk assessment methods (ASME/ANS, 2009), has created an environment that should permit a more timely implementation of the risk management framework described in Chapters 2 and 3, including the risk-informed and performance-based defense-in-depth concept.

In practice, the RMTF recognizes that operating reactors are, for the most part, well into the operations phase of their life cycle, meaning that their fundamental design was established many years ago and their licensing-basis occasionally has been updated to include the "patchwork" of additional requirements, such as station blackout and anticipated transients without scram.

With this background, the RMTF has examined its power reactor findings and recommendations and has developed several alternatives for implementation for operating reactors. These alternatives are discussed further in Appendix H and within the specific recommendations that follow.

Design-basis Accident Licensing Approach

As discussed above, the RMTF has found that the continued use of design-basis accidents established more than 30 years ago runs counter to one aspect of the NRC's Principles of Good Regulation—that regulations should be based on the best available knowledge from research and operational experience. However, this practice is consistent with another aspect of the NRC's principles—the value of stability in nuclear operational and planning processes. For operating reactors, the RMTF concludes that stability currently outweighs the potential value of changes to the design-basis accidents. However, should a new set of design-basis accidents be established for new reactor designs or an approach is developed by licensees or owners groups, the NRC staff should be amenable to its use by operating reactors at some future time.

> Recommendation OR-R-1: For operating reactors, the establishment of the design-enhancement category can be followed by a review of design-basis events and accidents and related revisions to anticipated operational occurrences and DBAs to integrate insights from operating history and more modern methods. The NRC need not impose such a requirement, but it should be amenable to related industry initiatives should they pursue revisions to design-basis events based on the introduction of the design-enhancement category.

Beyond-design-basis Accidents

In Appendix H, the RMTF defined a number of alternatives that could be used to implement the creation of the design-enhancement category. These alternatives included 1) the NRC defining additional specific events or conditions (including acceptance criteria) to be addressed for power reactors, 2) the NRC requiring licensees to perform risk assessments to identify design-enhancement events exceeding a defined threshold and reducing risks to levels as low as is reasonably achievable (ALARA), and 3) the NRC requiring licensees to perform risk assessments to identify design-enhancement events exceeding a defined threshold and reducing risks to levels below defined acceptance risk criteria. The RMTF does not recommend any of these alternatives, but provides them to assist in staff implementation.

> **Recommendation OR-R-2:** For operating reactors, the RMTF recommends that the NRC should establish through rulemaking a design-enhancement category of regulatory treatment for beyond-design-basis accidents.

External Hazards

As discussed above, the RMTF has found that the NRC's consideration of external hazards varies in the establishment of the design basis and the assessment of their risk. Although the Fukushima accident may not be directly relevant to all U.S. power reactors, it did highlight that the potential safety significance of external hazards needs to be addressed more systematically for U.S. reactors.

> **Recommendation OR-R-3:** The NRC should reassess the methods used to estimate the frequency and magnitude of external hazards and implement a consistent process that includes both deterministic and PRA methods. For operating reactors, the RMTF recommends that the design-enhancement category rulemaking include consideration of external hazards.
>
> **Recommendation OR-R-4:** For operating reactors, the RMTF recommends that the NRC develop and implement guidance for the collection and dissemination of external hazard information.

Defense-in-depth

As discussed above, operating power reactors have been the subject of sophisticated risk assessment analyses and currently have an infrastructure of methods and a large record of experience (both positive and negative). As such, an opportunity exists to expand the characterization of risk-informed and performance-based defense-in-depth to be more quantitative. This expansion could build upon and make use of existing performance monitoring approaches, such as the maintenance rule, reliability assurance programs, and the ROP's Mitigating System Performance Indicator (MSPI).

> Recommendation OR-R-5: The NRC should apply the risk-informed and performance-based defense-in-depth concept to power reactors in a more quantitative manner. For operating reactors, the RMTF recommends that this recommendation be implemented in the form of guidance to the NRC staff and in future requirements established for operating reactor licensees.

Security

As discussed above, the efficiency of the NRC's activities is reduced by differences in terminology and methods between security and safety assessments. Advancements being made by other organizations (DHS) may help remedy this inefficiency. Use of the RMTF characterization of risk-informed and performance-based defense-in-depth in both safety and security activities also could help.

> Recommendation OR-R-6: For operating reactors, the RMTF recommends that guidance be developed and implemented to better harmonize terminology and methods used for reactor safety and security.

4.2.1.2   Implementation Options—New Power Reactors

Background

The consideration of risk insights for new power reactors has built upon the requirements and initiatives associated with operating reactors. New reactors, for example, are required to have and update PRAs. Insights from these PRAs are used for activities such as the improvement of plant designs and the establishment of reliability assurance programs. New plants also incorporate severe accident mitigation features to provide some protection against scenarios involving significant core damage. The RMTF offers the following recommendations for new reactors, including both large and small light-water reactor designs.

Design-Basis Accident Licensing Approach

As with operating power reactors, the continued use of traditional DBAs requires consideration of both the use of the best available knowledge and regulatory stability. Given that the designs and regulatory reviews for new large light-water reactor designs are completed or well underway, the RMTF concludes that they should be treated similar to operating reactors. However, the NRC should be amenable to a more risk-informed approach to DBAs if it is proposed for a future design of a large light-water reactor. The iPWR designs being developed are currently at the conceptual stage and a certain degree of adjustment to DBAs will be necessary as a result of expected design features (e.g., no large reactor coolant piping). Designers and potential licensees want to move relatively quickly to complete the designs and begin the design and licensing reviews. Their current plans and schedules would be somewhat complicated by a significant change to the DBA requirements. The NRC should, however, be amenable to, and even promote, a more risk-informed and performance-based approach for iPWR DBAs, to the degree it could be done without unduly increasing the schedules or resources needed for design development and review.

> **Recommendation NR-R-1:** For new reactors, the RMTF recommends that the NRC be amenable to and promote, where practical, the adoption of more risk-informed and performance-based approaches for the selection of more relevant scenarios for design-basis events. Changes pursued for operating reactors (OR-R-1) should also consider applicability to new reactors.

Beyond-design-basis Accidents

There has been some consideration of beyond-design-basis accidents in the licensing of new reactors in terms of assessing PRA results, defining the regulatory treatment of non-safety systems for passive reactor designs, and establishing reliability assurance programs. The treatment of established events, such as station blackout and anticipated transients without scram, are handled basically the same way as they are for operating reactors, except where actual design differences warrant specific treatment (e.g., reduced reliance on alternating current power systems by the passive reactor designs). The similarities between new and operating reactor treatment of beyond-design-basis accidents result in a similar recommendation:

> **Recommendation NR-R-2:** Apply Recommendation PR-R-2 (design-enhancement category) to new reactors.

It should be recognized, however, that differences in plant designs and risk profiles are already reflected in the licensing and oversight processes for new reactors. For example, vendors and applicants have identified non-safety-related equipment to be included in reliability assurance programs based, in part, on PRA insights. Many new reactor designs are also subject to review by regulators in other countries, which may have defined design-enhancement requirements. For these reasons, as well as the existing requirements for new plant designs and facilities to have PRAs, the RMTF foresees easier implementation of this recommendation for new plants than might be anticipated for some operating plants.

External Hazards

The licensing of new reactors has benefited from updates to the guidance related to specific external hazards. However, the same inconsistency in handling various external hazards that was discussed for operating reactors is also applicable to new reactors. This leads the RMTF to suggest the same recommendations for new reactors as those proposed for operating reactors.

> Recommendation NR-R-3: Apply Recommendation PR-R-3 (include external events in design-enhancement category) to new reactors.
>
> Recommendation NR-R-4: Apply Recommendation PR-R-4 (periodically evaluate new information regarding external hazards) to new reactors.

Defense-in-depth

The new reactor designs have benefited from various studies and discussions of defense-in-depth. One example is efforts to incorporate design features to address the Commission's guidance in the Advanced Reactor Policy Statement (NRC, 2008) and a general reduction in the estimated risks associated with new reactor designs compared to operating reactors. Although the new reactor designs are somewhat improved in considering the features of risk-informed and performance-based defense-in-depth, adoption of the relevant recommendation from operating reactors could provide further improvements in areas such as the identification and resolution of cliff-edge effects.

> Recommendation NR-R-5: Apply Recommendation PR-R-5 (issue guidance to adopt risk-informed and performance-based defense-in-depth) to new reactors.

Security

New reactor designs and reviews have benefited from more recent requirements and guidance related to security. For example, although not incorporated into NRC regulations, the staff prepared guidance related to security assessments for new reactor designs and combined licensed applications that was used by both applicants and NRC reviewers. Applicants were able to incorporate features into reactor designs or site layouts to help address security concerns as well as new requirements to address aircraft impact and the loss of large areas due to fires and explosions. These activities have positioned new reactors to more easily incorporate risk insights into the security program. However, the RMTF recognizes that the adoption of common terminology and approaches for both safety and security areas would be beneficial.

> Recommendation NR-R-6: Apply Recommendation PR-R-6 (develop guidance and consistent approach between safety and security) to new reactors.

4.2.1.3   Implementation Options—Generation IV Power Reactors

Background

The primary distinctions between Generation IV and operating and new power reactors are: (1) they introduce significant changes in reactor technology, such as different fuel forms and reactor coolants, and (2) the designs and supporting analyses are not yet finalized. A high degree of uncertainty remains as to if, and, or when the NRC would be asked to review a Generation IV reactor design and what technology ultimately will be pursued. The NGNP program at the U.S. Department of Energy (DOE) continues to work on technical research and development and a licensing plan for a high-temperature gas-cooled reactor. DOE and private companies are also working on liquid-metal-cooled fast reactor designs and high-temperature fluoride salt-cooled designs. The NGNP activities and previous efforts related to Generation IV designs (e.g., PRISM and sodium advanced fast reactor (SAFR) preliminary designs, and modular high-temperature gas-cooled reactor (MHTGR) preliminary design) have incorporated risk assessments into the design and licensing processes to a larger degree than has been done for light-water reactors. This is due in large part to a recognition that existing NRC technical requirements in 10 CFR Part 50 were developed specifically for light-water reactor designs and the identification and resolution of issues would benefit from risk assessment techniques. For its part, the NRC developed and issued NUREG-1860 as part of its activities related to a potential "technology neutral" regulatory framework for future reactor designs (NRC, 2007). The technology neutral concept is to develop regulatory requirements that could be applicable to any advanced reactor design (e.g., gas-cooled reactors, liquid-metal cooled fast reactors). An alternative approach would be to develop regulations specific to each reactor technology. For either approach, the RMTF offers the following recommendations for Generation IV power reactors.

Design-Basis Accident Licensing Approach

The technology differences between current light-water reactor designs and Generation IV reactors require a fundamental reassessment of DBAs related to plant malfunctions. While some of the critical safety functions such as controlling reactivity and maintaining core cooling are similar, the design features and operating practices used to accomplish them will change. In general, Generation IV reactors have inherent features, such as higher system heat capacities, which will provide slower rates of temperature rises within the reactor core than what exists for light-water reactors. The Generation IV reactor designs also take advantage of passive safety features to reduce the reliance on active equipment, electrical power, and dedicated heat sinks. Differences in the fuel form and reactor coolant also have required alternate acceptance criteria for design-basis events since existing safety limits and regulatory acceptance criteria are defined specifically for light-water reactors with uranium dioxide fuel pellets within zirconium alloy cladding.

The NRC staff's interactions with NGNP and fast-reactor communities indicate that DBAs are under consideration for Generation IV technologies. These DBAs would share some of the attributes of current DBAs, such as reliance on safety-related equipment only. Other aspects of current practices, including the analysis of anticipated operational occurrences and the use of the single failure criterion, are being evaluated by the Generation IV reactor designers. The Generation IV activities have not proposed to go as far as NUREG-1860, "Feasibility Study for a Risk-Informed and Performance-Based Regulatory Structure for Future Plant Licensing," in terms of combining PRA and traditional methods to define special treatment requirements for specific SSCs. Based on information from NRC staff interactions with staff of Generation IV programs and general stakeholder feedback that DBAs are useful in establishing functional requirements and safety margins, the RMTF offers the following recommendation:

> Recommendation GIV-R-1: For Generation IV reactors, the RMTF recommends that the concept of design-basis accidents be maintained, but the NRC should be amenable to and promote, where practical, the adoption of more risk-informed approaches for the selection of relevant scenarios (e.g., alternatives to the single failure criterion) for design-basis accidents.

Beyond design-basis-Accidents

The NRC's experiences related to Generation IV reactor designs have included increased consideration of beyond-design-basis accidents. This increased focus is because future reactors have defined new events and placed them into categories using similar criteria as provided by NRC regulations and industry codes and standards developed for light-water reactors.

However, light-water reactor assessments used more operating experience and engineering judgment, since the need to define design-basis events predates the wide use of PRAs. Given that the Generation IV design and licensing activities began after the development of risk assessment techniques, PRAs were used to support plant design activities and licensing strategies for each of the subject Generation IV designs. The evaluation of beyond-design-basis events was also included in the licensing strategies for Generation IV designs in support of assessing plant design features and operational programs, such as emergency planning. As with design-basis events, alternative acceptance criteria usually are defined for Generation IV reactors, since the common surrogate measures for the NRC's safety goal (core damage frequency and large early release frequency) were developed for light-water reactor designs. NGNP and NUREG-1860 proposed acceptance criteria of dose at the site boundary, which is more directly related to the health objectives in the NRC safety goal.

NGNP and other Generation IV reactor initiatives have largely adopted the identification and disposition of what has been referred to as design-enhancement events within this report. Generation IV activities largely have been an international effort; therefore, this concept, as well as other features of the IAEA standards and guides, has been incorporated into the design and licensing programs (although sometimes using different terminology). Establishing the proposed Risk Management Regulatory Framework in a timely way for Generation IV reactors will support the design and pre-application interaction between designers and the NRC. Given the general alignment of Generation IV activities with the recommendations in this report, the RMTF recommends the following:

> Recommendation GIV-R-2: Apply Recommendation PR-R-2 (design-enhancement category) to Generation IV reactors.

External Hazards

The licensing of new reactors has benefited from updates to guidance related to specific external hazards, and some external events are being revisited in response to the Fukushima nuclear accident. However, the same inconsistency in handling various external hazards that was discussed for operating reactors and new reactors will, without NRC actions, probably become applicable to Generation IV reactors. This leads the RMTF to suggest the same recommendations for Generation IV reactors as those proposed for operating and new reactors.

> Recommendation GIV-R-3: Apply Recommendation PR-R-3 (include external events in design-enhancement category) to Generation IV reactors.
>
> Recommendation GIV-R-4: Apply Recommendation PR-R-4 (periodically evaluate new information regarding external hazards) to Generation IV reactors.

Defense-in-depth

Generation IV reactor designs and especially recent activities related to NGNP have benefited from various studies and discussions of defense-in-depth. Examples include specific white papers on the incorporation of defense-in-depth into the design and licensing approaches for NGNP and the increased use of IAEA standards and guides within specific Generation IV reactor technology groups. Although Generation IV reactor designs have largely embraced the previous discussions of risk-informed and performance-based defense-in-depth, the RMTF recommends formalizing the definition and overall Risk Management Regulatory Framework to support future interactions and ensure the identification and resolution of cliff-edge effects for advanced reactors.

> Recommendation GIV-R-5: Apply Recommendation PR-R-5 (issue guidance to adopt risk-informed and performance-based defense-in-depth) to Generation IV reactors.

Security

Generation IV reactor designs would be expected to benefit from NRC guidance related to security assessments for new reactor designs and combined licensed applications. Designers are considering how to better incorporate security features into reactor designs or site layouts to help address requirements, including analysis of aircraft impacts and the loss of large areas due to explosions or fires. As previously discussed for operating and new reactors, the RMTF recognizes that the adoption of common terminology and approaches for both safety and security areas would be beneficial for Generation IV reactor programs.

> Recommendation GIV-R-6: Apply Recommendation PR-R-6 (develop guidance and consistent approach between safety and security) to Generation IV reactors.

4.2.2 Nonpower Reactors

4.2.2a Background

In addition to the 104 commercial nuclear power plants licensed to operate in the United States, the NRC is also responsible for licensing and oversight of the nation's 42 Nonpower reactors (NPRs). NPRs, also called research and test reactors, are nuclear reactors primarily used for research, training, and isotope production. They contribute to almost every field of science, including physics, chemistry, biology, medicine, geology, archeology, and environmental sciences. Of the 42 NPRs, 31 have operating licenses and 11 have terminated their operating licenses and are either awaiting, or actively involved in, facility decommissioning.

The NRC's authority to license and regulate NPRs is provided in Sections 103 and 104 of the Atomic Energy Act (the Act) (NRC, 2011b). Section 103 of the Act pertains to the licensing of "industrial or commercial" reactors that may or may not be an NPR. Section 104 of the Act pertains to the licensing of NPRs for the purpose of "medical therapy and research and development." All NPRs currently licensed by the NRC are licensed under Section 104 of the Act. Unique to this authority are the provisions contained in paragraphs 104b and 104c of the Act. These paragraphs require the Commission to impose the minimum amount of such regulation and terms of license that will permit the Commission to fulfill its obligation under this Act to promote the common defense and security and to protect the health and safety of the public with the intent to permit the conduct of widespread and diverse research and development.

Nonpower reactors in the United States have been licensed using 10 CFR Part 50. Implementation of 10 CFR Part 50, as it applies to NPRs, has been achieved using only deterministic methods and acceptance criteria. Licensing decisions allowing construction and operation of NPRs have focused on assurance that worker and public doses are maintained within the limits contained in 10 CFR Part 20 for research reactors and 10 CFR Part 100, "Reactor Site Criteria," for test reactors. As was the case with power reactors, a set of licensing-basis events was established that was intended to ensure conservatism in design and protection from a wide spectrum of postulated events, up to and including design-basis accidents. Those accidents are highly stylized and do not consider multiple failures of safety systems. Qualitative approaches for ensuring reliable safety systems, such as the single failure criterion, were implemented. Testing plans and operational limits were established in technical specifications to ensure that if called upon safety systems would perform.

The licensing of NPRs includes an analysis of a maximum hypothetical accident (MHA). Analysis of the MHA is necessary because many NPRs are designed and operated so that an accident involving a radioactive release is not credible. The MHA assumes an incredible failure that results in consequences that bound all credible DBA consequences. The MHA assumes a radioactive release with radiological consequences that exceed those of any credible accident. Because the MHA is not expected to occur, only the potential consequences are analyzed and not the initiating event and scenario details.

4.2.2b Findings and Recommendations for Nonpower Reactors

The concept of defense-in-depth was developed early in NPR licensing and has been applied to compensate for the recognized limited knowledge of nuclear reactor operations and the consequences of potential accidents. A comprehensive defense-in-depth approach forms the foundation for all licensed NPRs. Nonpower reactor license application guidance (NUREG-1537, Part 1) (NRC, 1996a) specifically states that license applicants should provide a discussion in the safety analysis report of the multiple design features that comprise the facility's defense-in-depth. This guidance specifically identifies discussions of the restricted area surrounding the reactor to exclude and protect the public, confinement or containment designs for the control of radioactive releases, limitations on operation that will ensure thermal-hydraulic parameters remain well below the designed capabilities of the fuel and cladding, diversity and redundancy of instrumentation and control systems, and active or passive engineered safety features included to mitigate the consequences of accidents. Both experience with NPR operations and the knowledge base of potential consequences have grown over time. However, even with increased experience and knowledge, gaps in knowledge and uncertainties remain. These gaps and uncertainties continue to be effectively addressed by a defense-in-depth approach. As such, the concept of defense-in-depth remains relevant with NPRs.

> Finding NPR-F-1: The concept of defense-in-depth for NPRs remains relevant. Knowledge gaps and uncertainties continue to be effectively addressed by the defense-in-depth approach.

The applications of a defense-in-depth approach at both power reactors and NPRs share the common goal of preventing the radiological release to the environment. However, they differ significantly on potential accident consequences. These differences result from significantly different operating characteristics, including maximum power level, duration of operation, fission product inventories, and accumulation of spent fuel. To put these differences into perspective, consider the following: 1) the maximum licensed power level of the power reactor can be two to nine orders of magnitude greater than the Nonpower reactor, 2) the duration of operations of the power reactor is nearly continuous versus the periodic operation of the majority of NPRs, and 3) the onsite accumulation of significant quantities of spent fuel (20 years' to 40 years' worth) in fuel pools at power reactors as compared to little or no spent fuel at NPRs. Each of these factors contributes to a significantly smaller radioactive inventory and accident source term at the NPR when compared to the power reactor.

Emergency planning considerations for NPRs and power reactors are similar. The significantly smaller accident source term of the NPRs results in a reduction of potential consequences and emergency planning zones that are typically bounded well within ownercontrolled areas. The boundary of the room or building in which the NPR is housed often comprises the emergency planning zone. Additionally, of the four emergency classes defined in Appendix E, "Emergency Planning and Preparedness for Production and Utilization Facilities," to 10 CFR Part 50, the classification of general emergency is not included for any of the currently licensed NPRs since the MHA results for each NPR do not demonstrate a significant radiological impact at substantial distances from the reactor.

> Finding NPR-F-2: The analysis of design basis and the maximum hypothetical accidents based on conservative design limits, acceptance criteria, safety margins, and assumptions in conjunction with the application of a defense-in-depth philosophy continues to be a sound but highly conservative licensing approach to ensuring adequate safety of NPRs.

While significant conservatism has contributed to the demonstrated safety of NPRs, it is reasonable to assume that conservative design beyond some point does not yield an equivalent safety benefit. The imposition of excessively conservative NPR design and licensing criteria could be viewed as inconsistent with Section 104c of the Act. As presented previously, Section 104c requires the Commission to impose the minimum amount of such regulation and terms of license that will permit the agency to fulfill its obligation under this Act to promote the common defense and security and to protect the health and safety of the public with the intent of permitting the conduct of widespread and diverse research and development. The imposition of more stringent design requirements once an adequate level of safety or an acceptable level of risk has been achieved could be viewed as exceeding the requirements of the Act.

> Recommendation NPR-R-1: The proposed defense-in-depth framework should be applied to the NPR licensing process to ensure that the current amount of defense-in-depth is appropriate given the relatively small radioactive hazard. This application should include safety and security licensing matters.

Implementation of a Risk Management Approach

The assessment of risk at NPRs has been qualitative and based on conservative deterministic assumptions, traditional engineering analyses, and operational experience. To date, the operators of NRC-licensed NPRs have not used modern risk assessment methods in support of licensing activities. No reactor safety goals and objectives that parallel the Commission's Safety Goal Policy Statement (NRC, 1986) have been developed for NPRs.

The Department of Energy has sponsored PRAs on some of its NPR facilities. Eight international NPRs are known to have Level I PRA models. No Level 2 and 3 models are known to exist for NPRs.

The development of NPR risk assessment through the use of modern risk assessment methods and techniques may provide additional safety insight and benefit, especially for NPRs with maximum licensed power levels greater than 2 megawatts. However, the potential for additional safety insight and benefit will likely diminish rapidly as the facility power decreases. The greatest benefit of the development of a modern risk assessment would be a qualitative or quantitative measure of risk that is based on realistic information. Such information could provide a better understanding of the importance of facility systems and components to the overall risk presented by the facility, support the comparison of facility risk to an agency common risk management goal, and define the necessary barriers and controls to establish an adequate level of safety for the protection of individuals from radioactive materials. As discussed in Appendix B, the selection of appropriate analyses techniques depends on the radiological hazards, relevant scenarios, and associated uncertainties. The use or development of analytical tools less complicated than the detailed PRAs used for operating reactors may be appropriate given the simpler designs and lesser amounts of radioactive materials associated with NPRs. Recognizing the small staff sizes and operating budgets of NPRs, even less complicated risk assessments may require external financial support.

Excessive conservatism or the imposition of requirements that do not result in a proportional benefit to safety or only add minimally to safety beyond an already existing adequate level of safety can be contrary to an efficient and effective regulatory framework. The combination of the conservatisms introduced through the consideration of an incredible accident scenario (e.g., the MHA), the use of restrictive 10 CFR Part 20 standards for evaluation of the effects of a postulated accident at research reactors, and large safety margins associated with the traditional engineering analyses, may result in an overly conservative NPR regulatory framework. If that is the case, the expenditure of resources in the execution of licensing activities and oversight may not be providing a corresponding safety or security benefit. The performance of an NPR risk assessment using modern methods informed by the best and most realistic knowledge available would be valuable in identifying areas of risk previously unknown and not adequately addressed by the NPR licensing and oversight processes, or areas where licensing and oversight efforts are directed toward areas with little or no importance to facility risk. In either case, the risk assessment would be useful in providing information to a formal risk management process for a decision on appropriate changes to regulatory requirements.

Finding NPR-F-3: The application of modern risk assessment methods at NPRs could provide valuable insights into accident scenarios not previously identified by the earlier deterministic safety assessment and could be valuable in focusing the application of licensing and oversight resources on areas of risk importance. Risk assessment insights, in conjunction with a formal risk management decisionmaking process, could significantly contribute to the development of a more efficient and effective NPR regulatory framework. NPR PRA models developed by others could be used as a starting point for facility-specific PRA models at NRC-licensed NPRs. Even with this background, however, funding such assessments could be problematic for NPRs.

Recommendation NPR-R-2: The NRC should evaluate the utility of performing a pilot risk assessment, including consideration of external hazards, using modern risk assessment methods at an NPR. This evaluation would assess the value of the risk insights gained from the risk assessment on the basis of possible safety enhancements and possible contributions to a more efficient and effective risk-informed and performance-based regulatory framework for NPRs.

External Events

Nonpower reactors licensed by the NRC are required to demonstrate, in their design basis, reasonable assurance that external events would not preclude safe operation and shutdown of the reactor. They must also demonstrate that provisions are included to mitigate or prevent an uncontrolled release of radioactive material and the consequences of an external event are considered or bounded by analyzed accidents. The traditional licensing approach for NPRs considers documented historical averages and extremes, credible frequencies, and predictive potential for the specific external events. At a minimum, each research reactor facility is required to meet the local building codes for the specific type of event and test reactors are required to meet the requirements of 10 CFR Part 100.

The traditional NPR licensing approach attempted to establish conservative bounds on the magnitude of particular naturally occurring hazards using regional features. This approach is similar to that used for power reactors, and shares the same limitations. As such, power reactor recommendations PR-R-3 and PR-R-4 that call for a reassessment of the methods used to estimate the frequencies and magnitudes of external hazards, the implementation of a consistent process that includes both deterministic and PRA methods, the establishment of a program to systematically collect external hazard information, and the periodic evaluation of information for site-specific implications are also valid recommendations in the NPR case.

> Finding NPR-F-4: The traditional NPR licensing approach shares the same limitations as the power reactor approach for methods used to estimate frequencies and magnitudes of external events.

> Recommendation NPR-R-3: NPRs should be considered, to the extent practical, in the implementation of power reactor Recommendation PR-R-3 and Recommendation PR-R-4.

Security

Nonpower reactor security requirements are established to protect the same radioactive hazards as safety requirements. The physical protection measures at NPRs are established to address the current threat. As the threat changes significantly, appropriate changes are made to security requirements and physical protection measures. Security at NPRs uses a graded approach with a focus on the prevention of theft and diversion of materials for NPRs with a maximum licensed power level less than 2 megawatts and prevention of both radiological sabotage and theft and diversion of materials for NPRs 2 megawatts and greater. Current NPR security regulations are not risk-informed. The development of risk-informed security regulations or guidance would enhance efficiency by focusing resources on areas of importance.

Finding NPR-F-5: NPR security requirements are not risk-informed beyond the use of a graded approach based on NPR power levels.

Finding NPR-F-6: The development of risk-informed security regulations or guidance would enhance efficiency by focusing resources on identified areas important to facility security.

Recommendation NPR-R-4: If the NRC decides to develop and implement a risk-informed and performance-based defense-in-depth regulatory framework to ensure the safety of NPRs, then the agency should also develop guidance for use in its NPR security regulatory activities that uses a common language with safety activities and harmonizes methods with the risk-informed and performance-based defense-in-depth framework.

4.2.3   References

(ASME/ANS, 2009)	American Society of Mechanical Engineers/ American Nuclear Society, ASME/ANS RA-Sa-2009, "Standard for Level 1/Large Early Release Frequency Probabilistic Risk Assessment for Nuclear Power Plant Applications," Addendum A to RA-S-2008, ASME, New York, NY; ANS, LaGrange Park, IL.
(DHS, 2010)	U.S. Department of Homeland Security, "DHS Risk Lexicon, 2010 Edition," September 2010.
(IAEA, 2010)	International Atomic Energy Agency, "Draft—Safety of Nuclear of Power Plants: Design," Draft Safety Guide DS-414, September 2010.
(Johnson and Apostolakis, 2012)	B.C. Johnson and G.E. Apostolakis, "Seismic Risk Evaluation within the Technology Neutral Framework," *Nuclear Engineering and Design*, 242 (2012) 341– 352.
(NRC, 1975)	U.S. Nuclear Regulatory Commission, "Reactor Safety Study— An Assessment of Accident Risks in U.S. Commercial Nuclear Power Plants," NUREG-75-014 (WASH-1400), October 1975.
(NRC, 1985)	U.S. Nuclear Regulatory Commission, "Policy Statement on Severe Reactor Accidents Regarding Future Designs and Existing Plants," August 1985, published in the *Federal Register* on August 8, 1985 (50 FR 32138)

(NRC, 1986)   U.S. Nuclear Regulatory Commission, "Safety Goals for the Operation of Nuclear Power Plants," Final Policy Statement, published in the *Federal Register* on August 21, 1986 (51 *FR* 30028).

(NRC, 1988)   U.S. Nuclear Regulatory Commission, "Loss of All Alternating Current Power," Final Rulemaking, published in the *Federal Register* on June 21, 1988 (53 *FR* 23215).

(NRC, 1991)   U.S. Nuclear Regulatory Commission, "Monitoring the Effectiveness of Maintenance at Nuclear Power Plants," Final Rule, published in the *Federal Register* on July 10, 1991 (56 FR 31306).

(NRC, 1995)   U.S. Nuclear Regulatory Commission, "Use of Probabilisitic Risk Assessment Methods in Nuclear Regulatory Activities," Final Policy Statement, published in the *Federal Register* on August 16, 1995 (60 FR 42622).

(NRC, 1996)   U.S. Nuclear Regulatory Commission, "Status of the Integration Plan for Closure of Severe Accident Issues and the Status of Severe Accident Research," Commission Paper SECY–96–088, April 29, 1996.

(NRC, 1996a)   U.S. Nuclear Regulatory Commission, "Guidelines for Preparing and Reviewing Applications for the Licensing of Nonpower Reactors: Format and Content," NUREG-1537, Part 1, February 1996, Agencywide Documents Access and Management System (ADAMS) Accession No. ML042430055.

(NRC, 1996b)   U.S. Nuclear Regulatory Commission, "Guidelines for Preparing and Reviewing Applications for the Licensing of Nonpower Reactors: Format and Content," NUREG-1537, Part 2, February 1996, ADAMS Accession No. ML042430048.

(NRC, 2004)   U.S. Nuclear Regulatory Commission, "Regulatory Analysis Guidelines of the U.S. Nuclear Regulatory Commission," NUREG/BR-0058, Revision 4, September 2004, ADAMS Accession No. ML042820192.

(NRC, 2007)   U.S. Nuclear Regulatory Commission, "Feasibility Study for a Risk-Informed and Performance-Based Regulatory Structure for Future Plant Licensing," NUREG-1860, Vol. 1, December 2007, ADAMS Accession No. ML080440170.

(NRC, 2008)   U.S. Nuclear Regulatory Commission, "Policy Statement on the Regulation of Advanced Reactors," Final Policy Statement, October 7, 2008, ADAMS Accession No. ML082750370.

(NRC, 2011)	U.S. Nuclear Regulatory Commission, "An Approach for Using Probabilistic Risk Assessment in Risk-Informed Decisions on Plant-Specific Changes to the Licensing-basis," Regulatory Guide 1.174, Revision 2, May 2011, ADAMS Accession No. ML100910006.
(NRC, 2011a)	U.S. Nuclear Regulatory Commission, "Recommendations for Enhancing Reactor Safety in the 21st Century; The Near-Term Task Force Review of Insights from the Fukushima Dai-Ichi Accident," July 2011, ADAMS Accession No. ML112510271.
(NRC, 2011b)	U.S. Nuclear Regulatory Commission, "Nuclear Regulatory Legislation" NUREG-0980, Vol. 1, No. 9, January 2011.
(Okrent, 1981)	Okrent, David. "Nuclear Reactor Safety: On the History of the Regulatory Process," Madison, Wisconsin: University of Wisconsin Press, 1981.
(Walker, 2010)	Walker, S., "A Short History of Nuclear Regulation, 1946–2009," NUREG/BR-0175, September 2010.
(WENRA, 2008)	Western European Nuclear Regulators' Association, "WENRA Reactor Safety Reference Levels," Reactor Harmonization Working Group, January 2008.

4.3   Material Uses

4.3.1   Background

Reactor and accelerator-produced nuclear materials are used extensively throughout the United States for industrial applications, basic and applied research, manufacture of consumer products, academic studies, and medical diagnosis, treatment, and research. In addition, source materials are used in the production of processed uranium for nuclear fuel fabrication and a wide variety of other uses. The NRC materials regulatory programs are designed to ensure that licensees use these materials safely and securely and to present no undue risk to public health and safety and the environment. The regulatory framework for use of these materials is contained in the Commission's radiation protection standards in 10 CFR Part 20, in application-specific regulations in 10 CFR Part 30 through Part 39, and in related guidance and policies.

Under the provisions of Section 274 of the Atomic Energy Act, the NRC may enter into agreements with States under which the NRC discontinues its authority over certain radioactive materials and a State assumes that authority. The Agreement States, as they are known, now regulate more than 85 percent of the materials licensees in the United States. Presently, there are 37 Agreement States and, in concert with the NRC, they play a major role in the regulation of most materials uses.

The NRC materials regulatory program is the Agency's overall program that ensures safe and secure use of these materials by approximately 2,900 specific licensees. In addition, the program provides guidance and oversight to the Agreement States. Each year the NRC and Agreement States issue thousands of licensing actions (new applications, renewals, and license amendments) and conduct thousands of routine and reactive inspections. The program also regulates general licenses, licenses and inspects decommissioning activities and uranium recovery operations, implements the low-level waste program and conducts materials-related rulemakings.

The risks presented by the use of nuclear materials differ significantly from the risks presented by power reactors. The potential for releases to the environment are a major focus of the reactor program; that potential is not a significant part of the materials program because of the much smaller amounts of radioactive material possessed by most materials users and the lower risk posed by many of those specific uses. Therefore, the risk analysis for reactors focuses predominantly on the prevention of low-frequency, high-consequence scenarios, whereas the risk analysis for materials uses focuses primarily on higher-frequency, lower-consequence scenarios. Materials scenarios or events may include, for example, the loss of a portable gauge from the back of a truck, a well-logging device that is lost beneath the earth's surface, or a radiography camera whose source becomes stuck in an exposed position. These types of events are not unusual, but they generally do not pose substantial radiological hazard or risk. Furthermore, in the materials program, risk assessments are largely qualitative, based on operational experience; however, in the reactor program, such assessments are more often quantitative and probabilistic.

The materials program differs from the reactor program in other ways as well. Most of the technologies involved in materials uses are mature, having been implemented for decades, and are not undergoing fundamental change. New device and source development and uses are relatively infrequent. Costs, including regulatory costs and the difficulty in finding disposal options for radioactive sources, have led many private sector firms to seek nonradioactive technologies for their needs. All these factors have led to a relatively static or even shrinking number of materials licensees nationwide.

The primary concern in the materials program is radiation protection. Accordingly, the focus is on ensuring that worker and public doses are maintained within the limits contained in 10 CFR Part 20. There are no parallels to reactor safety goals in the materials program and, for the most part, there are no parallels to the General Design Criteria of 10 CFR Part 50 or emergency planning considerations because most materials users do not possess enough material to result in a credible accident scenario that might exceed EPA's Protective Action Guidelines (PAGs). Finally, the concept of defense-in-depth, which is a central part of reactor regulation, is more of an implicit rather than explicit part of the materials program.

Traditionally, the basis for the materials program has been largely a deterministic one, with rules and guidance developed over time and as a result of operational experience to provide reasonable assurance of adequate protection of public health and safety. Beginning in the 1990s, the NRC began a shift to a more risk-informed, performance-based approach for all of its regulatory programs. This shift was embodied in a series of Commission SECY papers, the "Commission's Final Policy Statement on Use of Probabilistic Risk Assessment Methods in Nuclear Regulatory Activities" and "Commission Direction-Setting and Policymaking Activities" papers that resulted from the agency's strategic assessment and rebaselining process. At that time, the NRC undertook a number of initiatives to better risk-inform and performance-base its nuclear materials (and other) regulatory programs. These led to fundamental changes in inspection frequency and approach as well as licensing policies and practices, and regulations and guidance.

Risk insights and performance considerations continue to be significant factors in materials program development and implementation, including rulemaking, guidance development, inspection, enforcement, and licensing. For example, in 2011, a working group of experienced Agreement State and NRC managers and staff conducted a re-evaluation of the nuclear materials inspection program to ensure its continued efficiency, effectiveness, and focus on safety and security. The group looked at various aspects of the inspection program, including risk insights. It considered several operational measures of risk—operational data, enforcement data, and expert elicitation through interviews with experienced NRC and Agreement State inspectors and program managers—in reaching its conclusion that the program is meeting its objective of ensuring the safe and secure use of nuclear material in a risk-informed, performance-based manner.

Medical Uses

The NRC and Agreement States issue licenses to hospitals and physicians for the use of radioactive materials in medical treatments. Medical uses fundamentally differ from all other activities regulated by NRC and Agreement States because it is the only area in which individuals are intentionally exposed to radioactive materials. Approximately one third of all NRC and Agreement State licenses in the United States are for medical uses. In addition, the NRC develops guidance and regulations for use by licensees and maintains a committee of medical experts (the Advisory Committee on Medical Uses of Isotopes, or ACMUI) to provide advice to the staff on the use of byproduct materials in medicine. The NRC regulations in 10 CFR Part 35, "Medical Use of Byproduct Material," require physicians and physicists to have special training and experience to practice nuclear medicine.

Industrial and Commercial Uses

The NRC and Agreement States license a wide variety of industrial and commercial applications, including industrial radiography, fixed and portable gauges, well logging, nuclear laundries, research and development, and manufacturing. These uses have varying risks associated with them, reflective of the type and amount of material involved, the environments in which they are used, and the operational experience of the users. For example, radiography uses high-activity gamma radiation sources to find structural defects in metallic materials and welds, in both fixed and field locations. Commercial irradiators are also relatively high-risk facilities that expose products, such as food, food containers, spices, medical supplies, and wood flooring, to radiation to eliminate harmful bacteria, germs, and insects, or for hardening or other purposes. The NRC and Agreement States license approximately 50 commercial irradiators nationwide that can contain upwards of 1 million curies of cobalt 60.

Fixed and portable nuclear gauges are used as nondestructive devices (e.g., to determine the thickness of paper products, fluid levels in oil and chemical tanks, and the moisture and density of soils and material at construction sites). In contrast to the other commercial uses described above, both fixed and portable gauges are relatively low-risk materials uses. They are regulated under the general requirements of 10 CFR Part 30, "Rules of General Applicability to Domestic Licensing of Byproduct Material," require little in the way of operator training, and are inspected at a 5-year frequency (the lowest inspection frequency for materials uses).

General Licensees

Generally licensed devices are devices that typically contain small amounts of radioactive material and are used to detect, measure, gauge, or control the thickness, density, level, or chemical composition of various items. Examples of such devices are gas chromatographs (detector cells), density gauges, filllevel gauges, tritium exit signs, and static elimination devices. Because of the small amounts of radioactive material contained in these devices, as well as their inherent robust design, an individual is not required to apply to the NRC or an Agreement State for a license. The devices may be received from any licensed vendor, and they carry basic accountability requirements. Most generally licensed devices are not subject to routine inspection oversight.

Security in the Materials Program

Over the past several years, the NRC and the Agreement States have increased security requirements on radioactive materials. These requirements were developed and implemented in a risk-informed manner, taking into account the form, quantity, and other aspects of the most risk-significant radionuclides. The NRC worked with its international partners in developing the International Atomic Energy Agency (IAEA) Code of Conduct on the Safety and Security of Radioactive Sources. The Code was published in 2004 and it lists 26 radionuclides and 3 activity thresholds for each (IAEA, 2004). Those activity levels are Category 1, 2, and 3, respectively, with Category 1 being the most risk significant. The NRC also issued orders, additional security measures, and increased controls to licensees that possessed Category 1 and 2 levels of material. The Commission recently approved a final rule, 10 CFR Part 37, which would place materials security requirements in the regulations. Working with other Federal agencies, such as the U.S. Department of Homeland Security, the NRC has also implemented a voluntary program of additional security improvements for materials licensees. In January 2009, the NRC deployed its National Source Tracking System (NSTS), by which the agency and the Agreement States track the manufacture, distribution, and ultimate disposal of Category 1 and 2 sources. Licensees use the NSTS, a secure Webbased system, to enter uptodate information on the receipt or transfer of tracked radioactive sources. In addition, the NRC is developing the Web-Based Licensing system and the License Verification System to track license information for materials licensees and plans to integrate these systems with the existing NSTS. Integration of the three systems will provide a Web-based solution to: enable accounting of the possession of the most risk-significant radioactive sources in the Nation, authenticate the validity of radioactive materials licenses, and modernize materials licensing. Together, these activities have made radioactive sources more secure and less vulnerable to potential terrorists.

Environmental Considerations in the Materials Program

The Commission's environmental protection regulations for implementing the requirements of the National Environmental Policy Act in the NRC's domestic licensing and regulatory functions are contained in 10 CFR Part 51, Environmental Protection Regulations for Domestic Licensing and Related Regulatory Functions." Section 51.22 lists those categories of actions that the Commission has determined are excluded from environmental review based on a determination by the Commission that they do not "…individually or cumulatively have a significant effect on the human environment." Subsection 51.22 (c)(14) identifies issuance, amendment, or renewal of materials licenses as one such category. As a result, environmental reviews in the form of environmental assessments or impact statements are not a routine part of the materials program. Accordingly, the RMTF did not consider environmental risks to materials.

4.3.2 RMTF Findings and Recommendations for Materials Uses

As noted above, the risk environment for materials uses is different from reactors and other regulated uses. Nonetheless, risk-informed, performance-based considerations have been built into the program for regulation of materials uses. Some of the more successful examples include:

- Programmatic Risk Assessments. The PRA Policy Statement stated that PRA should be used in the NRC's regulatory programs wherever appropriate. Staff in NMSS undertook a first step in doing that in NUREG/CR-6642, "Risk Analysis and Evaluation of Regulatory Options for Nuclear Byproduct Materials Systems," which performed a quantitative analysis of tasks, hazards, barriers, and doses (to workers and members of the public) to determine relative risk and regulatory options of various materials uses and systems (NRC, 2000). For purposes of this study, risk was defined in terms of the likelihood of workers or members of the public receiving doses of radiation that exceeded regulatory limits from normal operations or accidents. The study looked at more than 40 material uses or systems, including diagnostic nuclear medicine, high dose rate afterloaders, field radiography, pool irradiators, and portable nuclear moisture density gauges. This analysis accomplished its stated goals of providing the staff with regulatory options for these uses and systems that were informed by risk considerations, but noted that the diversity of materials uses and conditions posed significant challenges in terms of the lack of data and making comparisons to the use of quantitative probabilistic tools for regulatory purposes. The risk insights from NUREG/CR-6642 were used in the development of the NUREG-1556 series and to inform rulemakings, but for the aforementioned reason, they have not been used as a basis for individual materials licensing decisions.

- <u>Changes to Inspection approach and frequency.</u> The Phase II Byproduct Material Review was completed by a team of NRC and Agreement State managers and staff in 2001 (NRC, 2001). The team was charged with conducting a broad independent review of the nuclear byproduct materials program to, among other things, seek to provide a more rigorous risk basis to the program. The team applied risk insights from NUREG/CR-6642 as well as operational experience and enforcement data in making decisions on materials inspection approach and frequency. This work led to a fundamental restructuring of the inspection approach to focus inspection activities on risk significant aspects of licensed operations as opposed to only compliance and document review. It also led to significant changes in inspection frequency based on licensee performance and relative risk of licensed activities.

- <u>Materials Licensing Guidance.</u> The NUREG-1556 series, Volumes 1–21, "Consolidated Guidance About Materials Licensees," was developed in the late 1990s to pull together guidance documents written over the years for the wide variety of materials licensees. These documents allow license applicants to find the applicable regulations, guidance, and acceptance criteria used in granting a materials license, and help streamline the application and staff review processes. Operational experience (performance) and risk insights guided the development of these documents; higher risk activities with significant performance challenges have more prescriptive regulations and guidance than lower-risk activities. Over time, guidance in NUREG-1556 has been revised to further incorporate performance considerations and a new revision to the series is under development to address security issues.

The RMTF received a number of comments regarding materials users in response to the November 22, 2011 *Federal Register* notice, including the following:

- Due to the wide variety of licensed materials uses, there is not a common understanding of the terms risk-informed, performance-based, and defense-in-depth within NRC or with these licensees.

- The NRC's graded approach to radiation protection—focusing on higher dose, higher risk activities—has been one of the primary successes in the materials program.

- Performance-based inspections work well. They provide opportunities to recognize exemplary licensee performance and can help both the NRC and Agreement States to better focus their resources on risk-significant activities.

- Certain deterministic activities or concepts are necessary and effective—setting dose limits and possession limits, for example—but broader reliance on deterministic approaches can lead to ineffective use of licensee and regulatory resources.

- The NRC understands and attempts to build upon the different levels of risk associated with the various materials users and should build upon these efforts to establish and maintain a flexible regulatory approach that allows for and reflects the relative risks of these licensed activities.

- Regulation of highest risk, highest exposure activities—particularly medical—would likely benefit the most from transition to a risk management framework.

- The NRC should inform its deliberations on a holistic risk management structure by a thorough stakeholder engagement process in which specific input is solicited by the NRC from different categories of licensees, as was done in the development of the Safety Culture Policy statement and, in a more limited manner, in the development of the Commission's approach to security for cesium chloride irradiators.

- Major challenges to implementation of the framework include the need to work closely with the Agreement States; assuring that NRC management and staff support and can articulate the basis for the framework, and prioritizing ongoing regulatory initiatives to ensure that resources are available for transition to the risk management framework.

- The transition to a more holistic, risk management framework will likely take 5 to 10 years or more.

Implementation of the proposed risk management regulatory framework in the materials program does not pose a significant cultural challenge since many of the concepts are already in use. For example, NRC regional managers use risk management considerations as a routine part of the licensing and inspection oversight functions for materials licensees. The NRC Regions, as well as the Agreement States, issue licenses, conduct inspections to assure the safe and secure use of materials, issue violations, and take enforcement actions where necessary. Each of these areas provides for risk-informed decisionmaking, feedback on performance and appropriate adjustments to licensing, inspection and enforcement operations, and policy. Regional and Agreement State managers also routinely make risk management decisions in determining the appropriate level of followup for events involving radioactive materials. Based on health and safety considerations, as well as resources, events may warrant immediate dispatch of an inspector to the licensee's site, a note to the file for the next routine inspection to look into the matter, or no action at all. Irrespective of the level of risk and performance information already incorporated into NRC and Agreement State materials programs, risk enhancement of those programs can improve regulatory and licensee performance and further efficiency and effectiveness, provided it is done in a measured, systematic manner.

Finding M-F-1: The materials program has successfully developed and incorporated risk insights and performance considerations into its rulemaking, policy development, and routine licensing and inspection activities.

Finding M-F-2: Deterministic approaches such as dose limits and possession limits are useful concepts across the wide range of materials uses that should be retained, but broader use of deterministic approaches can lead to ineffective use of limited licensee, NRC, and Agreement State resources.

Finding M-F-3: Buy-in of the 37 Agreement States is essential to the success of risk management process implementation given their role in regulating more than 85 percent of the materials licensees in the United States.

Recommendation M-R-1: The NRC materials program should continue to apply risk insights and performance-based considerations, as appropriate, in rulemaking, guidance and policy development, and implementation in accordance with the proposed risk management framework.

Recommendation M-R-2: The development and rollout of the recommended Risk Management Policy Statement should be closely coordinated with the leadership of the Agreement States and a joint NRCAgreement State Working Group should be established to guide risk management implementation in the materials area.

The proposed risk management regulatory framework described in Chapters 2 and 3 could be helpful in better harmonizing approaches to safety and security risks in the materials program and improving efficiency and effectiveness. Presently, there are apparent disparities between these two areas, as exemplified by inspection frequency. Safety and security inspections are done separately, even though there may be benefits from doing them at the same time. The inspection frequency, for example, for self-shielded irradiators less than 10,000 curies is set at 5 years for the safety inspection and 3 years for the security inspection. A recent self-assessment of the materials inspection program recommended that "…safety and security inspections should be done together and, in the longer term, a revision to IMC [Inspector Manual Chapter] 2800 should be made to add/modify focus elements to fully integrate security and safety into one overall inspection."

Finding M-F-4: Differences in regulatory language and approaches between safety and security in the materials area may have reduced the efficiency and effectiveness of the NRC and Agreement State regulatory programs.

Recommendation M-R-3: The NRC should apply common risk approaches to safety and security based on the proposed risk management and defense-in-depth regulatory framework.

The proposed risk management regulatory framework described in Chapters 2 and 3 is very broad and represents an evolutionary, not revolutionary, approach to the agency's mission of protecting public health, safety, and the environment. While the framework is predicated on a defense-in-depth philosophy, that term is not commonly used within the materials program. However, the defense-in-depth concepts of hazards and barriers described above are implicit in the materials program. Considering the three primary components of materials licensing—specific licenses, general licenses, and exemptions—NRC and Agreement State regulations, licenses, and guidance provide for barriers to the hazard presented by radioactive material commensurate with the risk presented by the type and form of that material.

For example, licensing requirements for panoramic irradiators in 10 CFR Part 36, "Licenses and Radiation Safety Requirements for Irradiators," are arguably the most detailed requirements in the materials programs. The rule includes a system of defense-in-depth considerations that include physical barriers, engineered safeguards, access controls, and administrative and procedural controls designed to protect workers and members of the public from potentially significant exposure.

The licensing requirements for less hazardous uses, types, and amounts of radioactive materials can be and are correspondingly less prescriptive and reflect a less robust consideration of defense-in-depth. For example, portable and fixed gauges use small radioactive sources that are double encapsulated and contained within a relatively robust housing. The gauges can be used by individuals with a modicum of training that can be taken online.

Within 10 CFR Part 35 there are also defense-in-depth considerations to greater or lesser degrees based on the hazard or risk posed by the material or modality. For example, the requirements for therapeutic applications of byproduct material, particularly those involving highactivity sources, such as high-dose rate afterloaders or gamma stereotactic radiosurgery units, are more robust than those for diagnostic nuclear medicine and may include multiple physical barriers and administrative controls to protect workers, patients, and members of the public.

Defense-in-depth considerations are built into the design and manufacture of generally licensed devices so that an individual can possess and use such a device with no formal training or experience and only minimal requirements for accountability. For certain devices, which contain a sufficient amount of radioactive material that could pose a greater hazard, the NRC has required individuals to be registered (but not licensed).

> Finding M-F-5: The terminology of defense-in-depth is not used consistently across the agency's materials regulatory programs.

> Recommendation M-R-4: As part of the implementation of the proposed risk management regulatory framework, the RMTF recommends that the materials program should more explicitly consider the defense-in-depth philosophy in rulemaking, guidance, and program implementation, and modify appropriate parts of staff training to make these concepts a central part of such training.

So while there are numerous implicit applications of defense-in-depth consideration in the materials program, what is missing is explicit consideration of that philosophy as part of program development, implementation, and oversight.

As noted earlier, the risk management concept is largely embodied in the rulemaking, guidance, and inspection aspects of the materials program. It is lacking, however, in the budgeting process, particularly because of the amount of time between when budget formulation begins and when budget execution takes place, which is on the order of 18–24 months. During this time, operational factors that influenced the decisionmaking process of budget formulation can and often do change. Fact-of-life changes are considered during this interim period, but not in a systematic way.

The agency could benefit from a more structured application of the risk management process during this period. This would allow program managers to better apply resources to those areas of safety and security where risks are determined to warrant it. The annual nuclear materials events database report, the Agency Action Review Meeting, along with other operational data, would serve as the basis for decisions on changes in allocation of resources within major budget areas to reflect where priorities need to be adjusted to better focus on emerging safety and security risks. For example, if operational data indicated that a performance by a given class of licensees warrants greater regulatory attention, program managers at the NRC Headquarters or Regions would shift resources to provide that attention. Transition to this approach would be a challenge and could take considerable time and effort to change the current budget approach.

Finding M-F-6: The materials program could benefit from a more structured application of the risk management process in resource allocation. This process would allow program managers to more systematically apply resources to those areas where the safety or security risk warrants it.

Recommendation M-R-5: Headquarters and Region managers should undertake a more formal review of operational data and risk considerations to help determine if and where budgeted resources need to be adjusted. This option takes into account the fact that budgets are developed well in advance of budget implementation and that operational realities may change in the intervening time.

Implementation Options for Risk Management in Materials

In the introduction to Chapter 4, three options were identified for implementation of the proposed risk management regulatory framework for the NRC. Following are what those options might look like for the materials program.

Option A: Continue the current approach ("patchwork")

Under this option, the materials program would continue its current approach to risk-informed, performance-based programs. Insights from operational data and enforcement data would be considered in programmatic development, including rulemaking, guidance development, and inspection and enforcement. No significant new initiatives—including development of a Commission Policy Statement—would be undertaken based on risk considerations. Periodic self-assessments of specific program areas or lessons learned in response to events or other factors would continue to be performed as necessary but not under the direction provided by an overall risk management framework.

Option B: Implement the proposed risk management regulatory framework through selected guidance and rule changes

This option would emphasize specific rule and guidance changes to implement the proposed risk management regulatory framework contained in the proposed Commission Policy Statement. This would include early and substantive outreach to the Agreement States and formation of a Working Group to determine where risk management might offer the greatest opportunities. These might include changes to the materials inspection approach and frequency (including security), training on risk concepts, ongoing revisions to the NUREG-1556 series, and specific targeted rulemakings. Rulemakings might include revisions to the GL rule, changes to training and experience requirements in 10 CFR Part 35, systematic revision of 10 CFR Part 40, and the incorporation of ICRP Publication 103, "The 2007 Recommendations of the International Commission on Radiological Protection," into a revised 10 CFR Part 20. Implementation of this option could take up to 5 years.

Option C: Implement the proposed risk management regulatory framework through broad-scale regulatory framework changes

For materials, this option would involve a fundamental revision of the basic framework used for licensing and oversight for a specific materials area or areas. Such an option might begin with a specific area, such as industrial radiography, and taking into account risk management considerations and defense-in-depth, how that area might be regulated differently, while maintaining reasonable assurance of adequate protection of public health, safety, and the environment. That is, how might risk management and defense-in-depth lead to different licensing and oversight approaches in the regulations, guidance, and programmatic implementation As noted in Option B, the Agreement States would play a substantive role in the development of this option. Implementation of this option would take more than 5 years and could be as long as 10–15 years.

The RMTF recommends Option B. It provides a way to transition the agency to the risk management framework and process in a reasonable amount of time and reflects ongoing operational priorities and resource constraints. It will also ensure continuity and stability in the regulated community through the continued use of existing regulations and guidance while implementing the risk management framework and process.

4.3.3 References

(IAEA, 2004)	International Atomic Energy Agency, Code of Conduct on the Safety and Security of Radioactive Sources, January 2004.
(NRC, 2000)	U.S. Nuclear Regulatory Commission, "Risk Analysis and Evaluation of Regulatory Options for Nuclear Byproduct Materials Systems," NUREG/CR-6642, February 2000, Agencywide Documents Access and Management System (ADAMS) Accession No. ML003693052 (Not publically available).

(NRC, 2001)   U.S. Nuclear Regulatory Commission, Phase II Byproduct Material Review, August 2001, ADAMS Accession No. ML012270095.

(NRC, 2011)   U.S. Nuclear Regulatory Commission, Materials Inspection Working Group Final Report, May 2011, ADAMS Accession No. ML112020347.

4.4 Waste

4.4.1 Low-Level Waste

The NRC regulates the management and disposal of low-level radioactive waste (LLW) through regulations, licensing, inspection, enforcement, guidance, and policy development. The primary regulations in this regard are 10 CFR Part 20, Part 35, "Medical Use of Byproduct Material," (decay-in-storage provisions), and 10 CFR Part 61, "Licensing Requirements for Land Disposal of Radioactive Waste." The Agreement States also play a major role in LLW regulation since they regulate more than 85 percent of the materials licensees in the United States and the operating commercial LLW disposal sites. A comprehensive history of low-level radioactive waste management, regulation, and legislation is provided in a white paper prepared by the former Advisory Committee on Nuclear Waste (NRC, 2007).

4.4.1.1 Background

LLW may be generated by any licensee and includes a wide range of forms, activity, and associated hazard. LLW is as diverse as contaminated clothing, miscellaneous equipment, medical syringes, animal carcasses, ion-exchange resins, and decommissioning wastes. Similarly, the activity levels in LLW may range from material that is barely above background to material that could result in individual exposures well in excess of occupational limits and could pose operational challenges at the disposal site.

Before the development of 10 CFR Part 61, land disposal of radioactive wastes was governed by the general requirements of 10 CFR Part 20, as there were no specific regulations in place for such disposal. In response to operational issues at the existing LLW disposal facilities and a 1976 General Accounting Office review of those issues and other problems, the NRC began development of 10 CFR Part 61 in 1978 and the rule was completed in 1982. The agency performed many technical analyses to support the rulemaking. It sought to take into account the wide geographic and geologic variation in potential disposal sites by identifying four broad performance objectives—protection of the public from radioactive releases, protection against the inadvertent intruder, protection of disposal site workers, and site stability after closure—and specific technical requirements that needed to be met for granting a license for a disposal facility. The rule also took into account the variety in waste forms and activities by developing a waste classification system, in which waste was labeled Class A, B, or C based on radiological hazard. As the hazard increased, the disposal requirements for that class of waste became more robust. Certain wastes—commonly referred to as Greater than Class C— were identified as generally unacceptable for nearsurface disposal. Greater than Class C wastes are the responsibility of the U.S. Department of Energy (DOE) in terms of a long-term disposal alternative. In its final form, 10 CFR Part 61 contained a mix of prescriptive (technical requirements in Subpart D), risk-informed (waste classification requirements) and performance-based (performance objectives in Subpart C) requirements for the nearsurface disposal of radioactive waste. Although no quantitative risk analysis was performed, this rule was an early attempt to follow the risk-informed and performance-based approach supported by the best understanding of the LLW disposal scenarios and available methods and models at the time.

The only new LLW disposal sites licensed since the issuance of 10 CFR Part 61 have been in Agreement States—the Energy Solutions site in Clive, UT, and the Waste Control Specialists site in Andrews County, TX. In addition, the other LLW disposal sites—Barnwell, SC, and Richland, WA— and the major waste processors are also located in Agreement States. The

NRC's role has focused on development of regulations, guidance, and policy development with some inspection activity related to special nuclear material disposal licenses.

Since it was issued in the early 1980s, 10 CFR Part 61 has remained largely unchanged. Subsequently, however, there have been significant changes in the LLW environment. The Low-Level Radioactive Waste Policy Act of 1980 and amendments to that act in 1985 established that LLW disposal was a State responsibility. States were encouraged to enter into regional compacts that would develop common disposal facilities for use by member States of a compact. The Act has not been successful in developing a nationwide system of regional or State disposal facilities, and one of the longstanding operational sites—the Barnwell LLW facility in South Carolina—was closed to out-of-compact waste in 2008. A number of operating reactors have entered decommissioning, which created large volumes of radioactive waste that needed disposal homes. Waste processors began considering blending different classes of LLW to achieve a lower waste classification. Volume reduction practices continued in an effort to reduce overall disposal volumes and, thus, costs. New waste streams not envisioned at the time when 10 CFR Part 61 was written, such as large volumes of depleted uranium from commercial uranium enrichment operations, have appeared, and many licensees have had to consider onsite extended storage of LLW because of the lack of access to disposal sites.

The NRC staff has been proactive in recent years in responding to these events and identifying the changes needed to the national LLW program, as the following activities show:

- strategic assessment of the LLW program (SECY-07-0180)

- Commission paper on unique waste streams (SECY-08-0147)

- Commission paper on blending of LLW (SECY-10-0043)

- Commission paper on risk-informing 10 CFR Part 61 (SECY-10-0165)

- risk-informing Branch Technical Position on concentration averaging

- development of a Commission Policy Statement on volume reduction (SECY-12-0003)

In addition, the Commission issued a staff requirements memorandum (SRM) on January 19, 2012, on the staff's ongoing work on risk informing 10 CFR Part 61. The SRM directed the staff to expand its effort to bring a clearer risk-informed approach to 10 CFR Part 61. Specifically, the staff is to provide to the Commission within 18 months an expanded proposed rule that would:

- Allow licensees the flexibility to use International Commission on Radiological Protection dose methodologies in a site-specific performance assessment.

- Include a two-tiered approach that establishes a compliance period that covers the reasonably foreseeable future and a longer period of performance.

- Provide flexibility for disposal facilities to establish site-specific waste acceptance criteria based on the site's performance assessment and intruder assessment.

- Establish Agreement State compatibility criteria that ensure alignment between the States and the Federal Government on safety fundamentals while providing the States with flexibility.

A key part of LLW disposal licensing is performance assessment. The NRC staff developed the following description for the agency's Web site of the role of performance assessment in the regulation of waste disposal:

In the context of disposal of radioactive waste, a **performance assessment** is a quantitative evaluation of potential releases of radioactivity from a disposal facility into the environment, and assessment of the resultant radiological doses. The term performance assessment can refer to the process, model, or collection of models used to estimate future doses to human receptors. Typically, a performance assessment is conducted to demonstrate whether a disposal facility has met its performance objectives. In general, a performance assessment considers the following factors:

- Selected scenario (specific features and processes at the disposal facility and in the surrounding area, such as the location of the potential release, location and general characteristics of the receptors, and applicable transport pathways through which radionuclides might reach the environment and pose a threat to the selected receptor groups)

- Performance of the cask or other engineered barrier system used to store low-level waste, limit the influx of water, and reduce the release of radionuclides

- Release and migration of radionuclides through the engineered barrier system and geosphere (those deep-underground portions of the disposal facility where human contact is generally not assumed to occur)

- Radiological dose(s) to the selected receptor group(s)

Because it is not possible for computer models to precisely replicate all conditions of a realistic disposal facility, the staff of the U.S. Nuclear Regulatory Commission (NRC) uses abstraction to simplify the information to be considered in a performance assessment. The degree of abstraction normally reflects the need to improve reliability and reduce uncertainty, balanced with other practical considerations (such as making the model and its results easy for people to understand). Nonetheless, it is important for the model to be sufficiently detailed to ensure that it yields valid results for the performance assessment. Also, while traditional deterministic methods have been sufficient to ensure adequate safety, performance assessments can be more explicitly quantified through probabilistic approaches. In particular, a probabilistic performance assessment considers the risk triplet: "What can go wrong?" "How likely is it?" and "What are the consequences?" Use of performance assessment tools and methodologies aids the NRC in applying a risk-informed and performance-based approach to regulatory decisionmaking.

Because the existing 10 CFR Part 61 requirements were established when performance assessment was at an early stage of its development, a performance assessment is not required by 10 CFR Part 61 to demonstrate that the performance objectives have been met, nor is it discussed in the staff's standard format and content guide for LLW applications

(NUREG-1199, "Standard Format and Content of a License Application for a Low-Level Radioactive Waste Disposal Facility") and Standard Review Plan (NUREG-1200).

4.4.1.2 RMTF Findings and Recommendations for Low-Level Waste

The RMTF received responses from several commenters about LLW and risk management in response to the November 22, 2011, Federal Register notice. Among the comments provided were the following:

- The NRC should use caution in attempting to adopt a single approach to risk management in light of the diversity of activities that the agency regulates.

- 10 CFR Part 61 is one area of the regulations that is ripe for full conversion to a risk management regulatory approach.

- There is not a common understanding and usage of the terms risk-informed, performance-based, and defense-in-depth within the NRC, as well as outside the NRC.

- Deterministic requirements are inherently straightforward and provide for robust facility designs. But, in some cases, those requirements provide no commensurate benefit to public health and safety.

- Integration of deterministic and risk management approaches is both inevitable and appropriate given the time and resource constraints on the NRC.

- Challenges to accomplish a holistic risk management structure include resources (the NRC's as well as Agreement States'), difficulty in establishing performance-based measures of success, consistency in interpretation and approach; treatment of qualitative versus quantitative analyses; and availability of sufficient data.

- The NRC should engage stakeholders through public meetings specific to individual program areas on the risk management approach. The outreach approach that the agency took in developing the Safety Culture Policy Statement was offered as a model for consideration.

The RMTF considered these comments in light of the findings and recommendations in Chapter 2 on adoption of the Risk Management Regulatory Framework and concluded that there is no inherent conflict between the framework and the above views offered. The proposed risk management framework is particularly well suited to the regulation of LLW given its mix of prescriptive, risk-informed, and performance-based components.

The concept of defense-in-depth is implicit in the requirements and structure of 10 CFR Part 61, although the term itself is not explicitly used. The rule provides for a series of barriers or controls to assure that the performance objectives are met and that the public and the environment are adequately protected. For example, 10 CFR Part 61 requires that an applicant for a LLW disposal facility license to design disposal unit covers to minimize water intrusion into the disposal units. If water intrudes into the disposal units, other requirements in the rule on waste form, packaging, and placement serve as additional barriers or controls to minimize water coming into contact with the waste and serving as a transport mechanism for radionuclides. If somehow those radionuclides leach out of the waste, the rule requires additional barriers or controls in the form of a buffer zone between the disposal units and the disposal site boundary, which must be of sufficient size to allow mitigation measures to be taken.

As noted in Section 4.3, over the past several years, the NRC and the Agreement States have increased the security requirements on the most sensitive radioactive materials. The NRC issued orders, additional security measures, and increased controls to licensees that possessed Category 1 and 2 levels of material. LLW facilities were included in the group of licensees required to implement increased controls because they may possess Category 1 or 2 sources for disposal.

Finding LLW-F-1: The regulatory framework for LLW disposal has, for the most part, implicitly followed a risk-informed and performance-based approach. However, changes in the LLW environment and the maturing of the performance assessment method over the past 30 years have underscored the need to provide a stronger risk basis to the program. The staff is currently implementing Commission direction to make it more so and will be sending various products to the Commission over the next year or two. However, as noted in earlier sections of this report, the terms risk-informed and performance-based are not well understood and have not been consistently applied across NRC regulatory programs.

Finding LLW-F-2: Certain aspects of the LLW regulatory framework readily lend themselves better to the risk management approach, such as waste classification or concentration averaging. Applying the proposed risk management approach to comprehensive LLW licensing decisions, however, may be more challenging because it involves estimating facility performance due to events potentially far into the future.

Recommendation LLW-R-1: The NRC should adopt the concept of risk management to the LLW program, including any revisions proposed to 10 CFR Part 61 (including performance assessment requirements) and related guidance documents.

Finding LLW-F-3: The interlocking and reinforcing systems approach in 10 CFR Part 61 (site suitability, waste form and classification, intruder barrier, institutional controls, etc.) represents an implicit consideration of defense-in-depth features, based on the risk posed by various classes of waste.

Finding LLW-F-4: The NRC has not developed an explicit characterization of defense-in-depth considerations for the LLW program.

> Recommendation LLW-R-2: The NRC should develop an explicit characterization of how defense-in-depth within the proposed risk management framework applies to the LLW program and build this into current and future staff guidance documents and into training and development activities for the staff.

The NRC staff is required by Subsection 51.20 (b)(11) of the Commission's regulations in 10 CFR Part 51, "Environmental Protection Regulations for Domestic Licensing and Related Regulatory Functions," to prepare an environmental impact statement (EIS) or supplement to an EIS for issuance of a new license for land disposal of radioactive waste received from other persons. In addition, Subsection 51.20 (b)(12) requires an EIS or supplemental EIS to be prepared for an amendment to a LLW disposal license that authorizes: closure of a disposal site, transfer of the license for purposes of institutional control, or termination of the disposal site license. EIS preparation involves a consideration of impacts or risks to the environment against benefits from a proposed action or alternatives to that action. Renewal or significant amendments to the license other than those mentioned above may require an environmental assessment that can lead to preparation of an EIS. The requirements in 10 CFR Part 51 are Federal requirements that apply to the NRC. The Agreement States have varying environmental review requirements, but are not bound to those in 10 CFR Part 51.

> Finding LLW-F-5: Consideration of environmental risks as well as safety risks is a central part of the LLW regulatory program.

> Recommendation LLW-R-3: The NRC should include environmental reviews within the scope of its risk management framework.

Implementation Options for Risk Management in LLW

In the introduction to Chapter 4, three options were identified for implementation of the risk management framework for the NRC. Following are what those options might look like for the LLW program.

Option A: Continue the current approach

Under this option, the LLW program would continue its current approach to risk-informed, performance-based programs. No significant new initiatives, including development of a Commission Policy Statement, would be undertaken based on risk management or defense-in-depth considerations. Under this option, the agency would take the position that the risk-informed regulation and defense-in-depth concepts already in place in the LLW regulatory program are sufficient and no further action is needed to comport with the framework proposed by the Task Force. The LLW regulatory framework, including 10 CFR Part 61 and related guidance documents, would be revised in accordance with ongoing risk-informed, performance-based initiatives described earlier in this section. Periodic self-assessments or lessons learned in response to events or other factors would continue to be performed as necessary, but not under the direction provided by an overall risk management framework.

Option B: Implement the risk management framework through selected guidance and rule changes

Under this option, the agency would adopt RMTF Recommendation LLW-R-1, LLW-R-2, and LLW-R-3 to incorporate risk management and defense-in-depth into the basic philosophy of the LLW regulatory program. This option would emphasize specific rule and guidance changes to implement the risk management framework contained in the proposed Commission Policy Statement. NRC staff would undertake early and substantive outreach to the Agreement States (since all existing LLW disposal sites are in Agreement States) and form a working group to determine where risk management might offer the greatest opportunities. It would also help ensure that current and future regulatory initiatives had a stronger, more well-defined risk basis and would reduce deterministic, prescriptive regulation. Staff initiatives in progress, such as a revision to 10 CFR Part 61, consistent with the Commission's January 19, 2012 SRM, would be guided by the risk management approach, including any proposed revisions to the waste classification, requirements for performance assessment, periods of compliance, considerations of uncertainty in performance assessment, etc. Implementation of this option could take up to 5 years.

Option C: Implement the risk management framework through broad-scale regulatory framework changes

This option would involve redesigning the basic framework used for licensing and oversight for LLW disposal. In this case, that would mean a complete revision of 10 CFR Part 61 and its associated guidance documents to explicitly implement the Risk Management Regulatory Framework. As noted in Option B, the Agreement States would play a substantive role in developing this option. Implementation of this option would take more than 5 years and could be as long as 10 years to 15 years.

The RMTF recommends Option B. It provides a way to transition the agency to the risk management framework and process in a reasonable amount of time and reflects ongoing operational priorities and resource constraints. It would also ensure continuity and stability in the regulated community through the continued use of existing regulations and guidance while implementing the risk management framework and process.

4.4.2   High-Level Waste

4.4.2.1   *Regulatory Background and Framework on High-Level Waste Disposal*

U.S. policies governing the permanent disposal of high-level waste (HLW) are defined by the Nuclear Waste Policy Act of 1982, as amended (NWPA). This Act specifies that HLW will be disposed of underground, in a deep geologic repository, and that Yucca Mountain, NV, will be the single candidate site for characterization as a potential geologic repository. In light of the Federal Government's decision to discontinue activities directed at ultimate disposal of HLW at the Yucca Mountain site, the future direction and options for the NRC's regulation of HLW disposal are not certain at this time. The following paragraphs discuss how risk-informed and performance-based factors were considered in the development of the current regulations and guidance for HLW disposal and how the proposed risk management framework might interact with future regulatory development.

Under the NWPA Act, the NRC is one of three Federal agencies that has a role in the disposal of spent nuclear fuel and HLW from the Nation's nuclear weapons production activities:

- DOE is responsible for designing, constructing, operating, and decommissioning a permanent disposal facility for HLW, under NRC licensing and regulation.

- The U.S. Environmental Protection Agency (EPA) is responsible for developing site-specific environmental standards for use in evaluating the safety of a geologic repository.

- The NRC is responsible for developing regulations to implement EPA's safety standards and for licensing and overseeing the construction and operation of the repository.

In the late 1970s, the NRC began developing regulations for HLW disposal. Since that time, risk information and risk concepts have been used increasingly to assist and guide the development of the HLW program. Safety of an HLW geologic repository typically is divided into what is termed the "pre-closure" or "operational" period (the time when waste is emplaced in the repository and before the facility is permanently sealed) and the "post-closure" period (the time after the repository is permanently closed). Pre-closure safety is sufficiently similar to waste handling operations in other regulated activities (e.g., pool storage and spent fuel storage facilities); therefore, regulatory development in the pre-closure area has been guided principally by improvements implemented for similar activities for other operating facilities (e.g., security, emergency planning, worker safety and procedures, and operation of heavy-lift cranes). Conversely, the methods and approaches for evaluating post-closure safety were the principal subject of NRC development and improvement because of the unique aspect of evaluating the safety of a geologic repository thousands of years and longer into the future. Thus, this section primarily describes how post-closure performance assessment methods provided risk information that was used to 1) evaluate and improve the regulations for geologic disposal, 2) assist the resolution of technical issues and development of regulatory guidance, 3) address defense-in-depth requirements for the post-closure period, and 4) assist regulatory reviews of geological disposal.

Development of Risk-informed and Performance-based Regulations

Existing NRC regulations for geologic disposal are found in 10 CFR Part 60, "Disposal of HighLevel Radioactive Wastes in Geologic Repositories," (generic regulations for all sites other than Yucca Mountain, issued in 1983), and 10 CFR Part 63, "Disposal of HighLevel Radioactive Wastes in a Geologic Repository at Yucca Mountain, Nevada" (specifically for Yucca Mountain, issued in 2001). In the two decades between development of 10 CFR Part 60 and 10 CFR Part 63, a variety of technical assessment improvements and insights have taken place. The Commission stated that 10 CFR Part 63 allows more effective and efficient methods of analysis for evaluating conditions at Yucca Mountain than the NRC's existing generic criteria. These new methods were not envisioned when 10 CFR Part 60 criteria were established. It was thought that their implementation for Yucca Mountain would avoid the imposition of unnecessary, ambiguous, or potentially conflicting criteria that could result from the application of some of the Commission's generic requirements of 10 CFR Part 60 (NRC, 1999).

Perhaps the most significant change to the NRC regulations was the approach to defense-in-depth during the post-closure period of a geologic repository (i.e., implementation of the multiple barrier requirements). A longstanding principle of geologic disposal has been a reliance on multiple barriers to limit the release and transport of radionuclides. Engineered barriers (such as waste packages and waste forms) should complement and work with the geological or natural barriers so that safety does not depend solely on a single barrier or phenomenon.

During the late 1970s and early 1980s, when the NRC was first developing generic repository criteria for 10 CFR Part 60, quantitative techniques for assessing repository performance were in their infancy. This lack of experience with and confidence in the quantitative methods for addressing uncertainty was a key factor in selecting an approach for implementing the multiple barrier requirement. Therefore, for 10 CFR Part 60, the NRC elected to prescribe minimum performance standards for each major system (as they were envisioned at the time) and require the overall system to comply with the EPA standards. The standards for each major system, which became known as "subsystem requirements," were: 1) the length of time the waste package remains intact (300 to 1,000 years), 2) the rate of subsequent releases from the engineered system (release rate of 10^{-5} per year), and 3) the pre-emplacement ground water travel time to the accessible environment (1,000 years). Most of the bases for the NRC staff's recommendations of the numerical values for the subsystem criteria were generic judgments of what was thought to be protective of the public health and safety and feasible for consistent evaluation.

Some of the earliest applications of performance assessment methods for examining regulatory criteria was the examination of the relationship between the NRC's subsystem criteria (i.e., defense-in-depth or multiple barrier requirement) and compliance with the EPA's overall system standard (limits regarding allowable releases from the repository integrated over 10,000 years). These early examinations revealed that compliance with the subsystem criteria alone was not sufficient to ensure compliance with the EPA's overall release limit (64 FR 8648; February 22, 1999). Additionally, questions were being asked about the merits of the EPA releases limits for carbon-14 relative to overall safety of the repository, based, in part, on performance assessment calculations for a potential repository at Yucca Mountain (NRC, 1990; EPA, 1993; NAS, 1995).

In the 1980s and 1990s, work continued on the development of performance assessment methods for evaluating geologic disposal. This work significantly increased confidence in the technical ability to assess overall repository performance and to address and quantify the corresponding uncertainty (see for example Bonano, et al., 1989; Electric Power Research Institute [EPRI], 1990, 1992, and 1996; NRC 1992, 1995, and 1997; Sandia National Laboratories, 1992; Pacific Northwest Laboratory 1993; and DOE, 1995). Revisions to the standards and regulations for a potential repository at Yucca Mountain made significant use of risk information to support more effective and efficient standards and regulations. In particular, 10 CFR Part 63 reflected a risk-informed, performance-based approach that, for the post-closure period, focused on meeting a dose limit to a reasonably maximally exposed individual (i.e., dose limit provided in the EPA standard at 40 CFR 197).

4.4.2.2 *RMTF Findings and Recommendations for High-Level Waste*

Regulations in 10 CFR Part 63 impose criteria for performance assessment that must be used for demonstrating compliance with the overall safety standard (i.e., dose limit). As a result, the NRC found no need for added quantitative subsystem requirements. The quantitative criteria for repository subsystems limited the flexibility of the repository developer, DOE, in achieving compliance with the overall performance goal. Instead, the NRC adopted an approach to regulation that provided DOE with the flexibility to select and defend the barriers it decides to rely on for demonstrating the safety of the repository. The 10 CFR Part 63 approach for implementing the multiple barrier approach for Yucca Mountain is tied to the performance assessment used to demonstrate compliance with the overall performance standard. The NRC stated that when it finalized its regulations for Yucca Mountain, it was "…confident that evidence for the resilience, or lack of resilience, of a multiple barrier system will be found by examining a comprehensive and properly documented performance assessment of the behavior of the overall repository system" (NRC, 2001a). Such an approach allows "…DOE to use its available resources effectively to achieve the safest repository without unnecessary constraints imposed by separate, additional subsystem performance requirements" (NRC, 2001a).

Similar to the approach used for demonstrating compliance with the post-closure performance objectives (i.e., the use of a performance assessment), the NRC also prescribed a system analysis for evaluating compliance with the pre-closure performance objectives (i.e., dose limits applicable during the operational period when waste is being emplaced in the repository). Regulations in 10 CFR Part 63 identify the need for, and general scope of, the Pre-Closure Safety Analysis (PCSA) to be done to demonstrate compliance with the performance requirements for the operational phase of the repository. The PCSA was intended to refer to a broad category of analyses to be used by DOE in its evaluation of repository operations and design in meeting the pre-closure performance objectives. The regulations also prescribe the need for certain design measures to ensure the availability of safety systems (e.g., means to prevent criticality, radiation alarms, means to provide reliable emergency power, explosion and fire detection systems, and appropriate suppression systems). This approach is consistent with ensuring safety at an operating facility where actions could be taken to prevent or minimize accidents and their consequences. This is distinct from the post-closure period, where it is assumed actions will not be taken to ensure safety due to thousands of years and longer covered by the post-closure period.

As experience with performance assessment methods and results increased over time, so did the use of risk information to resolve technical issues and to assist the development of regulatory guidance. In particular, the NRC and DOE conducted regular technical exchanges on performance assessments for a potential repository at Yucca Mountain. The technical exchanges considered the data and science supporting the abstracted models used in performance assessment. They also considered the analysis of the results and interpretation of the risks insights derived from the analysis. These technical exchanges integrated technical work (e.g., data collection, confirmatory research) with results from the performance assessment that assisted prioritization of the technical work and improved performance assessment models. For example, geosphere flow and transport models in the performance assessment were appropriately simplified (i.e., risk-significant attributes of flow and transport could be sufficiently represented with simple models); further detail was added to previous models for waste package degradation and release of radionuclides from the repository; significantly more detail was added to models for disruptive events (i.e., seismic and volcanic events) specific to the Yucca Mountain site; detailed analyses and data collection on rainfall and infiltration were conducted to provide an infiltration rate in the performance assessment; and field investigations were initiated to better understand redistribution and mass loading of volcanic ash.

The increased understanding of system performance and the increased development of scientific support and information for risk-significant aspects of repository performance were used to resolve technical issues and to develop the Yucca Mountain Review Plan (YMRP) (NRC, 2003). The post-closure portion of the YMRP was developed based on the attributes needed to develop a credible performance assessment.

Finding HLW-F-1: The development of regulations for geologic disposal of HLW at Yucca Mountain was based significantly on risk information developed from performance assessments and closely followed the proposed Risk Management Regulatory Framework.

Finding HLW-F-2: The NRC's regulatory philosophy of defense-in-depth is reflected in the multiple-barrier requirement for post-closure in 10 CFR Part 63. Compliance with the multiple barrier requirements is demonstrated through the performance assessment.

Finding HLW-F-3: As performance assessment capabilities and experience increased at the NRC during the past 30 years, so did the use of risk insights to help guide the HLW program. Risk insights and performance assessment capabilities have been used to improve the efficiency and effectiveness of guidance documents; inform pre-licensing interactions with DOE; and help identify and direct data needs and experimental activities.

Recommendation-HLW-R-1: Any future revisions to the regulatory framework for geologic disposal of HLW should be done in accordance with the proposed risk management framework to ensure that risk information continues to be appropriately considered in the development of requirements and appropriately reflect any future HLW disposal paradigm.

Preparations for Regulatory Reviews

A fundamental aspect of the risk-informed, performance-based approach in 10 CFR Part 63 was that staff would focus its technical and regulatory reviews consistent with safety significance. During the staff's technical review of DOE's license application for Yucca Mountain, the NRC encountered some inefficiencies in applying its risk-informed approach partly because of the approach used to develop the YMRP and partly because of the somewhat more "open-ended" nature of the risk-informed, performance-based approach adopted in 10 CFR Part 63.

First, the YMRP was developed well before DOE submitted its license application as a way to provide DOE and other stakeholders with information on how the NRC would review DOE's license application. The NRC provided significant detail in the YMRP about how the staff would evaluate all the topics that "could" be relevant during the review. However, the NRC staff always understood and explained in public meetings on the YMRP that although the YMRP covered the full spectrum of topics, the regulatory review would focus on aspects that were significant to safety (i.e., a risk-informed approach). Despite the intended risk-informed approach, there was a strong desire (based on legal and adjudication concerns) to address "all" of the topics mentioned in the YMRP, simply because they appeared in the YMRP. Thus, the details included in the YMRP increased the topics that had to be addressed in the staff review regardless of their safety significance. It is appropriate to document how certain topics are determined to be not significant to safety; however, the selection of such topics to be discussed in technical evaluation reports should be based on the technical merits instead of the appearance of the topic in the YMRP. In hindsight, staff development of guidance (i.e., YMRP) that attempted to provide an identification of 'all' topics that might be significant unintentionally increased the burden of the regulatory review and potentially limited the effectiveness of the risk-informed approach.

Secondly, the risk-informed, performance-based approach in 10 CFR Part 63 was intended to focus technical reviews on those aspects of the geological repository important to safety. As such, 10 CFR Part 63 prescribed in broad terms how the performance assessment of overall safety was to be performed (e.g., include features, events, and processes that have a significant effect on the overall dose estimate; consider events with an annual probability greater than 10^{-8}), which provided the applicant flexibility to develop its safety case and design the facility as it deemed appropriate. This approach, however, also leaves open a wide range of topics that could potentially need to be adjudicated in the licensing hearing. For example, contentions can raise a topic related to performance; however, any detailed discussion on the quantitative impact on safety (i.e., risk significance) is generally considered a discussion addressed in the hearing. Therefore, a risk-informed, performance-based approach that provides flexibility to applicants in developing a safety case for a geological repository may also provide a potential for increased topics to be adjudicated in the licensing hearing. Although this may be an inevitable consequence of a risk-informed, performance-based approach for geologic disposal regulations, further development of regulations and guidance should consider how it affects the efficiency and effectiveness of the NRC's licensing process.

> Finding HLW-F-4: The risk-informed, performance-based approach for regulating geologic disposal (YMRP and regulations) was intended to provide flexibility to the applicant in developing the best approach in meeting the regulatory requirements, and allow the use the risk information to focus regulatory review on safety-significant issues.

4.4.2.3   Implementation Options for High-Level Waste

The nature and future direction of a national program for HLW disposal within the United States remains to be determined. The recent Blue Ribbon Commission report was the first step in determining that direction in light of the cessation of activities for the Yucca Mountain site. The NRC has implemented an orderly closeout of its review of DOE's license application for the Yucca Mountain site. However, the NRC must be prepared to exercise its regulatory responsibility for whatever decision is made on the national HLW program. The recommendations outlined above are intended to assist the agency in its future development of regulations and guidance, but because the timing and direction of the national HLW program are uncertain, the RMTF is not identifying any implementation options.

4.4.3 References

(Bonano, E.J., et al. 1989) E.J. Bonano, et al., "Demonstration of a Performance Assessment Methodology for HighLevel Waste Disposal in Basalt Formation," NUREG/CR-4759, U.S. Nuclear Regulatory Commission, Washington, DC, 1989.

(DOE, 1995) U.S. Department of Energy, "Total System Performance Assessment—1995: An Evaluation of the Potential Yucca Mountain Repository," prepared by TRW Environmental Safety Systems, Inc., for DOE, Office of Civilian Radioactive Waste Management, B-01717-2200-00136, Revision 01, 1985.

(EPRI, 1990) Electric Power Research Institute, "Demonstration of a Risk-Based Approach to HighLevel Waste Repository Evaluation," EPRI NP-7057, Palo Alto, CA, 1990.

(EPRI, 1992) Electric Power Research Institute, "Demonstration of a Risk-Based Approach to HighLevel Waste Repository Evaluation: Phase 2," EPRI TR-100384, Palo Alto, CA, 1992.

(EPRI, 1996) Electric Power Research Institute, "Yucca Mountain Total System Performance Assessment, Phase 3," EPRI TR-107191, Palo Alto, CA, 1996.

(EPA, 2008) U.S. Environmental Protection Agency, "Public Health and Environmental Radiation Protection Standards for Yucca Mountain, Nevada; Final Rule; 40 CFR 197," published in the *Federal Register*, on October 15, 2008, (73 FR 61256).

(IAEA, 2006) International Atomic Energy Agency, "Geological Disposal of Radioactive Waste," Safety Requirements No. WS-R-4, 2006.

(NRC, 1983) U.S. Nuclear Regulatory Commission, "Disposal of HighLevel Radioactive Wastes in Geologic Repositories: Technical Criteria; Final Rule; 10 CFR 60," published in the *Federal Register* on June 21, 1983, (48 FR 28194).

(NRC, 1992) U.S. Nuclear Regulatory Commission, "Initial Demonstration of the NRC's Capability to Conduct a Performance Assessment for a HighLevel Waste Repository," NUREG-1327, May 1992, Agencywide Documents Access and Management System (ADAMS) Accession No. ML012980272.

(NRC, 1995) U.S. Nuclear Regulatory Commission, "NRC Iterative Performance Assessment Phase 2—Development of Capabilities for Review of a Performance Assessment for a HighLevel Waste Repository," NUREG-1464, October 1995, ADAMS Accession No. ML040790450.

(NRC, 1997)   U.S. Nuclear Regulatory Commission, "NRC High-Level Radioactive Waste Program Annual Progress Report: Fiscal Year 1996," prepared for NRC by the Center for Nuclear Waste Regulatory Analyses, NUREG/CR-6513, January 1997, ADAMS Accession No. ML012750518.

(NRC, 1999)   U.S. Nuclear Regulatory Commission, "Disposal of HighLevel Radioactive Wastes in a Proposed Geologic Repository at Yucca Mountain, Nevada," Proposed Rule; 10 CFR 63, published in the *Federal Register* on February 22, 1999, (64 FR 8640).

(NRC, 2001a)   U.S. Nuclear Regulatory Commission, "Disposal of HighLevel Radioactive Wastes in a Proposed Geologic Repository at Yucca Mountain, Nevada; Final Rule; 10 CFR 63, published in the *Federal Register* on November 2, 2001, (66 FR 55732).

(NRC, 2001b)   U.S. Nuclear Regulatory Commission, "Preliminary Performance-based Analyses Relevant to Dose-Based Performance Measures for a Proposed Geologic Repository at Yucca Mountain," NUREG-1538, October 2001, ADAMS Accession No. ML020100312.

(NRC, 2003)   U.S. Nuclear Regulatory Commission, "Yucca Mountain Review Plan; Final Report," NUREG-1804, Revision 2, July 2003, ADAMS Accession No. ML032030389.

(NRC, 2007)   U.S. Nuclear Regulatory Commission, "History and Framework of Commercial Low-level Radioactive Waste Management in the United States: ACNW White Paper," NUREG-1853, January 2007, ADAMS Accession No. ML070600684.

(NRC, 2009)   U.S. Nuclear Regulatory Commission, "Implementation of a Dose Standard After 10,000 Years; Final Rule; 10 CFR 63," published in the *Federal Register* on March 13, 2009, (74 FR 10811).

(NRC, 2011)   U.S. Nuclear Regulatory Commission, "Technical Evaluation Report on the Content of the U.S. Department of Energy's Yucca Mountain Repository License Application, Post-closure Volume: Repository Safety after Permanent Closure," NUREG-2107, August 2011, ADAMS Accession No. ML11223A273.

(NAS, 1983)   National Academy of Science, "A Study of the Isolation System for Geologic Disposal of Radioactive Wastes," Waste Isolation Systems Panel, Board on Radioactive Waste Management, National Academy Press, 1983.

(NAS, 1995)   National Academy of Science, "Technical Basis for Yucca Mountain Standards," Committee on Technical Basis for Yucca Mountain Standards, National Academy Press, 1995.

(PNL, 1993)   Pacific Northwest Laboratory, "Preliminary Total-System Analysis of a Potential High-Level Nuclear Waste Repository at Yucca Mountain," PNL-8444, January 1993.

(SNL 1992)   Sandia National Laboratory, "TSPA 1991: An Initial Total-System Performance Assessment for Yucca Mountain," SAND-91-2795, July 1992.

4.5   Uranium Recovery

4.5.1   Background

Uranium recovery is the first step in the nuclear fuel cycle. It involves extracting uranium from its parent ore and processing it into a physical and chemical form (yellowcake) that will allow additional processing and fabrication to become nuclear fuel. Although natural uranium is found throughout the United States, economically recoverable uranium ore reserves are found primarily in the western United States. The concentration of uranium in the parent ore is highly variable based on a number of geological and geographic factors. In general, the uranium concentrations in U.S. ore deposits are much lower (approximately 0.1 percent by weight) than those found in ore deposits in Canada and Australia (as high as 20 percent by weight).

Because of this low concentration and variability, a variety of recovery methods and technologies have been employed in the United States. These include conventional mining and milling; insitu recovery (ISR), and heap leaching.

Conventional uranium recovery involves mining of the uranium bearing ore, either by surface or underground mining methods, and milling (crushing, grinding, solvent extraction, concentration, drying, and product packaging and shipment). The U.S. Nuclear Regulatory Commission's (NRC's) regulatory authority under the Atomic Energy Act, the Uranium Mill Tailings Radiation Control Act of 1978 (UMTRCA), Title 10 of the Code of Federal Regulations (10 CFR) Part 20, "Standards for Protection Against Radiation," and 10 CFR Part 40, "Domestic Licensing of Source Material," do not extend to the mining of uraniumbearing ore by conventional methods. That activity is regulated primarily by States and other Federal agencies. Rather, the NRC regulates the processing of uranium ore to concentrate uranium and the disposal of the large amounts of tailings resulting from that processing. Conventional methods dominated U.S. uranium production through the mid1980s.

As uranium prices dropped, reflecting a lack of demand beginning in the mid-1980s, other processes came into greater use that were less expensive and made lower quality or geologically deeper ore bodies economically attractive. The primary one is ISR. Unlike conventional mining and milling, ISR does not involve removal of ore from its place in nature. Rather, ISR is based on injection of a chemical solution—lixiviant—into the ore body, which releases uranium and the resultant loaded solution is brought to the surface through a recovery well. Uranium is removed from the solution through an ion exchange process and is then concentrated, purified, and dried at a processing plant in a manner similar to that used at a conventional mill.

Heap leach is a third method used to extract uranium. This method involves placing ore over a lined pad that is overlain with perforated piping. An acid solution is then applied to the ore to release uranium, which percolates through the heap into the piping beneath it. The uraniumbearing solution is then collected and, as with ISR facilities, the uranium is removed from solution for processing into yellowcake.

Uranium prices began to rebound beginning around 2003 and continued to increase into 2007, peaking at nearly $140 per pound, reflecting increased worldwide interest in nuclear power and other factors. This economic incentive has led to a resurgence of interest in new uranium recovery facilities, primarily ISRs, although the NRC has also received letters of intent for conventional mills and heap leach facilities. Presently, the NRC has received five applications for new facilities and expects more than 20 additional applications for new, expanded, or restarted uranium recovery facilities. Uranium prices have declined since their peak in 2007 to less than $60 per pound, which has led to deferral or cancellation of a number of applications, but overall interest in new facilities remains high.

As mentioned above, the NRC's regulatory authority over uranium recovery is limited to the concentration of uranium as a result of processing and the disposal of tailings produced from that process. Before 1978, the NRC did not have specific authority over the disposal of tailings. Congress passed UMTRCA to ensure that there would be a comprehensive Federal and State program to deal with the environmental problems mill tailings had created. UMTRCA gave the U.S. Department of Energy (DOE) authority and responsibility for the cleanup of abandoned uranium mill processing sites and tailings under Title I of the Act. Title II gave the NRC authority and responsibility for regulating tailings at licensed processing sites and the U.S. Environmental Protection Agency (EPA) was given authority and responsibility for setting generally applicable environmental protection standards for uranium mills and tailings sites. NRC standards to implement the agency's responsibilities under UMTRCA are required to conform to standards set by the EPA. In this regard, EPA's uranium recovery standards are contained in Title 40, "Protection of Environment," CFR Part 192, "Health and Environmental Protection Standards for Uranium and Thorium Mill Tailings."

Agreement States may also regulate uranium recovery activities. Authority to regulate uranium recovery in Agreement States is based on an explicit request by a State to have 11(e)(2) byproduct material authority as part of their Agreement. Presently, the States of Colorado, Texas, Utah, and Washington have such authority and have active licensing and regulatory programs in place.

The NRC's regulatory framework for uranium recovery is contained in 10 CFR Parts 20 and 40, and in 10 CFR 40 Appendix A, "Criteria Relating to the Operation of Uranium Mills and the Disposition of Tailings or Wastes Produced by the Extraction or Concentration of Source Material from Ores Processed Primarily for Their Source Material Content." Appendix A is the vehicle that NRC uses to conform its uranium recovery standards—primarily as they apply to the management of conventional uranium mills and mill tailings—to those of the EPA in 40 CFR Part 192. These regulations (both those of the NRC and EPA) are largely deterministic and do not reflect risk-informed, performance-based considerations. The Commission considered a revision to the regulatory framework for uranium recovery in 2000 that would have updated and risk-informed the regulations but chose not to pursue it.

The NRC does not have specific regulations governing the licensing of ISRs. Licensing decisions are made based on staff guidance contained in NUREG-1569, "Standard Review Plan for In-Situ Uranium Recovery Licensing" (SRP) and implemented through license conditions (NRC, 2003). The SRP is in the process of being revised to reflect risk insights and licensing experience. Licenses for ISR facilities also include a performance-based license condition that allows licensees to make certain changes without requesting NRC approval in the form of a license amendment. This licensing approach recognizes the unique process, geological, and hydrological characteristics at ISR sites. The EPA has been working on an ISR rule for several years and the NRC staff expects that it could be issued as a proposed rule in the second half of 2012. The NRC would then need to conduct a conforming rulemaking to make its regulations track those of EPA. In the absence of these regulations, the staff plans to continue its current licensing approach of using the SRP and performance-based license conditions.

Security Considerations in the Uranium Recovery Program

Uranium is not one of the radionuclides contained within the International Atomic Energy Agency's "Code of Conduct on the Safety and Security of Radioactive Sources." (IAEA, 2004). Accordingly, uranium recovery licensees were not included as part of the group of licensees that received Increased Controls Orders by the NRC. Security of licensed material at these facilities is governed under the requirements of 10 CFR 20.1801 and 10 CFR 20.1802.

4.5.2 RMTF Findings and Recommendations for Uranium Recovery

The risk environment for uranium recovery is different from power and Nonpower reactors as well as many materials uses. The environment was characterized in a risk assessment prepared by NRC staff and contractors in NUREG/CR-6733, "A Baseline Risk-informed, Performance-Based Approach for In Situ Leach Uranium Extraction Licensees," issued in September 2001, which influenced development of the Standard Review Plan (NRC, 2001). The risk assessment concluded that ISR facilities are of inherently low risk, noting that they "...contain no operating reactors, no fission products and no high radiation areas requiring extensive shielding; and they have operating records that confirm low exposures to workers and members of the public." The only significant operational radiological risk is posed by the yellowcake dryer which, under certain accident scenarios, could lead to occupational exposures in excess of NRC limits in 10 CFR Part 20. Radiological risk to members of the public is low. Uranium recovery facilities are not required to have offsite emergency plans since there are no credible accident scenarios that might exceed EPA's Protective Action Guidelines. The risks associated with uranium recovery have largely been environmental ones, in terms of potential or actual ground water contamination, long-term tailings management, and radon emissions.

Finding UR-F-1: Uranium recovery facilities are of low radiological risk to workers and members of the public under normal operational conditions and most accident scenarios.

Finding UR-F-2: Although the NRC staff has made inroads to risk-informed, performance-based licensing of uranium recovery facilities, the regulatory framework is largely a deterministic one. NRC regulations in 10 CFR Part 40 and Appendix A to 10 CFR Part 40 reflect the requirements of UMTRCA and EPA's rules in 40 CFR Part 192, which pre-date the Commission's move to a risk-informed, performance-based framework in the mid-1990s. Similarly, program guidance, especially for conventional mills, could benefit from a greater risk basis.

Finding-UR-F-3: The exact nature of the ISR rule under development by the EPA is not clear at this time; therefore, presents an uncertainty to adoption of the proposed risk management regulatory framework.

> Recommendation UR-R-1: Notwithstanding the current uncertainty associated with the EPA rulemaking, the NRC should adopt the proposed risk management regulatory framework to the uranium recovery program to provide greater efficiency, effectiveness, and predictability in policy development and regulatory decisionmaking.
>
> Recommendation UR-R-2: The NRC should work closely with the Agreement States and the regulated community to guide implementation of risk management in the uranium recovery program.

Earlier sections of this report have discussed the concept of defense-in-depth as it applies to different program areas within the NRC. Similar to the finding in Section 4.3 for the materials program, the concept of defense-in-depth is not commonly used as an explicit consideration in the NRC's regulation of uranium recovery. In large measure, this reflects the fact that uranium recovery is a relatively low-risk activity. There are instances, including design features and regulatory review of mill tailings impoundments, as well as the arrangement of injection, recovery and monitoring wells at ISR facilities that reflect defensein-depth considerations.

The NRC staff is required by 10 CFR 51.20 (b)(8) to prepare an environmental impact statement (EIS) for issuance of new source material license authorizing uranium recovery. EIS preparation involves a consideration of impacts or risks to the environment against benefits from a proposed action or alternatives to that action. Renewal or significant amendments to a uranium recovery may require an environmental assessment that can lead to an EIS. These requirements and practices reflect the current and future potential risks posed by uranium recovery in terms of potential ground water contamination, long-term tailings management, radon emissions, and other factors. The NRC staff has been forward-looking in addressing these risks from increased interest in ISR licenses by preparing a "Generic Environmental Impact Statement [GEIS] for In-Situ Leach Uranium Milling Facilities (NUREG-1910)" (NRC, 2009). The GEIS assesses environmental impacts of common environmental issues associated with the construction, operation, and decommissioning of ISL facilities, as well as the ground water restoration at such facilities, if they are located in particular regions of the western United States. In addressing environmental issues common to the ISL process, the NRC staff will use the GEIS as a starting point for its site-specific environmental review of license applications for new ISL facilities and applications to renew or amend existing ISL licenses.

> Finding UR-F-4: Consideration of environmental risks is a central part of the uranium recovery regulatory program.

> Recommendation UR-R-3: The NRC should include environmental reviews within the scope of its risk management framework.

The RMTF received input from several commenters about uranium recovery and risk management in response to the November 22, 2011 Federal Register notice. Among the comments provided were the following:

- Section 84 of the Atomic Energy Act mandates that the Commission consider risk to public health, safety, and the environment in its management of 11(e)(ii) byproduct material, to the extent practicable.

- Risk-informed, performance-based approaches are good public policy because they promote efficient use of already-limited agency, licensee, and other stakeholder resources.

- Uranium recovery is by its nature a low-risk activity, as noted in NUREG/CR-6733, the GEIS, and other studies. Moreover, the location of uranium recovery facilities in remote, low-population areas in the western United States also contributes to low risk to the public. Risks of physical injury to workers at uranium facilities are far greater than radiological risks from uranium.

- There is a common understanding of the terms risk-informed and performance-based in the uranium recovery industry.

- Deterministic approaches limit the flexibility of both the NRC and the industry to respond to lessons learned from operational experience and should be eliminated to the greatest extent possible.

- A key characteristic of a holistic risk management regulatory structure is a ranking of radiological risks from uranium recovery along with the risks associated with other NRClicensed activities so that it is clear where they fit into the overall risk picture.

- There are many opportunities to identify and prioritize activities that are amenable to a risk management approach: regulations, licensing actions, policy development, and inspection and enforcement prioritization. Specific examples from the uranium recovery perspective include remediation or restoration of UR facilities, applying timeliness in decommissioning to uranium recovery wellfields, and contaminated soil cleanup at uranium recovery facilities.

The RMTF considered these comments in light of the findings and recommendations in Chapter 2 on adoption of the proposed risk management regulatory framework and concluded that there is no inherent conflict between the framework and the views offered by commenters.

4.5.3 Implementation Options for Risk Management in Uranium Recovery

In the introduction to Chapter 4, three options were identified for implementation of the proposed risk management regulatory framework for the NRC. Following are what those options might look like for the uranium recovery program.

Option A: Continue the current approach

Under this option, the uranium recovery program would continue its current approach to risk-informed, performance-based programs. Insights from licensing reviews, environmental reviews, and inspection findings would be considered in programmatic development, including rulemaking, guidance development, inspection, and enforcement as resources permit. No significant new initiatives, including development of a Commission Policy Statement, would be undertaken based on risk considerations. Assuming that EPA issues an ISR rule, the NRC would conform its regulations to the rule. Periodic self-assessments of specific program areas or lessons learned in response to events or other factors would continue to be performed as necessary but not under the direction provided by an overall risk management framework.

<u>Option B: Implement the proposed risk management regulatory framework through selected guidance and rule changes</u>

This option would emphasize specific rule and guidance changes to implement the proposed risk management regulatory framework contained in the proposed Commission Policy Statement. This would include, assuming that EPA issues an ISR rule, making conforming changes based on a risk management framework. Development of updated guidance for ISR facilities, such as the revision to NUREG-1569, also would be based on risk management considerations, as would any guidance updates for conventional milling operations. It would also involve early and substantive outreach to the Agreement States and formation of a working group to determine where risk management might offer the greatest opportunities. Implementation of this option could take 5 years or more.

<u>Option C: Implement the proposed risk management regulatory framework through broad-scale regulatory framework changes</u>

The NRC's ability to make broad scale regulatory framework changes in the uranium recovery program is somewhat limited because of the requirement that NRC rules must conform to those generally applicable environmental standards developed by EPA. However, the agency could consider how the proposed risk management regulatory framework might lead to different licensing and oversight approaches within the constraints of the EPA rule. As noted in Option B, the Agreement States would play a substantive role in developing this option. Implementation of this option would take more than 5 years and could be as long as 10 to 15 years.

The RMTF recommends Option B. It provides a way to transition the agency to the proposed risk management regulatory framework and process in a reasonable amount of time and reflects ongoing operational priorities and resource constraints. It will also assure continuity and stability in the regulated community through the continued use of existing regulations and guidance while implementing the risk management framework and process.

4.5.4   References

(IAEA, 2004)   International Atomic Energy Agency, Code of Conduct on the Safety and Security of Radioactive Sources, January 2004.

(NRC, 2001)   U.S. Nuclear Regulatory Commission, "A Baseline Risk-informed, Performance-Based Approach for In Situ Leach Uranium Extraction Licensees," NUREG/CR-6733, September 2001, Agencywide Documents Access and Management System (ADAMS) Accession No. ML012840152.

(NRC, 2003)   U.S. Nuclear Regulatory Commission, "Standard Review Plan for In-Situ Uranium Recovery Licensing," NUREG-1569, June 2003, ADAMS Accession No. ML031550272.

(NRC, 2009)   U.S. Nuclear Regulatory Commission, "Generic Environmental Impact Statement for In-Situ Leach Uranium Milling Facilities," NUREG-1910, May 2009, ADAMS Accession No. ML091480244.

4.6   Fuel Cycle

4.6.1   Background

The NRC regulates major fuel cycle facilities, including those involved in conversion of uranium ore to uranium hexafluoride (UF_6), gaseous centrifuge and diffusion enrichment, reactor fuel fabrication, plutonium processing, and UF_6 deconversion. Reactor fuel fabrication facilities include those that produce lowenriched uranium, highenriched uranium, and mixedoxide (Pu+U) fuels. The NRC also regulates possession of small amounts of special nuclear material (SNM), usually for research purposes.

The regulations applicable to licensing of fuel cycle facilities vary because of differences in the nature and amounts of the materials licensed and the historical origins of the facilities. Hazards at these facilities include those from radioactive materials, toxic chemicals, and inadvertent nuclear criticality accidents. Regulation of risk to workers is a major focus for these facilities since these workers typically are in close proximity to the hazards.

Regulations under 10 CFR Part 70, "Domestic Licensing of Special Nuclear Fuel," apply to licensees authorized to possess SNM, which consists of specified fissile materials. In terms of fuel cycle facilities, 10 CFR Part 70 applies to major low-enriched uranium fuel fabrication facilities, high-enriched uranium processing facilities, and new enrichment facilities (gaseous centrifuge). For these major fuel cycle facilities, Subpart H, "Additional Requirements for Certain Licensees Authorized To Possess a Critical Mass of Special Nuclear Material," to 10 CFR Part 70, requires licensees to perform an integrated safety analysis (ISA). Regulations under 10 CFR Part 70 also apply to licensees possessing amounts of SNM less than that required to perform ISAs. Including the ISA requirement, 10 CFR Part 70 also applies to the Mixed-Oxide Fuel Fabrication Facility (MOX FFF) as a plutonium processing facility.

As part of an ISA, Subpart H to 10 CFR Part 70 requires licensees or applicants to identify all accident sequences that could lead to "high" or "intermediate consequences" (as defined in the rule) to workers or the public. The rule also requires items relied on for safety (IROFS) to be applied to credible sequences sufficient to make high-consequence sequences "highly unlikely," and intermediate-consequence sequences "unlikely".[1] The rule allows considerable flexibility to licensees for ISA methods and definitions of terms. For this reason, ISAs vary in methods and acceptance criteria. Some use considerable quantitative analysis and quantitative criteria. It is worth noting that for the purposes of the rule, ISAs may be conservative rather than realistic estimates. Unlike a Probabilistic Risk Assessment (PRA), no ISA sums risk from multiple sequences to individuals.

Regulations under 10 CFR Part 76, "Certification of Gaseous Diffusion Plants," apply to certification of pre-existing DOE gaseous diffusion enrichment plants. The regulation does not require an ISA or PRA, but it does require technical specifications. There is just one operating 10 CFR Part 76 facility at this time.

The regulations in 10 CFR Part 40 address domestic licensing of source material (uranium and thorium ores). In terms of fuel cycle facilities, 10 CFR Part 40 applies to conversion

---

[1]   Discussions and guidelines on which accident sequences are considered "highly unlikely" and "unlikely" are provided in NUREG-1520, "Standard Review Plan for the Review of a License Application for Fuel Cycle Facility" (NRC, 2010). In addition to the qualitative guideline, the quantitative guideline for "unlikely" is less than 10^{-4} per year and less than 10^{-5} per year for "highly unlikely."

and deconversion facilities as licensees authorized to possess source material. The safety requirements for 10 CFR Part 40 uranium conversion licenses are general, not prescriptive, and do not require performance of any risk assessment. However, the staff is working on a proposed amendment to 10 CFR Part 40 that would require licensees authorized to possess more than 2,000 kilograms (kg) of UF_6 to perform ISAs.

Prior Risk-Informed/Performance-Based Efforts for Fuel Cycle

As directed in the staff requirements memorandum (SRM) to SECY-99-100, the NRC established a Risk Task Group (RTG) to define a framework for risk-informing certain nonreactor activities, including fuel cycle facilities (NRC, 1999). The RTG developed methods and criteria for risk-informing these activities, which were published in "Risk-informed Decisionmaking for Nuclear Material and Waste Applications" (NRC, 2008). This guidance, as directed, included guidelines on how safe is safe enough, analogous to the reactor safety goals, as well as decision algorithms for certain situations.[2]

As applicable to fuel cycle facilities, Subpart H to 10 CFR Part 70 is a risk-informed, performance-based rule that addresses all three elements of the risk triplet: What can go wrong? What are the consequences? What is the likelihood? The regulation includes qualitative performance requirements (10 CFR 70.61, "Performance Requirements") formulated in terms of consequences and likelihoods of accident sequences. For example, "highconsequence" events are required to have IROFS to be applied until they are "highly unlikely". This rule became final in 2000. As such, the industry and staff have more than 10 years of experience in implementing this ISA rule. As a result, risk concepts are now more widely used in this industry. Lessons based on this experience were recently incorporated into Revision 1 of the fuel cycle Standard Review Plan (NRC, 2010).

ISAs are required for new fuel cycle facilities, including the recent gaseous centrifuge plants (GCP). Information on consequences and riskindex likelihood results from two of the GCP ISAs were used to prioritize IROFS for the operational readiness reviews of these plants.

Ongoing Risk-Informed/Performance-Based Activities for Fuel Cycle

Issues relating to implementation of the ISA rule (Subpart H to 10 CFR Part 70) continue to be addressed by developing additional guidance. These issues include defining what plant features must be IROFS and clarifying reporting requirements. These definitions and clarifications relate to whether a given process design is acceptable for compliance with 10 CFR 70.61 performance requirements; therefore, they relate to controlling risk. ISA analyses are also being systematically revisited by certain licensees in response to specific findings.

The NRC inspection program and licensee corrective action programs inherently result in reevaluation, on a continuing basis, of parts of the ISAs in response to identified deficiencies or potential improvements. Revision of the fuel cycle oversight process (F-COP) to be more risk-informed, transparent, and predictable was proposed to the Commission in SECY-11-0140, "Enhancements to the Fuel Cycle Oversight Process," and approved with added direction in a Staff Requirements Memorandum (NRC, 2011, and 2012). The proposed approach includes

---

2   Detailed algorithms were provided for decisions related to new or revised safety requirements. These
     algorithms, in part, made use of pre-existing NRC guidance for regulatory analysis of rulemaking and backfit.
     However, detailed algorithms for other regulatory applications, such as license reviews and inspections,
     were not provided in this document.

a significance determination process, cornerstones, and an action matrix; all to be developed within 5 years.

4.6.2 RMTF Findings and Recommendations for Fuel Cycle

Most fuel cycle facilities consist of a large number of separate process nodes. For example, about 60 to 80 process steps are required to produce a light-water reactor fuel bundle, given enriched UF_6 as input. The safety controls and designs of processes are very diverse. Safety controls typically are not complex automatic controls requiring power to complete their safety mission. Safety or process controls usually are manual and governed by procedure and training. Process interactions may occur because of output and input relationships, or violent events, but processes are otherwise independent. Reliance on power or other motive forces for safety functions is limited. Confinement by ventilation is one exception for the MOX FFF. Chemical and nuclear criticality hazards vary substantially in magnitude between processes and licensees. Radiological hazards are not a dominant hazard except for the MOX FFF.

ISAs share some elements in common with PRAs; namely, all significant accident scenarios must be evaluated and the consequences estimated. Consequences of fuel facility chemical and criticality accidents are acute effects, such as death or irreversible health effects. These effects have thresholds; therefore, the relevant risk metric is the probability of the discrete health effect instead of an expected value of dose. While most, but not all, ISAs have developed some form of quantitative or semiquantitative evaluation of accident sequence consequences and frequencies, these quantitative evaluations may be highly conservative and may not have the same consequence end point. As such, and as noted above, none of the ISAs sums accident sequence frequencies to obtain a metric of risk to specific individuals or of collective risk. Quantification of overall risk is further hampered by the fact that technologies for assessing quantitative risk to individuals onsite or offsite for the collection of processes in a fuel cycle facility have not been fully developed.[3]

The RMTF received the following comments regarding fuel cycle facilities in response to the November 22, 2011 *Federal Register* notice:

- Due to the wide variety of fuel cycle and other licensed operations, there is not a common understanding of the terms risk-informed, performance-based, and defense-in-depth within NRC or with these licensees.

- The NRC understands and attempts to build upon the different levels of risk associated with the various fuel cycle and other materials users and should build upon these efforts to establish and maintain a flexible regulatory approach that allows for and reflects the relative risks of such licensed activities.

- The NRC could and should improve its inspection of fuel cycle facilities to be more risk-informed and performance-based using the results of the facility-specific ISAs and required annual updates. Inspectors should spend less time evaluating strict compliance and focus more on evaluating the actual safety significance of facility operations.

---

3   For example, an analytical code for calculating potential chemical or radiological risk to an individual worker because of potential accidents at multiple locations in a plant, for a single accident scenario, or sum across all accident scenarios, does not exist. A computer code similar to MACCS, a probabilistic offsite consequence analysis code for potential power plant accidents, with a heavy gas model applicable to UF6 releases also would be needed to estimate risk to individuals outside the controlled area to account for location and meteorology variations.

- The licensing process, particularly amendments and renewals, could be improved by better ensuring that requests for additional information reflect a true safety benefit or need as opposed to nonsafety-significant matters.

- The NRC should inform its deliberations on a holistic risk management structure by a thorough stakeholder engagement process in which specific input is solicited by the NRC from various categories of licensees, as was done in the development of the Safety Culture Policy statement.

- The NRC should consider a more risk-informed approach to the annual license fee rule, given that license fees for a Category I fuel cycle facility are higher than for a reactor when the overall risk from fuel facilities is orders of magnitude lower than reactors.

- As a class of licensees, fuel cycle facilities might benefit most from a more risk-informed, performance-based approach since they are primarily chemical processing plants that safely and securely handle licensed radioactive materials.

The proposed risk management regulatory framework described in Chapters 2 and 3 applies to managing risk through the NRC fuel cycle regulatory program. Similar general concepts of risk-informing decisions in the nonreactor areas of NRC, including fuel cycle facilities, were laid out in the document "Risk-informed Decisionmaking for Nuclear Material and Waste Applications" (NRC, 2008).

Events with adverse consequences to persons can occur that involve fuel cycle facilities regulated by the NRC. Decisions need to be made concerning long-term policies, requirements, practices, and resource allocations to manage these potential adverse events. This is true if the adverse events result from routine plant operations, accidents, or malevolent acts. Hence, a common risk management situation confronts those who must make decisions about radiation health, safety, and security. NRC regulatory practices for both safety and security apply structured reasoning concerning consequences, likelihood, and other stable aspects about potential adverse events to set long-term policies, regulatory requirements, and allocation of resources. Therefore, it would be useful if both disciplines recognized the commonalities of their risk management frameworks and facilitated communication by a joint refinement of their risk management approach and terminology.

The requirement for and definition of defense-in-depth in safety of fuel cycle facility processes is explicit in 10 CFR 70.64(b). That definition is identical to the one contained in SECY-98-144, "White Paper on Risk-Informed and Performance-Based Regulation," which defined "risk-informed," "defense-in-depth," and related concepts (NRC, 1998). In addition, the double contingency principle has been an industry standard in the nuclear criticality safety field for decades and is also mandated by 10 CFR 70.64(a)(9). Thus, defense-in-depth is applied in regulation of fuel cycle facilities consistent with Commission guidance. However, unlike power reactors, where more permanent barriers and controls such as a containment are built into the design and operation, defense-in-depth for each fuel cycle unit process is different. As new processes are added or existing ones are changed, the design and maintenance of defense-in-depth at these facilities are based on the characteristics of the most current operations. Therefore, defense-in-depth is a continuing process at fuel cycle facilities, not one permanently established by the initial design.

Finding F-F-1: The current fuel cycle regulatory approach incorporates several elements of the proposed risk management regulatory framework, such as the use of ISAs to identify safety significant items, and the implementation of a revised fuel cycle oversight program as directed by the Commission.

Finding F-F-2: The concept of defense-in-depth, as embedded in fuel cycle regulatory requirements and practices, is consistent with Commission guidance. Its implementation changes as the processes change at the fuel cycle facilities.

Recommendation F-R-1: The fuel cycle regulatory program should continue to evaluate the risk and the associated defense-in-depth protection by using insights gained from ISAs. ISAs should continue to evolve to support regulatory decisionmaking.

4.6.3   Implementation Options for Risk Management in Fuel Cycle Facilities

In the introduction to Chapter 4, three options were identified for implementation of the proposed Risk Management Regulatory Framework for the NRC. Listed below are what those options might look like for fuel cycle programs.

Option A: Continue the current approach

Under this option, the fuel cycle programs would continue current approaches to risk-informed, performance-based programs. Insights from operational data and enforcement data would be considered in programmatic development, including rulemaking, guidance development, and inspection and enforcement. This approach would include the development of a more risk-informing performance-based fuel cycle oversight process, as authorized by the SRM to SECY-11-0140. No other major new initiatives, including development of a Commission Policy Statement, would be undertaken based on risk considerations. Periodic self-assessments of specific program areas or lessons learned in response to events or other factors would continue to be performed as necessary, but not under the direction provided by an overall risk management framework.

Option B: Implement the proposed risk management regulatory framework through selected guidance and rule changes

This option would emphasize specific rule and guidance changes over time to implement   the proposed risk management regulatory framework as contained in the proposed Commission policy statement. This would include outreach to fuel cycle stakeholders and formation of an NRC Fuel Cycle Working Group to determine where risk management might offer the greatest opportunities. These new opportunities might include guidance and tools development to support risk-informing for various regulatory activities, including licensing, inspection, and policy development, with particular emphasis on revision of the F-COP and evolution of ISAs. Other new actions may include enhancement of riskrelated staffing and training, particularly in support of a risk-informed F-COP, development and implementation of methods to support prioritizing regulatory activities, and development of risk assessment tools applicable to fuel cycle facilities.

Specific risk-informed changes to licensee safety programs may be identified and implemented as a result of the proposed revision of the F-COP.

Practical Considerations

The staff would need to formulate objectives; then develop, test, and implement guidance. The NRC's approach to involving stakeholders in the development of specific changes to regulatory practices is especially important for this program area because of the large number and variety of fuel cycle unit operations. Often, even at one facility, detailed knowledge about their risk is dispersed among licensee staff members instead of a few NRC experts. Therefore, these stakeholders have knowledge needed in the development of the risk management process. Likewise, developing applicable risk assessment tools would require a certain amount of external technical support because of the limited availability of qualified staff. Training would need to focus on practical applications of risk methods, in particular, to support a more risk-informed and performance-based F-COP with a significance determination process.

This option, as in the case of the fuel cycle oversight process revision, faces the technical challenge of risk-informing regulatory activities without expending large resources. The above developments in support of an overall risk management approach could be implemented in a phased manner over a period of years. Rulemaking or backfit justification would be necessary if a particular change in regulatory practice so required.

Option C: Implement the proposed risk management regulatory framework through broad-scale regulatory framework changes

This option would involve a revision of the basic framework and regulations used for licensing and oversight for fuel cycle facilities. Current regulations do not require quantitative risk assessment of fuel cycle facilities; therefore, rulemaking would be needed.

Practical Considerations

This option would require comprehensive development and application of quantitative risk assessment technology for all fuel cycle facilities. This would first require development and testing of technical PRA tools for such facilities. The rulemaking would have to follow the process and schedule required.

The RMTF recommends Option B. It provides a way to transition the agency to the proposed risk management regulatory framework in a reasonable amount of time and reflects ongoing operational priorities, such as revision of the F-COP. It will also ensure continuity and stability in the regulated community through the continued use of existing regulations and guidance while implementing the proposed risk management regulatory framework.

4.6.4   References

(NRC, 1998)   U.S. Nuclear Regulatory Commission, SECY-98-144, "White
                     Paper on Risk-Informed and Performance-Based Regulation,"
                     June 22, 1998, Agencywide Documents Access and Management
                     System (ADAMS) Accession No. ML992880068.

(NRC, 1999)   U.S. Nuclear Regulatory Commission, Staff Requirements
                     Memorandum Regarding SECY-99-100, "Framework for
                     Risk-informed Regulation in the Office of Nuclear Material
                     Safety and Safeguards," June 28, 1999, ADAMS Accession
                     No. ML003751901.

(NRC, 2000)   U.S. Nuclear Regulatory Commission, Letter from B.J. Garrick and
                     D.A. Powers to R.A. Meserve, "Use of Defense-in-depth in Risk
                     Informing NMSS Activities," ACRS Letter, May 25, 2000, ADAMS
                     Accession No. ML003718610.

(NRC, 2001)   U.S. Nuclear Regulatory Commission, "Integrated Safety Analysis
                     Guidance Document," NUREG-1513, May 2001, ADAMS
                     Accession No. ML011440260.

(NRC, 2008)   U.S. Nuclear Regulatory Commission, "Risk-Informed
                     Decisionmaking for Nuclear Material and Waste Applications,
                     Revison 1," February 2008, ADAMS Accession No. ML080720238.

(NRC, 2010)   U.S. Nuclear Regulatory Commission, "Standard Review
                     Plan for the Review of a License Application for a Fuel Cycle
                     Facility," NUREG-1520, Rev. 1, May 2010, ADAMS Accession
                     No. ML101390110.

(NRC, 2011)   U.S. Nuclear Regulatory Commission, SECY-11-0140,
                     "Enhancements to the Fuel Cycle Oversight Process,"
                     October 7, 2011, ADAMS Accession No. ML111180708.

(NRC, 2012)   U.S. Nuclear Regulatory Commission, Staff Requirements
                     Memorandum Regarding SECY-11-0140, "Enhancements to
                     the Fuel Cycle Oversight Process," January 5, 2012, ADAMS
                     Accession No. ML120050322.

4.7   Interim Spent Fuel Storage

4.7.1   Background

The NRC has regulatory responsibility for the storage of spent nuclear fuel (SNF) and reactor-related greater than Class C (GTCC) waste in an independent spent fuel storage installation (ISFSI); and the storage of SNF, solid highlevel radioactive waste (HLW) and reactor-related GTCC waste (in solid form) in a monitored retrievable storage installation (MRS). The NRC carries out its responsibilities in this area through rulemaking, licensing, and inspection.

Regulatory background and framework

The need for storage of commercial power reactor spent fuel stems from the Federal Government's 1977 decision to forego reprocessing. While power reactor facilities include pools for storing SNF, they were not designed to hold all of the SNF generated by reactors during their operating life. Dry storage in ISFSIs was selected as a way to expand a reactor facility's SNF storage capacity. ISFSIs were licensed under the agency's general regulation 10 CFR Part 70, "Domestic Licensing of Special Nuclear Material," until late November 1980, when a new regulation specifically for ISFSIs was added as 10 CFR Part 72, "Licensing Requirements for the Storage of Spent Fuel in an Independent Spent Fuel Storage Installation." This rule covers both wet and dry storage and licenses individual ISFSIs, the specific-license process.

The regulation has changed over time to address needs arising from policy and legislative requirements and those identified through petitions for rulemaking. To meet the requirements set out in the Nuclear Waste Policy Act (NWPA) of 1982, as amended, 10 CFR Part 72 was twice modified to introduce a licensing mechanism for MRSs and a mechanism for licensing SNF dry storage technologies, a general-license option coupled with approval of dry storage system designs. In addition to storage of SNF and solid HLW (MRS only), 10 CFR Part 72 now includes ways to license the storage of solid reactor-related GTCC waste at ISFSIs and MRSs.

Other relevant regulations for storage include the security requirements for licensees in 10 CFR Part 73, "Physical Protection of Plants and Materials," as well as the Commission's environmental protection requirements in 10 CFR Part 51, "Environmental Protection Regulations for Domestic Licensing and Related Regulatory Functions." Also, U.S. Environmental Protection Agency (EPA) regulations contain public dose limits with which licensees must comply (e.g., 40 CFR 191.03(a)).

Licensing

The NRC reviews applications for specific licenses and specific-license amendments for ISFSIs and MRSs and certificates of compliance (CoC) and CoC amendments for SNF dry storage system designs. These reviews are conducted using guidance in the applicable Standard Review Plans (SRPs), NUREG-1567, "Standard Review Plan for Spent Fuel Dry Storage Facilities" (NRC, 2000), and NUREG-1536, Revision 1, "Standard Review Plan for Spent Fuel Dry Storage Systems at a General License Facility" (NRC, 2010). Applicable interim staff guidance (ISG) documents are also used, in addition to regulatory guides (RGs) and industry standards, to the extent appropriate. To address license and CoC renewal, the NRC recently published NUREG-1927, "Standard Review Plan for Renewal of Spent Fuel Dry Cask Storage System Licenses and Certificates of Compliance" (NRC, 2011). For obligations under the

National Environmental Policy Act, the NRC prepares environmental impact statements and environmental assessments, as warranted, for each approved license, CoC, and amendment.

A general license is granted by regulation (i.e., 10 CFR Part 72) to 10 CFR Part 50 and 10 CFR Part 52 power reactor licensees. To make use of an approved SNF dry storage system, the general licensee completes an evaluation in accordance with the requirements in 10 CFR 72.212 to determine that the selected system is appropriate for its site. These evaluations are not reviewed by the NRC licensing staff; instead, they are reviewed as part of an inspection of the general license ISFSI (e.g., pre-operational or dry-run inspection). Conservative requirements in dry storage system safety analyses (e.g., design-basis events and margins of safety), the evaluation of siting factors, the application of quality assurance to cask design and fabrication, and physical protection at the site provide the NRC with reasonable assurance that this regulatory oversight approach will ensure SNF is stored in an adequately safe manner under the general license.

The above review guidance does not apply to one facility: the GE Morris wet-storage, or pooltype, ISFSI in Illinois. This facility is the only wet-storage ISFSI licensed in 10 CFR Part 72. In 2004, the NRC renewed the license for this facility using appropriate guidance from its other regulatory programs (NRC, 2004).

Inspection

The NRC inspects 10 CFR Part 72 licensees, CoC holders, and system component fabricators to ensure safe and secure storage of SNF. Inspections focus on risksignificant activities, such as fabrication of important-to-safety components by fabricators. Baseline inspection frequencies differ for the types of 10 CFR Part 72 entities listed above. For ISFSIs, there is no set frequency; instead, inspections are eventdriven (e.g., before loading begins at a new ISFSI). For the other entities, frequencies are based on factors such as available resources, the population of the entity type, the level of production, and the frequency of design modifications. Other factors may result in increased inspection frequency or additional reactive inspections.

Security

Physical security plans for all licensees are designed to protect against sabotage and to promote the common defense and security. After September 11, 2001, security assessments were performed on four ISFSI/dry storage system designs (or casks) considered representative of the current fleets of dry storage systems. The purpose of the studies was to identify if licensees needed to take immediate actions to mitigate potential security concerns above and beyond the requirements in the ISFSI security orders issued in 2002. None was identified. Risk was considered as one factor in the selection of security scenarios to be analyzed as part of the assessments. As part of any licensing review for a specific-license ISFSI, an assessment of the ISFSI's security plans is performed to ensure that the licensee can provide high assurance of adequate protection of public health and safety and the common defense and security. For general-license ISFSIs, the licensee updates its security plans to reflect the presence of the ISFSI and verifies that the ISFSI does not decrease the effectiveness of the reactor security plans. These revised plans are subject to onsite inspection by the NRC staff. While the assessments did not indicate the need for immediate actions, they did indicate there was a need to consider changes to the ISFSI security regulations. The staff is currently engaging with stakeholders and gathering further information to support the regulatory basis for an ISFSI security rulemaking.

Previous and Current Risk-Informed, Performance-Based Efforts

The NRC has updated and is currently updating its guidance in this area to make it more risk-informed and performance-based. These efforts include prioritization of the review procedures in the SRPs, NUREG-1536, and NUREG-1567. The prioritization method is based on a qualitative assessment, based on staff judgment, of the risk and impact on defense-in-depth, where applicable, of each review procedure. For improvements to guidance related to burnup credit, the NRC produced a report investigating the probabilities of improper loading of contents into a dry storage system, known as misloads, to inform guidance addressing these events with respect to nuclear criticality safety (NRC, 2011a). Staff judgment regarding risk is being used to evaluate the appropriate margins for criticality safety for misloads. The NRC is also assessing the level of review effort needed for dry storage system shielding and radiation protection evaluations and the types of CoC and license conditions necessary to adequately capture the system design and operations requirements. While this assessment was initiated based on concerns derived from a particular licensee's and dry storage CoC holder's activities, interactions with other stakeholders indicated the need to evaluate staff guidance in shielding and radiation protection. Guidance related to SNF condition (e.g., intact, damaged) recently was modified from a prescriptive set of definitions to a framework for defining the SNF condition in terms of fuel-specific or system-related functions that must be performed, if any, to ensure compliance with applicable regulatory requirements (NRC, 2007).

Changes have been made to 10 CFR Part 72 based on licensing experience and scientific and technological advancements. A notable example is the 2003 modification of the requirements addressing the geological and seismological site characteristics for specific-license facilities storing SNF in dry casks. For facilities licensed before the regulation change in October 2003, probabilistic techniques were allowed for use as a site selection criterion but not for determining the design earthquake (DE). The DE for sites in areas with known potential seismic activity was to be equal to the safe shutdown earthquake for a nuclear power plant. Considering the approach to the dry storage system (i.e., dry cask) designs—particularly their robustness against events that challenge cask integrity more than a seismic event as well as the parameters and conditions that lead to lower radiological consequences of a release versus a release from a power plant—the NRC believes the seismically induced radiological risk associated with facilities using dry casks is significantly less than the risk associated with a nuclear power plant. Therefore, the regulation was changed to allow use of a DE level more commensurate with the risk associated with an ISFSI or MRS. Requirements for determining the appropriate DE level include addressing the uncertainties of the DE estimate using a probabilistic seismic hazard analysis or suitable sensitivity analyses.

The NRC published NUREG-1864, "A Pilot Probabilistic Risk Assessment of a Dry Cask Storage System at a Nuclear Power Plant" in 2007 (NRC, 2007a). The study's focus was limited in the way it was conducted and in its scope of applicability; it was geared toward developing and demonstrating an assessment method. Several aspects of a risk assessment, such as human factors and sensitivity and uncertainty analyses, were not performed. The study evaluated a particular cask design at a specific site and relied on several assumptions, including the storage system or cask being fabricated as designed and operated correctly (i.e., no operator error). Since the quantitative results were quite low (given in terms of latent cancer fatality due to radioactive release), further studies were not pursued. However, the results have been considered by inspection staff to reinforce staff's understanding as to those aspects of operations that are more risksignificant, and the results are used to inform review evaluations of fuel rod and storage canister integrity under accident conditions.

A separate probabilistic risk assessment (PRA) entitled "Probabilistic Risk Assessment of Bolted Storage Casks: Quantification and Analysis Report," was performed by the Electric Power Research Institute (EPRI) and published in December 2003 (EPRI, 2003). The method of evaluation, the storage system design, and the site evaluation parameters differ from the NRC study, but the quantitative results for it are also quite low.

In addition to these two PRAs, a human reliability analysis (HRA) study was recently completed and published as NUREG/CR-7016, "Human Reliability Analysis—Informed Insights on Cask Drops" (NRC, 2012), and NUREG/CR-7017, "Preliminary, Qualitative Human Reliability Analysis for Spent Fuel Handling" (NRC, 2012a). The goals of this study were: to identify what should be included in a qualitative HRA for SNF and cask handling operations, to demonstrate that "A Technique for Human Event Analysis" (ATHEANA) HRA methods can be used for these types of operations, and to serve as a starting point for building a technical basis for potential improvements to procedures and practices in SNF handling operations. Though the studies were not done in the context of a larger or plant-specific PRA, they use the NRC's pilot PRA study and EPRI's PRA to provide information about SNF and cask handling operations. The HRA study attempted to be generic in scope; it was not tied to a specific nuclear power plant. The study did focus on two SNF cask designs, but this was done to encompass the scope of complexity in handling operations. The study included consideration of cask misloads in addition to cask drops and how human performance can plausibly lead to radiological consequences that affect plant personnel and, to a lesser extent, the public and the environment.

4.7.2 RMTF Findings and Recommendations for Spent Fuel Storage

To a certain extent, 10 CFR Part 72 requirements are informed by risk insights. For example, 10 CFR 72.48, "Changes, Tests, and Experiments," allows some changes to be made without NRC review and approval. The requirement for radioactive releases and direct radiation exposure to be maintained "as low as is reasonably achievable" implies the use of risk to determine if effluent and direct radiation levels are sufficiently low. The Backfit Rule in 10 CFR Part 72.62 requires that the increase in protection of personnel and the public health and safety be sufficient to justify the cost of implementing the backfit.

SNF dry storage systems are designed to be robust, passive systems. The systems are designed to withstand the effects of "worst case" events or designbasis events and phenomena while maintaining the capabilities to provide adequate shielding and confinement of radioactive contents and to prevent nuclear criticality. The systems are designed to perform these functions while requiring minimal maintenance or repair. This design approach is the basis for the current regulatory approach, with requirements aimed at ensuring the systems are robust and that consequences from any radiological release are low. For specific-license ISFSIs, where the characteristics of the site and its surroundings are known, consideration of these characteristics can allow for certain events and phenomena not to be considered. Dry storage systems are designed, however, to be useable at as many potential sites as possible. Exclusion of consideration for certain events and phenomena is not permitted unless the design CoC includes a condition limiting the design's use to those sites where the excluded events and phenomena are not credible. Further, specific-license ISFSIs may use a dry storage design that significantly differs from the approved (by CoC) dry storage system designs. An example of this scenario is the ISFSI at Ft. St. Vrain, which is a vault-type design.

The Commission has directed the NRC staff to revisit the paradigm for SNF storage and transportation to include evaluating the dry storage of SNF for periods significantly in excess of those envisioned previously. Therefore, the licensing and inspection programs are being reevaluated to determine the enhancements needed to provide adequate regulatory oversight of extended storage. Any risk assessment and risk-informing efforts should adequately account for these activities.

Consideration also should be given to the nature of the dry storage industry. This industry basically consists of four major vendors in competition to provide dry storage systems, which sets up a dynamic in which design margins are reduced to increase system capacity and reduce costs. In response, the NRC has used a conservative approach in its SFS regulations, guidance, and licensing practices to minimize the likelihood of adverse consequences to public health and safety.

The current situation and trends should influence the direction taken to implement the proposed approach to risk-inform the interim SNF storage program. The following summarizes the situation and trends relevant to risk-informing considerations:

- No additional wet storage applications are expected; thus, licensing activity for wet storage is expected to be very limited and no specific guidance has been developed for licensing wet storage facilities.

- While no applications have been submitted for a MRS, the future of MRS is tied to the U.S. policy under deliberation.

- The trend in dry storage is toward the general license option; therefore, few, if any, specific-license applications are expected beyond amendment applications and license renewal of the few existing facilities.

- The trend toward general licenses could change based on national policy changes or any transition of general license to specific license by decommissioning 10 CFR Part 50 licensees or other factors.

- Amendment applications for specific licenses are few and limited in scope; therefore, most of the review is for dry storage system designs.

- Guidance for specific-license reviews refers substantially to guidance for dry storage systems.

- Dry storage system design CoC and CoC amendment applications (predominantly amendment applications) continue to be submitted frequently.

- Licensees and Coc holders have started to submit applications for license and CoC renewals as the initial license and CoC terms have begun to expire.

As such, the implementation of the approach should be influenced by the potential for benefit in the oversight of the different interim storage entities.

Beyond the current situation and trends in SNF storage, there are other factors that require consideration. Some factors may be driven by changes in technology in other aspects of the fuel cycle. For example, changes in fuel designs—whether for current or future reactor design generations—may influence what may be necessary to store such fuel after its use. Reactions to current events, such as national or regulatory policy, also should be considered. At present, there is some discussion on the need to move SNF more quickly from SNF pools into dry storage. This discussion was prompted by the recent events associated with the Fukushima Dai-ichi nuclear station. Also, at some point, the SNF will have to be transported. Additional discussion on transportation of nuclear material is provided in Section 4.8.

In making its findings and recommendations, the Risk Management Task Force also considered comments on spent fuel storage submitted in response to its November 22, 2011, Federal Register notice. These comments included the following:

- The risk presented by dry cask storage is much lower than for power reactors, but the regulatory framework for SFS calls for a relatively high degree of information and involves an unusually protracted approval process.

- PRAs performed to date—both by the NRC and EPRI—reached the same conclusion: the risks of dry storage ranged from very low to extremely low. However, no subsequent action was taken to review the regulatory framework based on these findings.

- Regulations for SNF storage and transport need to be cohesively linked since many of the casks are dual purpose. Ideally, this would be done under one set of regulations.

- The regulations in 10 CFR Part 71 and Part 72 are not risk-informed. As a result, they are not conducive to efficient and timely movement of SNF from pools to storage. The ability to move fuel from spent fuel pools to storage, or ideally a final repository, is a critical lesson learned from the event at Fukushima Dai-ichi in Japan.

- The regulations in 10 CFR Part 71 and 72 are performance-based. As such, RGs are highly dependent on identifying methods acceptable to the NRC. Presently, however, many requirements are implemented through ISG, which lacks the rigor, reliability, and predictability of RGs.

- The NRC should apply lessons learned from PRAs performed to date in its comprehensive review of the regulatory framework for spent fuel storage and transportation.

As already noted, the SNF dry storage industry is a fairly small and competitive industry with basically four major vendors. Their focus is to provide dry storage systems that meet the needs of current and potential customers with the most costeffective designs. This means there are great incentives to find ways to meet regulatory requirements that allow the greatest flexibility. Additionally, any flexibility in the regulatory requirements that might be achieved through risk-informing of the regulations would be of interest to the industry. Activities such as the EPRI PRA are indicative of industry interest in and objectives for pursuing risk-informed regulations and regulatory practices in the interim SNF dry storage area.

The regulations in 10 CFR Part 72 include performance-based requirements. The annual dose limits and accident dose limits in 10 CFR 72.104 and 72.106, "Controlled Area of an ISFSI or MRS," are examples of these performance-based requirements. However, other NRC programs have developed specific riskacceptance goals. For example, the reactors program uses core damage frequency as a goal in RG1.174, "An Approach for Using Probabilistic Risk Assessment in Risk-Informed Decisions on Plant-Specific Changes to the Licensing-basis." In a letter to the Chairman in May 2000, the Chairman of the Advisory Committee on Reactor Safeguards (ACRS) and the Advisory Committee on Nuclear Waste stated that implementation of regulations within a risk-informed framework requires the establishment of risk-acceptance goals (NRC, 2000a). No such goals have been developed for regulation of dry storage activities, although probabilities have been used in specific licensing actions to inform decisions on the need to consider certain events and phenomena. Development of a consensus on an appropriate goal and the supporting basis will require a significant effort and will have to be set by a Commission Policy Statement.

The NRC revises its guidance periodically or as other circumstances (e.g., development of new information) warrant. With these revisions, engineering judgment and insights from experience related to risk inform decisions on guidance changes. Also, changes to 10 CFR Part 72, such as a recent rulemaking to extend the license and CoC term limits, involve consideration of risk. Often, however, this process is not guided or informed by risk information gained from risk assessments. This was pointed out by the ACRS in reference to the prioritization of the review procedures in the latest revision of NUREG-1536 when it was presented to the Committee. As previously stated, some risk studies have recently been performed, but they are few in number and limited in scope, applicability, and purpose. However, these studies could aid in the identification of risk-information needs. Some information useful to the development of risk information is available, but it has not been collected in a systematic way geared toward risk-informing. This information includes the staff's technical knowledge and operational, licensing, and inspection experience.

Finding S-F-1: The regulatory approach for SNF storage is largely based on meeting applicable industry consensus standards and conservative guidance to ensure adequate safety margins in the facility and cask designs and operations. More recently, insights from a limited number of risk studies have been gradually factored into this regulatory approach. Furthermore, though qualitative, a systematic approach that parallels answering the risk triplet was used in the latest revision of the Standard Review Plan.

Recommendation S-R-1: While elements of the proposed risk management approach have been used in the SNF storage regulatory approach to evaluate the acceptable level of risk and the sufficiency of defense-in-depth (physical barriers, controls or margins) more consistently, the NRC should develop the necessary risk information, the corresponding decision metrics, and numerical guidelines. This is important in guiding further changes to the existing SNF storage regulatory approach and the evaluation of strategies for extended SNF storage activities.

As noted in earlier portions of this report, defense-in-depth is an important part of the NRC's regulatory program. The concept is most notably incorporated in 10 FR 72.124(a), the double contingency principle to prevent nuclear criticalities. In addition to the current licensing approach, defense-in-depth may also be inherent in the designs and operations of the various dry storage systems. However, these aspects are not explicitly identified or recognized as defense-in-depth considerations. Therefore, while there are implicit applications of defense-in-depth consideration in the SNF storage regulatory program, more explicit consideration and application of that philosophy is warranted.

> <u>Finding S-F-2</u>: The concept of defense-in-depth is not explicitly or consistently applied in the SNF storage regulatory program.

> <u>Recommendation S-R-2</u>: As part of the implementation of the proposed risk management regulatory framework, the NRC should more consistently consider the concept of defense-in-depth explicitly and evaluate its proper use in the SNF storage regulatory program. The NRC should also improve appropriate parts of staff training to make this concept a central part of such training.

4.7.3 Implementation Options for Risk Management in Interim SNF Storage

The introduction of Chapter 4 identifies three options for implementation of the proposed risk management regulatory framework. The following are what those options might look like for the interim SNF storage program.

<u>Option A: Continue the current approach</u>

The interim storage program would continue its current approach to risk-informed, performance-based programs under this option. Insights from operational and enforcement data would be considered in programmatic development, including rulemaking, guidance development, and inspection and enforcement. In instances where it is deemed necessary, the data may be evaluated within the bounds of a very narrow scope to inform specific, targeted actions, such as in the case of the study of misload probabilities for informing the revision of guidance related to burnup credit implementation. No significant new initiatives, including development of a Commission Policy Statement, would be undertaken based on risk considerations. Periodic self-assessments of specific program areas or lessons learned in response to events or other factors would continue to be performed as necessary but not under the direction provided by an overall risk management framework. An example of such assessments is the current work on extended storage and transportation of SNF, which has been initiated at the direction of the Commission because of the evolving national policy regarding final disposal of SNF.

Option B: Implement the proposed risk management regulatory framework through selected guidance and rule changes

This option would emphasize specific rule and guidance changes to the existing SNF storage regulatory approach to implement the Risk Management Regulatory Framework contained in the proposed Commission Policy Statement developed to adopt this framework. It would also apply the proposed risk management regulatory framework to the Commission-directed review of the paradigm of spent fuel storage. This would include substantive outreach to interim SNF storage stakeholders. It might involve forming a working group to determine where risk management might offer the greatest opportunities. These opportunities might include staff training on risk concepts, DID and practical applications of risk methods (including development of qualified risk assessment staff), development of guidance and tools to support risk-informing regulatory activities, and guidance and rule changes that may result from activities related to extended storage and transportation.

The actions taken as part of this option would necessitate implementation of the previously noted recommendations. Development of risk assessment tools and methods would require a significant amount of external technical support because of the lack of qualified risk assessment staff in the storage program area. This development should use the previously noted efforts already expended toward development of risk assessment tools appropriate for interim SNF storage activities. Feasibility and benefit also would have to be considered in determining which opportunities to pursue.

Option C: Implement the proposed risk management regulatory framework through broad-scale regulatory framework changes

This option would involve a fundamental revision of the basic framework and regulations used for licensing and oversight for interim SNF storage facilities and dry storage systems. Appropriate consideration of risk management concepts and defense-in-depth—a potentially different approach to licensing and oversight that maintains reasonable assurance of adequate public health and safety and protection of the environment—could be identified. This effort would require quantitative risk assessments to support such an approach. Only two such studies have been performed to date, and they are are limited in scope and applicability. The considerations described for implementing Option B would also apply to this option. Significant outreach to stakeholders also would be an important aspect of this option.

The RMTF recommends Option B. It provides a way to transition the agency to the proposed risk management regulatory framework and process in a reasonable amount of time and reflects ongoing operational priorities and resource constraints. It will also ensure continuity and stability in the regulated community through the continued use of existing regulations and guidance while implementing the risk management framework and process.

4.7.4   References

(EPRI, 2003)   Electric Power Research Institute, "Probabilistic Risk Assessment (PRA) of Bolted Storage Casks: Quantification and Analysis Report," EPRI, Palo Alto, CA: 2003. 1002877.

(NRC, 2000)   U.S. Nuclear Regulatory Commission, "Standard Review Plan for Spent Fuel Dry Storage Facilities," NUREG-1567, March 2000, Agencywide Documents Access and Management System (ADAMS) Accession No. ML003686776.

(NRC, 2000a)   U.S. Nuclear Regulatory Commission, Letter from B.J. Garrick and D.A. Powers to R.A. Meserve, "Use of Defense-in-depth in Risk Informing NMSS Activities," ACRS Letter, May 25, 2000, ADAMS Accession No. ML003718610.

(NRC, 2004)   U.S. Nuclear Regulatory Commission, "Encl 2 (Safety Evaluation Report) to 12/21/04 Ltr J E Ellis, General Electric Co, Issuance of Renewed Materials License No. SNM-2500, Morris Operation Independent Spent Fuel Storage Installation (72-1) (L23091)," December 21, 2004, ADAMS Accession No. ML043630514.

(NRC, 2007)   U.S. Nuclear Regulatory Commission, "SFST ISG-1, Rev. 2, Division of Spent Fuel Storage and Transportation Interim Staff Guidance No. 1, Revision 2, 'Classifying the Condition of Spent Nuclear Fuel for Interim Storage and Transportation Based on Function,'" May 11, 2007, ADAMS Accession No. ML071420268.

(NRC, 2007a)   U.S. Nuclear Regulatory Commission, NUREG-1864, "A Pilot Probabilistic Risk Assessment of a Dry Cask Storage System at a Nuclear Power Plant," March 2007, ADAMS Accession No. ML071340012.

(NRC, 2010)   U.S. Nuclear Regulatory Commission, "Standard Review Plan for Spent Fuel Dry Storage Systems at a General License Facility," NUREG-1536, Revision 1, July 2010.

(NRC, 2011)   U.S. Nuclear Regulatory Commission, "Standard Review Plan for Renewal of Spent Fuel Dry Cask Storage System Licenses and Certificates of Compliance," NUREG-1927, March 2011, ADAMS Accession No. ML111020115.

(NRC, 2011a)   U.S. Nuclear Regulatory Commission, "Estimating the Probability of Misload in a Spent Fuel Cask," November 30, 2011. ADAMS Accession No. ML113191144.

(NRC, 2012)   U.S. Nuclear Regulatory Commission, "Human Reliability Analysis—Informed Insights on Cask Drops," NUREG/CR-7016, February, 2012, ADAMS Accession No. ML110610673.

(NRC, 2012a)   U.S. Nuclear Regulatory Commission, "Preliminary, Qualitative Human Reliability Analysis for Spent Fuel Handling," NUREG/CR-7017, February 2012, ADAMS Accession No. ML110590883.

4.8   Transportation

4.8.1   Regulatory Background and Framework

The transportation of radioactive materials within the United States is regulated jointly by the U.S. Department of Transportation (DOT), the NRC, the U.S. Department of Energy (DOE), and State and local governments. In general, DOT has primary responsibility for carrier safety, hazardous communications, the definition of radioactive materials used for transportationrelated regulations, and highway routing. State and local governments have a significant role in overseeing shippers' and carriers' compliance with DOT regulations for carrier safety. They also have lead responsibility for responding to transportation emergencies within their jurisdictions. Local and State governments also can impose additional requirements on the shipment of radioactive materials within their jurisdictions, provided such requirements are compatible with applicable DOT requirements.

The approval or certification of shipping package designs for radioactive materials is shared jointly between the NRC, DOT, and DOE. NRC and DOT responsibilities for certification of shipping package designs are delineated in a 1979 Memorandum of Understanding (MOU) between DOT and the NRC (NRC, 1979). The MOU assigns the NRC lead responsibility for certification of shipping package designs used to transport Type B quantities (i.e., quantities requiring accidentresistant packaging) and fissile material. DOT oversees safety standards for packages not required to be certified as accident resistant (e.g., Type A and industrial packages). These packages generally are self-certified by shippers or package vendors to withstand normal conditions experienced during shipments. DOE has the authority, under DOT regulations, to approve transportation package designs for DOEowned material, except where congressional legislation has specified otherwise (e.g., packages used to ship transuranic wastes to the Waste Isolation Pilot Plant and to a national repository that require NRC certification under the Nuclear Waste Policy Act).

Since the NRC, DOT, and DOE all have authority to approve Type B shipping packages, it is important that the safety standards used by these agencies be nearly identical. The United States, a participating International Atomic Energy Agency (IAEA) member state, has also endorsed the concept that its domestic transportation regulations in Title 49, "Transportation," of the Code of Federal Regulations (49 CFR) Part 173, "Shippers General Requirements for Shipments and Packages," need to be compatible with the IAEA's transportation regulations to the greatest extent practicable. The practical result is that the NRC and DOT periodically undertake a joint rulemaking effort to revise their respective regulations to be compatible with the latest revision to IAEA transportation regulations.

Licensing

The NRC's primary licensing role in transportation safety is the review and certification of Type B and fissile material shipping package designs under 10 CFR Part 71, "Packaging and Transportation of Radioactive Material." Under 10 CFR Part 71, NRC reviews individual shipping package designs submitted by an applicant to certify that the design can withstand a series of specified hypothetical accident conditions and subsequently meet specific safety criteria dealing with criticality safety, containment, and package dose limits. In making this certification, the package design must consider both the packaging hardware and the contents. The tests include a 30-foot drop onto an unyielding surface, a puncture test consisting of a 40 inch drop onto a metal spike, a half-hour fire test at 1,475 degrees Fahrenheit and an immersion test for fissile material. After the test, the package (packaging and contents) must remain subcritical, have a dose rate of less than 1 rem per hour at 1 meter from the package surface, and a release rate not exceeding an A_2 quantity per week. An A_2 quantity is the quantity below which an accident-resistant package is not required. If the package design complies with the safety criteria, it is issued an NRC certificate of compliance (COC).

In general, individual shipments by NRC licensees do not need specific approval. The NRC grants its possession licensees a general license pursuant to 10 CFR 71.17, "General License: NRC Approved Package," to deliver material to a private or contract carrier for shipment. The NRC licensee is responsible under the general license for ensuring that the material is properly prepared for shipment, including the selection of an appropriate DOT- or NRCapproved package. NRC licensees also may obtain a specific license for transportation under 10 CFR Part 71. However, this option has rarely been used (not since the mid-1980s). The NRC COC enables the package to be used under the general license in 10 CFR 71.17 as well as under DOT regulations for non-NRC licensees.

Inspection

The NRC inspects 10 CFR Part 71 certificate holders and package component fabricators to ensure that transportation casks are fabricated and tested in accordance with the package specifications in the NRC certificate. These inspections are conducted under Inspection Manual Chapter 2690 using inspection procedures applicable to those entities. Inspections focus on risk-significant activities, such as fabrication of important-to-safety components. The frequency of inspections is based on factors that include the population of the entity type, level of production, frequency of design modifications, and available resources. Other factors, including performance history, allegations, significant licensing issues, or previous significant inspection findings may result in increased inspection frequency or additional, reactive inspections.

Previous and Current Risk-Informed, Performance-Based Efforts

The current framework for certifying Type B and fissile material shipping packages has evolved from the efforts undertaken in the 1960s at the IAEA to achieve a set of internationally acceptable safety standards that could be applicable for both domestic and transboundary shipments. In theory, acceptance of common safety standards would ensure that bordering Nations were using a minimum set of safety standards for radioactive material shipments. It would also ensure that package approvals by one Nation could be accepted by other Nations through which a package may travel.

In 1961, the IAEA issued safety standards for transportation requiring packages that carry over certain quantities of radioactive material (Type B) to withstand "worst-case credible accidents" (IAEA, 1961). Not surprisingly, there was considerable disagreement among IAEA member states on what constituted a worst-case credible accident. By 1964, IAEA member states defined a worst-case credible accident in terms of a series of physical tests that would bound, but not mimic, the potential forces seen in real-life accidents (IAEA, 1964). The physical tests were defined to be easily reproducible in all member states. These were the origin of the 30foot drop, punctures, and fire tests used to certify shipping packages today. Because there was not much information available on the effects that occur during transportation accidents, the tests were based on the collective engineering judgment of transportation experts representing IAEA member states. Implementation of the physical tests was accompanied by a recommendation for member states to acquire more accident data and revise the IAEA Safety Standards as needed.

The basic physical tests implemented in the 1964 IAEA regulations are still the primary tests used today for approving Type B and fissile material shipping packages. The continued use of these physical tests has been supported by numerous risk studies done in the United States, as well as in other IAEA member states (AEC, 1972; NRC, 1977; NRC, 1988; NRC, 2000). All of the studies have shown that the risk of shipping spent fuel is very low. In addition, the United Kingdom (1984), Germany (1999), and the United States (1970s) have conducted fullscale demonstration tests by staging real-life accidents for spent fuel casks. In addition, the IAEA continually reviews and updates its transportation safety regulations and has repeatedly concluded that major changes in the core physical tests were unnecessary.

Consequently, most revisions to IAEA regulations over the years have dealt with areas other than the core test requirements for Type B and fissile material shipping packages. A notable example was the inclusion of new requirements for Type C shipping packages for shipments of large quantities by air in 1995. The development of Type C standards was based on a risk analysis of data on aircraft accidents. The impact speed that a Type C package would have to withstand was based on a "knee of the curve"-type analysis, a point where a large increase in impact speed would capture an increasingly small percentage of air crashes. At the time, the United States argued for a much larger impact speed based on its implementation of a Federal law in the 1980s that required air shipments of plutonium to withstand the crash of a highflying aircraft. The United States has not adopted Type C standards.

In 1995, the United States presented a working paper to IAEA's Transportation Unit proposing a risk-informed approach to future revisions of its Transportation Safety Regulations. The proposal advocated accepting the current version of the regulations (1996) as a consensual baseline and using risk information and safety goals to screen future revisions. The intent of the paper was to slow down the adoption of amendments to regulations that were not risk-informed. The paper was considered at an IAEA Technical Committee Meeting, but it was not pursued further.

In 2003, the NRC adopted a provision for special package approvals, similar to the provision for special arrangements in IAEA regulations. The provision allows the NRC to approve packages for domestic shipments that do not meet some of the prescribed Type B tests if an applicant can prove equivalent safety.

4.8.2   RMTF Findings and Recommendations

While the term "defense-in-depth" is not explicitly used, the current regulatory approach for approving and inspecting radioactive shipping packages follows the risk-informed and performance-based defense-in-depth approach in a general sense. For example, the safety requirements for different types of shipping packages become more stringent with the quantity (radioactivity), or hazard, contained. The threshold for an accidentresistant package is based on an A_1 (special form or encapsulated material) or A_2 (normal form) quantity. In turn, the A_1 and A_2 quantities are based on accident models that keep the anticipated dose to first responders below the occupational exposure limit of 5 rem. If a package contains greater than an A_1 or A_2 quantity (i.e., has a potential to cause an exposure greater than 5 rem), it is required to meet Type B accident conditions. The current system also allows shipments of quantities that would normally require Type B packages to be made in less robust packages that take credit for the low, specific activity of the material being shipped.

The current regulatory framework for the certification of radioactive material shipping packages is based on a 50-year consensus of IAEA member states. That consensus incorporates the acceptance of a very-low risk based on the assumption that "every transportation accident occurs offsite" and that transportation packages often are used by third parties not under the nuclear regulator's control. In addition, the need to maintain compatibility with IAEA member states and among U.S. regulatory agencies makes changes to regulations resource intensive and requires the consent of multiple parties. The current regulatory framework has resulted in an outstanding safety record.

In making its findings and recommendations, the RMTF also considered comments on transportation submitted in response to its November 22, 2011 Federal Register notice. These comments included the following:

- In the case of transportation, the NRC's regulatory framework appears to be graded in an inverse manner from what the risk insights would suggest.

- The risk presented by transportation is much lower than that of power reactors; however, the regulatory framework calls for a relatively high degree of information and involves an unusually protracted approval process.

- Regulations for storage and transportation of spent fuel need to be cohesively linked since many casks are dual purpose. Ideally, this would be done under one set of regulations.

- The regulations in 10 CFR Parts 71 and 72 are not risk-informed. As a result, they are not conducive to efficient and timely movement of spent fuel from pools to storage and ultimate disposal. The need to be able to move fuel from spent fuel pools to storage, or ideally a final repository, in an efficient manner is a critical lesson learned from the accident at Fukushima Daiichi in Japan.

- The regulations in 10 CFR Parts 71 and 72 are performance-based. As such, there is a large dependence on regulatory guides to identify methods acceptable to the NRC.

- The NRC should apply the lessons learned from probabilistic risk assessments performed to date in its comprehensive review of the regulatory framework for spent fuel storage and transportation.

Much of the conservatism in approving shipping packages under 10 CFR Part 71 stems from assumptions used in implementing the regulations rather than in the regulations themselves. Risk-informing guidance for implementing current regulations may be desirable in some cases. While some aspects of the staff guidance are prescriptive, other aspects have been made more performance-based (e.g., definition of fuel condition). Some areas of the guidance and the overall review process can be revised to incorporate risk insights based on staff experience and judgment. For example, risk-informing to allow packages designed to current standards to ship other types of materials (e.g., high-burnup spent fuel) may be desirable.

In addition to operational history, since 10 CFR Part 71 applies to all Type B shipping packages in order to more systematically implement the proposed risk management framework for transportation, risk analyses of shipping radioactive materials other than spent fuel should be developed to complement the existing spent fuel transportation risk studies.

Finding T-F-1: While the U.S. transportation regulatory approach is governed by the IAEA transportation regulations, the current NRC transportation regulatory approach uses several elements of the proposed risk management framework.

Finding T-F-2: Risk assessments have been conducted on the safety of transportation of spent fuel. However, there is a lack of risk information on the transportation of other radioactive materials.

Recommendation T-R-1: Considering the strong international regulatory basis for transportation and the need to conform U.S. standards to those of the IAEA and other member states, application of the proposed risk management regulatory framework should focus on implementation guidance.

Recommendation T-R-2: The risk management process should be used to influence the future outcome of IAEA deliberations on proposed changes in international transportation regulations.

Recommendation T-R-3: The NRC should explore the value of using risk insights to justify regulations different from the IAEA's for domestic use only, such as regulations dealing with domestic storage and transportation of high burnup fuel. Risk information could be used to develop a more flexible approach toward implementing and making gradual changes to current transportation regulations.

4.8.3 Implementation Options for Risk Management in Transportation

The introduction of Chapter 4 identifies three options for implementation of the proposed risk management regulatory framework. The following are what those options might look like for the approval of transportation shipping containers.

Option A: Continue the current approach

The transportation program would continue its current approach of using consentbased standards. These standards are based on compatibility with DOT and IAEA regulations. Risk studies would be used to verify that current standards and guidance continue to protect against severe real-life accidents. Risk studies also could be used to influence the future outcome of IAEA deliberations on proposed changes in international regulations. Risk concepts could be used to justify regulations different from IAEA's for domestic use only, such as regulations dealing with domestic storage and transportation of high burnup fuel.

Option B: Implement the proposed risk management regulatory framework through selected guidance and rule changes

This option would emphasize specific guidance changes to implement the proposed risk management regulatory framework contained in the proposed Commission Policy Statement. This would include substantive outreach to industry, State, and local governments and the public. It might involve the formation of a working group to determine where risk management might offer the greatest opportunities. It would focus on using risk information to develop a more flexible approach toward implementing and making gradual changes to international standards and, eventually, to domestic regulations. It could be based on accepting the current version of the regulations as a consensual baseline and using risk information and safety goals to screen future revisions.

Option C: Implement the proposed risk management regulatory framework through broad-scale regulatory framework changes

This option would involve a fundamental revision of the basic framework and regulations used for licensing and oversight of transportation package approvals. This effort would require quantitative risk assessments to support such an approach. Significant outreach to stakeholders would also be an important aspect of this option. The greatest challenge would be to overcome resistance to abandoning a worldwide consent-based system perceived to work fairly well and that has an outstanding safety record.

The RMTF recommends Option B. The transportation regulations are unique in their international conformance. As such, changes to the regulations need to be done in a manner consistent with IAEA standards and those of member states. Option B would allow the NRC to focus its risk management efforts, in the near term, on improving the guidance used in implementation of the current transportation regulations. In the long term, it would allow the NRC to apply risk insights to international standards development. Option B provides a way to transition the agency to the proposed risk management regulatory framework and process in a reasonable amount of time and reflects ongoing operational priorities and resource constraints. It will also ensure continuity and stability in the regulated community through the continued use of existing regulations while implementing the proposed risk management regulatory framework and process to improve the transportation guidance.

4.8.4 References

(AEC, 1972)   U.S. Atomic Energy Commission, "Environmental Survey of Transportation of Radioactive Materials to and from Nuclear Power Plants," WASH-1238, December 1972.

(IAEA, 1961)   International Atomic Energy Agency, "Regulations for the Safe Transport of Radioactive Materials," Safety Series No. 6, 1961.

(IAEA, 1964)   International Atomic Energy Agency, "Regulations for the Safe Transport of Radioactive Materials," Safety Series No. 6, Revised Edition, 1964.

(NRC, 1977)   U.S. Nuclear Regulatory Commission, "Final Environmental Statement on the Transportation of Radioactive Material by Air and Other Modes," NUREG-0710, December 1977.

(NRC, 1979)   U.S. Nuclear Regulatory Commission, "Transportation of Radioactive Materials, Memorandum of Understanding," published in the *Federal Register* on July 2, 1979, (44 FR 38690).

(NRC, 1988)   U.S. Nuclear Regulatory Commission, "Shipping Container Response to Severe Highway and Railway Accident Conditions, NUREG/CR-4829," Lawrence Livermore National Laboratory, October 1988, Agencywide Documents Access and Management System (ADAMS) Accession No. ML070810403.

(NRC, 2000)   U.S. Nuclear Regulatory Commission, "Reexamination of Spent Fuel Shipments Risk Estimates," NUREG/CR-6672, Sandia National Laboratory, February 2000, ADAMS Accession No. ML003698324.

5. CONCLUSIONS

The NRC has made progress in its efforts to implement risk-informed and performance-based approaches into its regulation of the various uses of byproduct, source, and special nuclear materials. Nevertheless, it is necessary to re-assess that progress and the underlying strategic vision from time to time.

The Risk Management Task Force (RMTF) has found that the NRC's programs do not require radical or revolutionary changes, but they could benefit from continuing the evolution that has been embraced throughout the agency's history. To that end, a Risk Management Regulatory Framework, as depicted in Figure 5-1, is being recommended as the next logical step for the NRC. This proposed framework uses a disciplined risk management process to identify and evaluate issues and make decisions about appropriate defense-in-depth protections for various radiological hazards. The risk-informed and performance-based defense-in-depth protections provide sufficient barriers, controls, and personnel to prevent, contain, and mitigate the exposure of workers or the public to radioactive materials. The appropriate barriers, controls, and personnel are based on the hazards present, the relevant scenarios leading to possible exposures, and the associated uncertainties to ensure that the risks resulting from the failure of some or all of the established barriers are maintained acceptably low.

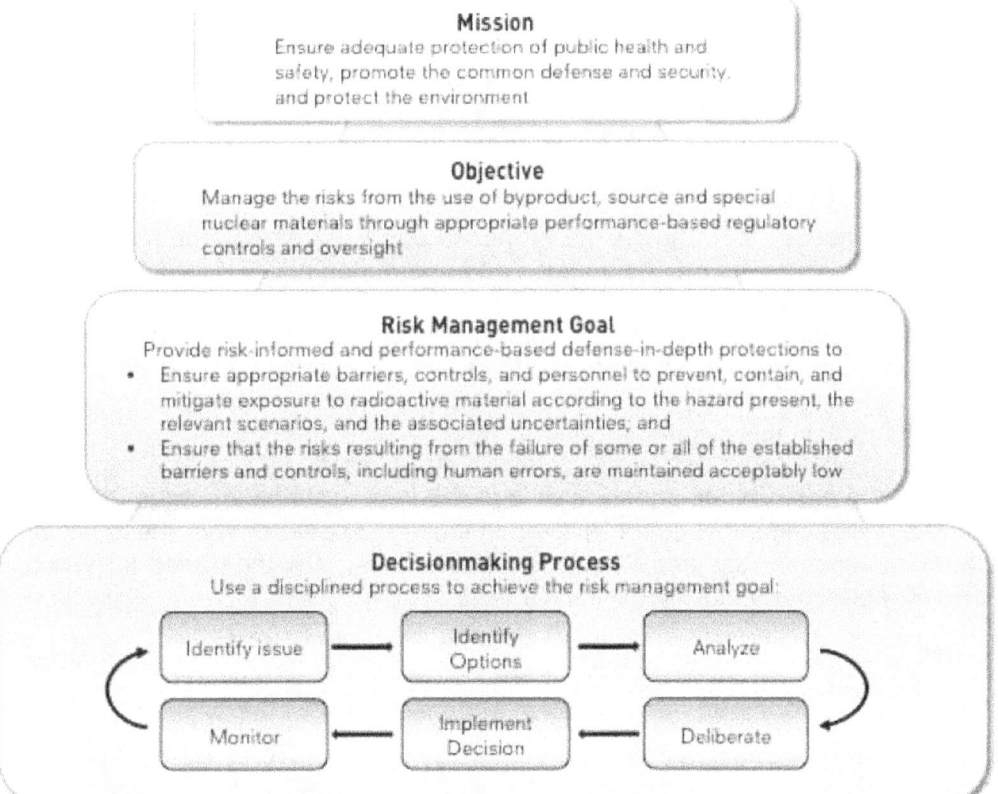

Figure 5-1 A Proposed Risk Management Regulatory Framework

Chapter 5: Conclusions   Risk Management Task Force | 5-1

In addition to the findings and recommendations in Chapter 4, the RMTF proposed options for implementation. Generally, the RMTF recommends an approach of systematically implementing the risk management framework through guidance and rule changes. This option is an evolutionary one that focuses on specific guidance and rule changes, as well as related activities (such as training) to implement a proposed Risk Management Regulatory Framework that would be rooted in a Commission Policy Statement. The RMTF recognizes the considerable work the NRC has done in risk-informing its programs as well as the need to do more.

The first step in implementation of the recommendations in this report, if adopted, would be to initiate a thorough stakeholder engagement process. The process would begin with the issuance of a draft Commission Policy Statement on risk management for review and comments. Interactions on the draft policy statement could be supported by a series of workshops similar to those held as part of the Commission's Safety Culture policy statement development. As reflected in the summary of comments provided in Appendix I, several stakeholders stated a strong preference for workshops as a way to better understand and help shape Commission direction in implementing the framework. The workshops would afford licensees, nongovernmental organizations, industry groups, States, and the public at large an opportunity for early involvement in shaping the agency's risk management approach.

The results of this stakeholder engagement would be considered by the NRC staff in preparing a final Commission Policy Statement on risk management and developing a long-term action plan to shape future direction of implementation. The staff would also need to consider existing commitments or initiatives, budget and resource constraints, and impacts to licensees and other stakeholders.

During the period needed to formulate the proposed policy statement, the NRC's regulatory practices will be changing, as will the designs and operations of U.S. power reactors, as a result of the lessons learned from the Fukushima Dai-ichi accident in Japan. It is important that these shorter-term actions are compatible with the longer-term changes envisioned by the Commission, whether they are related to the RMTF recommendations or a different course of action. Consideration of the proposed design-enhancement category for power reactors is particularly important in this regard.

None of this will be easy, as the agency's experience with past risk-informed initiatives has demonstrated. However, subsequent NRC and industry experience with those initiatives, such as the Maintenance Rule and Regulatory Guide 1.174, and the Phase II Byproduct Material Review, has shown their net positive value.

Challenges that would have to be addressed to implement the vision proposed in this report are broadly summarized below:

- **A change would be required within the agency and externally to increase understanding of the value and use of risk concepts and risk management language.**

  The proposed framework involves change for the agency, its licensees, and other stakeholders. While it is considered to be evolutionary, it is nonetheless a change that would involve agency and licensee programs and resources. Accordingly, the NRC would need to assure buy-in across the agency, develop a communications strategy, and prepare key messages. The proposed stakeholder workshops would be an important part of that communication strategy.

- **The proposed risk-informed and performance-based concept of defense-in-depth may require the development of additional decision metrics and numerical guidelines.**

  The NRC has developed a working set of risk-informed decision metrics for power reactors. Examples include measures on an overall acceptable level of risk (safety goals) and guidance to the NRC staff on addressing incremental risk increases. Similar metrics do not exist in other program areas, but they may be needed to fully implement the proposed defense-in-depth framework.

- **The approach would likely require developing new or revised risk-assessment consensus codes and standards.**

  The increased use of consensus standards is a Governmentwide goal intended to improve the efficiency of agency decisionmaking. The practical difficulties in achieving this goal have been well demonstrated as NRC has worked to implement risk assessment standards for power reactor applications.

- **Consideration of cost in the proposed design-enhancement category in the power reactor regulatory program would necessitate a reconsideration of the agency's tools for performing cost-benefit analysis.**

  Under some implementation alternatives considered by the RMTF, cost-benefit analyses would be performed (perhaps by licensees) on a potentially larger set of accident scenarios and on a more frequent (periodic) basis. The current set of tools used by agency staff for cost-benefit analyses was not designed with such uses in mind and would require improvements to make them easier to use, if such alternatives were chosen. In addition, the NRC staff is currently assessing possible updates to the actual criteria and factors incorporated into the NRC's cost-benefit analyses.

- **A long-term commitment from the Commission and senior agency management would be required for implementation.**

  Development and issuance of a Commission Policy Statement on risk management would signal the commitment of the agency to the proposed framework. However, transitioning to a risk-management framework would take time. To successfully transition to this framework, the NRC would need to continue finding ways to bring risk concepts into the daily activities of the staff and investigate how risk analysis works in concert with the traditional deterministic analysis, such as relying on engineering margins, to achieve the risk management goal. This would begin with a top-down commitment to the concept, would be reinforced with training, and would be made a part of routine practices. Development of guidance, agencywide training, and a consistent emphasis from the highest levels of the agency to the risk management framework are key activities that the agency would need to undertake.

Despite these challenges, there are substantial benefits to be realized by transitioning to the risk management framework proposed in this report. These include:

- **Updated knowledge from contemporary studies, such as risk analysis, would be incorporated into the regulations and guidance, thereby improving their realism and technical basis.**

  While stability in a regulatory authority is generally considered to be a positive attribute, that authority also has a responsibility to examine new information and, when appropriate, make changes. Important aspects of the NRC's regulatory practices were established more than 30 years ago, before an extensive base of operating experience and methods existed and risk assessments were developed and successfully applied. The proposed Risk Management Regulatory Framework and the program-area-specific recommendations are intended to build upon the successes of the past 30 years while improving the efficiency and effectiveness of the NRC's regulatory program.

- **Implementation of a systematic approach would foster a consistent regulatory decisionmaking process throughout the agency and improve resource allocation.**

  The proposed Risk Management Regulatory Framework and process are intended to be used universally for NRC actions involving the regulation and oversight of licensed activities. They take into account that the risk environment varies substantially across the wide variety of uses of radioactive materials that the NRC regulates. This common process would be used to identify and evaluate issues, make decisions, and allocate the appropriate amount of resources to implement actions to manage these diverse risks. While the details of licensing decisions, inspections, and other regulatory actions would continue to be different, a consistent decisionmaking process would be used.

- **Consistency in language and communication would be improved across the agency and externally.**

  Several stakeholders have noted there is not a common understanding and use of terms such as risk-informed, performance-based, and defense-in-depth within the NRC and by external stakeholders. This makes effective communication a challenge. By focusing on the common use and applicability of these terms across the agency and the difference in approaches that various activities entail, the proposed framework would help improve consistency and understanding of the agency's internal and external communications.

- **Support of issue resolution would be achieved in a systematic, consistent, and efficient manner.**

  The proposed Risk Management Regulatory Framework provides a vehicle to systematically integrate relevant recent information into the decisionmaking process. Under this framework, a common decisionmaking process and the risk management goal for risk-informed and performance-based defense-in-depth would support developing program-specific criteria for the identification and disposition of issues. This common approach, with specific guidance for different types of licensees, could help ensure a more systematic, consistent, and efficient resolution of issues across the NRC's regulatory programs.

- **The design-enhancement category proposed for the power reactor regulatory program would clarify the attributes of all requirements established as substantial safety (beyond-design-basis) improvements. This approach may contribute to the resolution of the "patchwork" issue identified by the Fukushima Near-Term Task Force.**

  It should not be surprising that requirements, such as the rules for station blackout and aircraft impact, which emerged at different times and were considered "additional" protection, were not established with a consistent approach. The RMTF recommendation for a design enhancement category is intended to remedy this "patchwork" approach. Although there would be some costs incurred to establish this rule, these costs would be more than balanced by the long-term benefit of having a clear, consistent description of treatment for beyond-design-basis events and accidents.

To a certain degree, some resistance to change is natural and perhaps even desirable for the NRC to maintain a clear and stable regulatory environment. In addition, the ongoing activities within the agency and limited resources can impede the development and implementation of acknowledged process improvements. However, a patchwork of regulatory requirements has been created as a result of addressing problems on a case-by-case basis over many years. The RMTF has concluded that cost-effective changes are possible and recommends they be undertaken in a holistic manner across the NRC's programs. With this in mind, the implementation of the proposed Risk Management Regulatory Framework can be pursued in a planned and deliberate manner so that it does not disrupt the NRC's mission, but ensures that the NRC continues to improve on how it protects the public health and safety, promotes the common defense and security, and protects the environment.

APPENDIX A

RISK MANAGEMENT SYSTEMS

A.1   Introduction

As discussed in Chapters 2 and 3 of the report, "A Proposed Risk Management Regulatory Framework," the objective of the U.S. Nuclear Regulatory Commission (NRC) is to manage the risks from the use of byproduct, source, and special nuclear materials by establishing and maintaining appropriate performance-based regulatory controls and oversight. In its simplest form, this objective is met by ensuring that sufficient barriers (supported by associated controls and personnel) are placed between radioactive materials and members of the workforce and public to minimize the risks of adverse effects on health and safety. These barriers, controls, and personnel must provide protection during routine operations, as well as for challenges such as equipment failures and external hazards (e.g., fires and floods). The NRC has longstanding positions and guidance related to the consideration of risk-insights and performance-based concepts in its decisionmaking processes. As discussed in a March 1999 (NRC, 1999) guidance document issued by the Commission, the NRC's risk definition takes the view that when one asks, "What is the risk?" one is really asking three questions:

- What can go wrong?

- How likely is it?

- What are the consequences?

These three questions can be referred to as the "risk triplet" (Kaplan and Garrick, 1981). In simple terms, when one deals with risk, one must consider both consequences *and* likelihoods. The traditional definition of risk—that is, probability times consequences—is fully embraced by the "triplet" characterization of risk, although it is more limited. The consequences of concern to the NRC are related to the use of byproduct, source, and special nuclear materials and to the possible exposure of workers or the public to radiation from the normal use of those materials or as a result of accidents. Although the NRC's regulatory programs range from the control of simple sealed sources to complex nuclear power plants, the RMTF described in the main body of this report the way in which most of the agency's regulatory decisions involve asking the above questions and determining an appropriate course of action based on the answers derived from analyses or judgment. The RMTF defined a risk-informed and performance-based defense-in-depth concept to serve as the risk management goal for NRC-regulated activities. In addition, a move to a common, structured decisionmaking model would ensure consistency and improved communications throughout the NRC's regulatory programs.

Before focusing on risk management as the logical evolution of the NRC's current risk-informed, performance-based approach to regulation, the task force considered a number of approaches to accomplishing its charter. In doing this, the RMTF considered the agency's historic approach to consideration of risk, researched and reviewed literature across the field of risk, and held discussions with internal and external stakeholders to develop options for a more comprehensive and holistic approach for its regulatory programs. These included: 1) no action, 2) a purely deterministic approach, 3) a single risk-based numerical criterion, and 4) an approach based on the concept of risk management.

The no-action option would continue the agency's current risk-informed, performance-based approach. This approach has been used successfully since the late 1990s to greater or lesser degrees in the various regulatory programs of the NRC. Its attributes are by and large known to the regulated community and are therefore predictable. The approach has provided the NRC the ability to make its regulatory decisionmaking more systematic, more objective, more consistent, and more transparent. However, it does not position the agency to address the challenges of the future, specifically the prospect of flat or declining budgets for most, if not all, Federal agencies. Nor does it specifically address the direction of Executive Order 13563, "Improving Regulation and Regulatory Review" (Exec. Order, 2011).

The second option considered by the RMTF was a purely deterministic one. Under this approach, the agency would discontinue its risk-informed, performance-based approach and return, over time, to deterministicbased regulations and guidance for its regulatory programs. Risk insights and performance considerations would no longer be considered in developing regulatory standards and guidance. This approach would arguably be clearer and easier to comply with than standards that allow licensees to choose one of a number of ways to meet a performance-based regulation. The approach would also be less flexible than the current approach, not allowing for site or facility-specific considerations, and would represent a departure from the trend of international standardssetting approaches, which have an increasingly robust risk basis. Finally, such an approach would more likely than not drive up the costs of regulatory compliance with no compelling increase in safety or security.

The third option considered by the task force was the concept of a risk-based numerical criterion to be used across all agency regulatory programs. This approach is used by some other countries and is attractive from the prospect of providing a single unifying basis for protection of public health and safety and the environment. It could also be useful in communicating to the public the primary basis for the NRC's various regulatory programs. However, it would not take into account the wide variety of hazards posed by various uses of nuclear materials and could incur unnecessary and perhaps unacceptable costs to the regulated community. It would also represent a significant change in the NRC's approach to regulation that would necessitate an overhaul of the regulations and guidance for reactors, materials, fuel cycle, waste, and transportation programs at a substantial cost to the agency.

The fourth option, risk management, is being widely used in various sectors, including Government agencies, financial institutions, and technology companies, to address the kinds of challenges the NRC faces and that the RMTF was tasked to address. Risk management allows for various approaches to address risks as part of the decisionmaking process, including the use of both quantitative and qualitative tools, which is essential in the broad range of NRC regulatory programs. It represents a logical evolution from the risk-informed, performance-based philosophy that has governed NRC activities for many years. It may also provide program managers with a more systematic approach to resource allocation, whether in budget formulation, response to events, or licensing decisions. The RMTF believes that risk management offers the potential for an improved regulatory framework, which is described generally and for specific regulatory programs in other parts of this report.

Given the NRC's mission and the fact that most regulatory decisions involve an evaluation of and control of the risks posed by the use of radioactive materials, the RMTF recommends (Recommendation 2.2) that "risk management" be used as the framework for a comprehensive and holistic risk-informed, performance-based regulatory approach for reactors, materials, waste, fuel cycle, and transportation. Risk management concepts, process descriptions, and decisionmaking models are widely used in various sectors, including Government agencies, financial institutions, and technology companies. The U.S. Department of Homeland Security Risk Lexicon (DHS, 2010) defines risk management as follows:

> *Risk management is the process for identifying,*
> *analyzing, and communicating risk and accepting, avoiding,*
> *transferring, or controlling it to an acceptable level considering*
> *associated costs and benefits of any actions taken.*

A.2   Risk Management Frameworks

There are many general descriptions of risk management processes or methodologies that include similar key points. A review of many of the frameworks and processes for risk management reveals that most share common steps or phases. Not surprisingly, these risk management frameworks are a variation of traditional decisionmaking models. The steps to a basic decisionmaking model are as follows:

- define the problem and the desired outcome

- research and identify options

- analyze alternatives

- make a decision, (i.e., choose an alternative)

- implement the decision

- monitor the results

Descriptions of several risk management frameworks and how they support the decisionmaking process are described in the following sections.

A.2.1   Framework for Environmental Risk Management
The Presidential/Congressional Commission on Risk Assessment and Risk Management; Final Report 1997

In the 1990 Clean Air Act Amendments, Congress mandated that a Commission on Risk Assessment and Risk Management be formed to:

> *…make a full investigation of the policy implications and*
> *appropriate uses of risk assessment and risk management in*
> *regulatory programs under various Federal laws to prevent*
> *cancer and other chronic human health effects which may*
> *result from exposure to hazardous substances.*

A Proposed Risk Management Regulatory Framework

The Commission, consisting of specialists in various fields of science and public policy, was assembled in May 1994. Following its deliberations and interactions with various stakeholders, the Commission issued its final report in 1997, which included a framework for incorporating risk management concepts into the decisionmaking processes for the U.S. Environmental Protection Agency (EPA) (Commission, 1997). The Commission's Framework defines a clear, sixstage process for risk management that can be scaled to the importance of a public health or environmental problem and that:

*   Enables risk managers to address multiple relevant contaminants, sources, and pathways of exposure so that threats to public health and the environment can be evaluated more comprehensively than is possible when only single chemicals in single environmental media are addressed.

*   Engages stakeholders as active partners so that different technical perspectives, public values, perceptions, and ethics are considered.

*   Allows for incorporation of important new information that may emerge at any stage of the risk management process.

The basic structure of the framework developed by the Presidential/Congressional Commission is shown in Figure A-1.

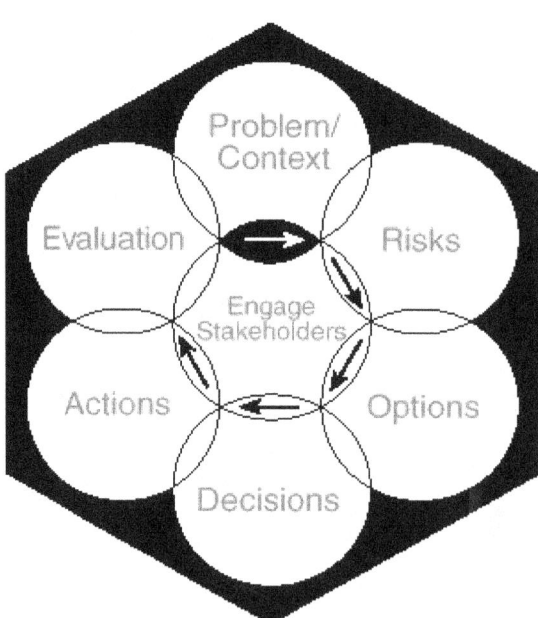

Figure A-1   Framework for Environmental Health Risk Management
The Presidential/Congressional Commission on Risk
Assessment and Risk Management. Final Report 1997

A.2.2   Understanding Risk: Informing Decisions in a Democratic Society
        National Research Council, 1996

The National Resource Council has performed several studies related to risk assessments and risk management as they apply to various government functions and public health concerns. In 1996, the Council documented (National Research Council, 1996) a study it had undertaken to address the following task statement:

> *"Risk characterization" is a complex and often controversial activity that is both a product of analysis and dependent on the processes of defining and conducting analysis. The study committee will assess opportunities to improve the characterization of risk so as to better inform Decisionmaking and resolution of controversies over risk. The study will address: technical issues such as the representation of uncertainty; issues relating to translating the outputs of conventional risk analysis into non-technical language; and social, behavioral, economic, and ethical aspects of risk that are relevant to the content or process of risk characterization."*

The report offers seven principles to increase the likelihood of achieving sound and acceptable decisions. These principles are as follows:

1.   *Risk characterization should be a decisiondriven activity, directed toward informing choices and solving problems.*

2.   *Coping with a risk situation requires a broad understanding of the relevant losses, harms, or consequences to the interested and affected parties.*

3.   *Risk characterization is the outcome of an analyticdeliberative process. Its success depends critically on systematic analysis that is appropriate to the problem, responds to the needs of the interested and affected parties, and treats uncertainties of importance to the decision problem in a comprehensible way. Success also depends on deliberations that formulate the decision problem, guide analysis to improve decision participants' understanding, seek the meaning of analytic findings and uncertainties, and improve the ability of interested and affected parties to participate effectively in the risk decision process. The process must have an appropriately diverse participation or representation of the spectrum of interested and affected parties, of decisionmakers, and of specialists in risk analysis at each step.*

4.   *The analytic-deliberative process leading to a risk characterization should include early and explicit attention to problem formulation; representation of the spectrum of interested and affected parties at this early stage is imperative.*

5.   *The analytic-deliberative process should be mutual and recursive. Analysis and deliberation are complementary and must be integrated throughout the process leading*

to risk characterization: Deliberation frames analysis, analysis informs deliberation, and the process benefits from feedback between the two.

6.   *Those responsible for a risk characterization should begin by developing a provisional diagnosis of the decision situation so that they can better match the analyticdeliberative process leading to the characterization to the needs of the decision, particularly in terms of level and intensity of effort and representation of parties.*

7.   *Each organization responsible for making risk decisions should work to build organizational capability to conform to the principles of sound risk characterization. At a minimum, an organization should pay attention to organizational changes and staff training efforts that might be required, to ways of improving practice by learning from experience, and to both costs and benefits in terms of the organization's mission and budget.*

The Council's report provided the following schematic representation of a risk decision process:

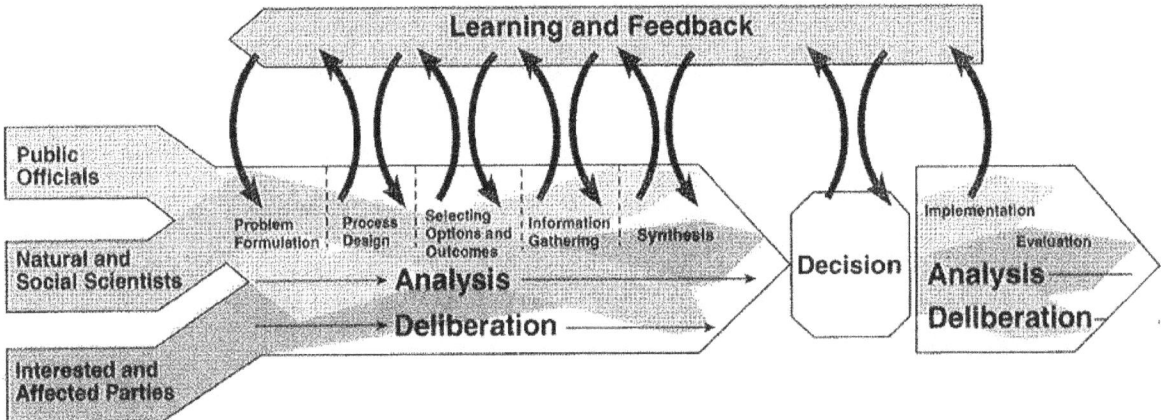

Figure A-2   Understanding Risk: Informing Decisions in a Democratic Society, National Resource Council, 1996

A.2.3   Risk Governance Framework
International Risk Governance Council

Risk and Regulatory Policy – Improving the Governance of Risk
OECD Reviews of Regulatory Reform, 2010

The International Risk Governance Council (IRGC) is a private, independent, notforprofit foundation based in Geneva, Switzerland. It was founded in 2003 with a mission to support governments, industry, nongovernmental organizations, and other organizations in their efforts to deal with major and global risks facing society and to foster public confidence in risk governance. The IRGC was established to address widespread concern within the public sector, the corporate world, academia, the media, and society at large that the complexity and interdependence of an increasingly large number of risk issues was making it ever more difficult for risk managers to develop and implement adequate risk governance strategies.

They produced a white paper, entitled "Risk Governance – Towards an Integrative Approach" (IRGC, 2005) to describe a framework for an integrated, holistic, and structured approach for improving the ways risk is identified, assessed, managed, monitored, and communicated. The IRGC framework is depicted in Figure A-3.

The IRGC risk governance framework and its sequence of pre-assessment, risk appraisal, risk characterization, risk evaluation and risk management is also discussed in a report by the Organization for Economic Cooperation and Development (OECD). The report, entitled "Risk and Regulatory Policy – Improving the Governance of Risk" (OECD, 2010) is one of a series of OECD reports on regulatory reforms. A premise of the report is that there is a gap between the level of risk that is aspired to by policymakers and the level that is achievable through regulation. In addition, the report acknowledges that since not all risks can be reduced to zero, tradeoffs in risk reduction measures are inevitable. The OECD studied areas for the improvement of risk governance through an analysis of the legal, procedural, and practical challenges for risk regulation.

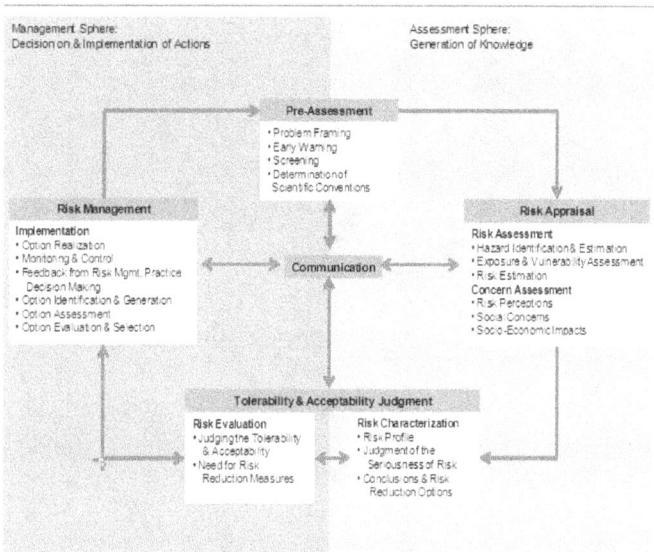

Figure A-3   International Risk Governance Council (IRGC)
                Risk Governance Framework

A.2.4   Strengthening the Use of Risk Management Principles in Homeland
        Security; Government Accountability Office, June 2008

        Risk Management: Further Refinements Needed to Assess Risks and
        Prioritize Protective Measures at Ports and Other Critical Infrastructure;
        Government Accountability Office, December 2005

The U.S. Government Accountability Office (GAO) convened a forum of 25 national and international experts on October 25, 2007, to advance a national dialogue on applying risk management to homeland security. Participants included Federal, State, and local officials and risk management experts from the private sector and academia. Forum participants identified (1) what they considered to be effective risk management practices used by organizations from the private and public sectors, and (2) key challenges to applying risk management to homeland security, and actions that could be taken to address them.

The report (GAO, 2008) refers to the risk management framework developed during a previous GAO activity on assessing risks related to ports and other critical infrastructure (GAO, 2005). The framework was prepared based on industry best practices and other criteria. This framework, shown in Figure A-4, divides risk management into five major phases: (1) setting strategic goals and objectives and determining constraints, (2) assessing risks, (3) evaluating alternatives for addressing these risks, (4) selecting the appropriate alternatives, and (5) implementing the alternatives and monitoring the progress made and results achieved.

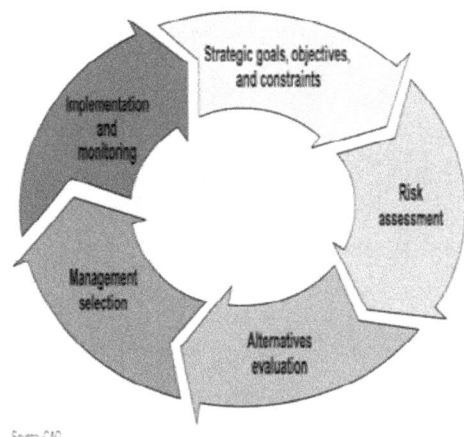

Figure A-4 GAO Risk Management Framework

A.2.5   National Aeronautics and Space Administration Risk-Informed Decisionmaking Handbook, 2010

The preface to the U.S. National Aeronautics and Space Administration (NASA) handbook observes the following:

> *Risk management (RM) is an integral aspect of virtually every challenging human endeavor, but well-defined RM processes have only recently begun to be developed and implemented as an integral part of systems engineering at NASA, given the complex concepts that RM encapsulates and the many forms it can take. However, few will disagree that effective risk management is critical to program and project success.*

The NASA risk-informed decisionmaking (RIDM) handbook (NASA, 2010) was prepared to provide a framework and guidance for risk-informed decisionmaking at that agency. RIDM at NASA is intended to ensure decisions between alternatives are made with an awareness of the risks associated with each, thereby helping to prevent late design changes, which can be key drivers of risk, cost overruns, schedule delays, and cancellation. The framework is shown below:

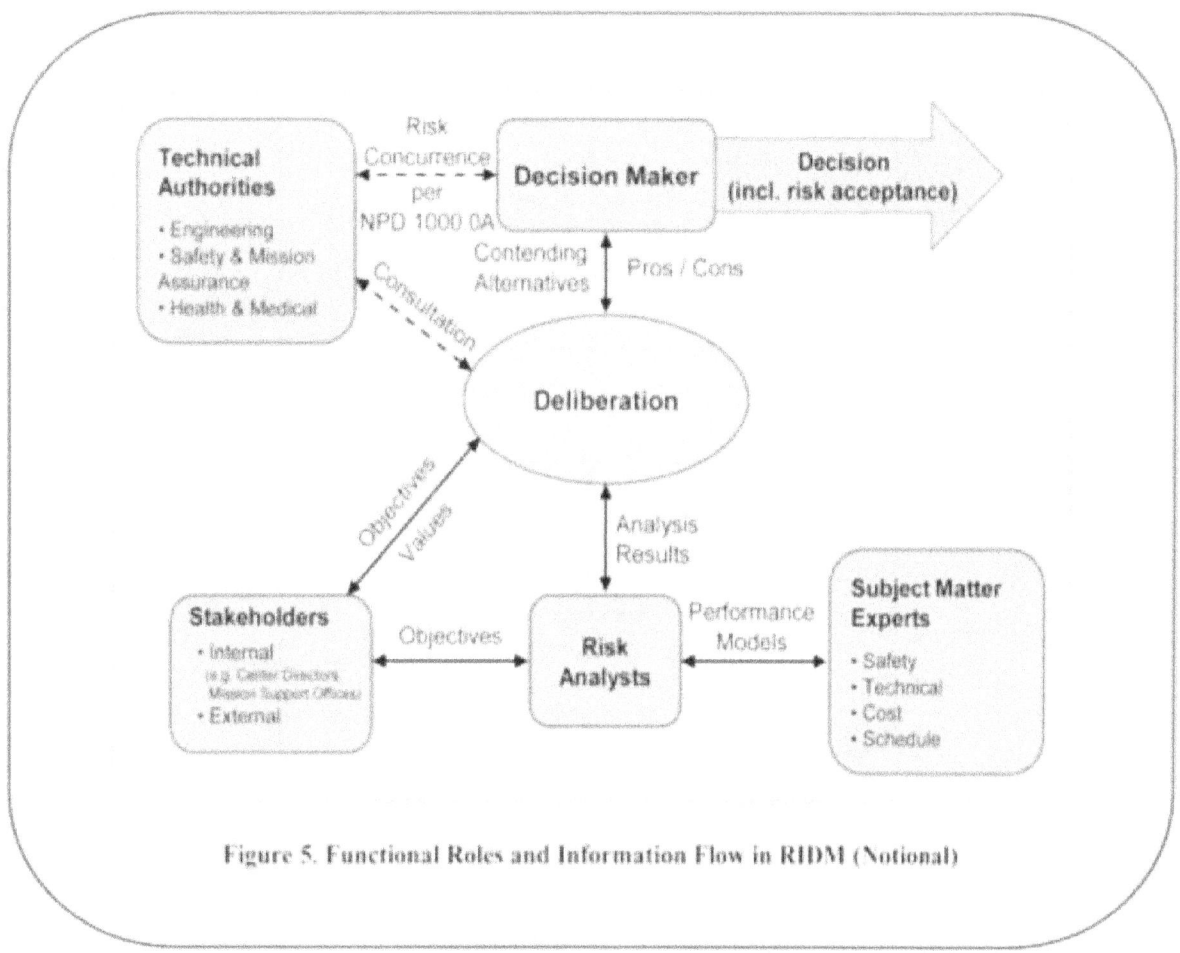

Figure 5. Functional Roles and Information Flow in RIDM (Notional)

Figure A-5   NASA Framework

A.2.6   Commandant Instruction 3500.3
        Operational Risk Management, U.S. Coast Guard

In response to various mishaps and related investigations, the U.S. Coast Guard prepared training material to emphasize risk management principles that outline a systematic process to continuously assess and manage risks: the operational risk management (ORM) process (USCG, 1999). The ORM process is a continuous, systematic process of identifying and controlling risks in all activities according to a set of preconceived parameters by applying appropriate management policies and procedures. This process includes detecting hazards, assessing risks, and implementing and monitoring risk controls to support effective, risk-based decisionmaking. The ORM process can be represented by the seven steps shown in Figure A-6.

Figure A-6 U.S. Coast Guard Framework

A.2.7   International Organization for Standardization, Standard 31000,
        "Risk Management – Principles and Guidelines"

The International Organization for Standardization (ISO) is a worldwide federation of national standards bodies. ISO prepared its 31000 standard, "Risk Management – Principles and Guidelines" (ISO, 2009) and supporting standards in recognition that all types of organizations need to address factors and uncertainties that challenge their objectives (i.e., risks). The standard defines various principles regarding the integration of risk management into organizations' processes and the expectations for risk management evaluations. The consistent use of risk management requires a management framework to embed, support, and improve the processes throughout the organization. The implementation of a risk management approach is, in turn, supported by a structured process that includes establishing the context (e.g., the specific objectives and decisions, internal and external factors to consider, and decisionmaking criteria), assessing the risks, selecting options, monitoring results and revising actions and processes, and communicating with various stakeholders. The relationship between the principles, framework, and risk management process are shown in the following figure.

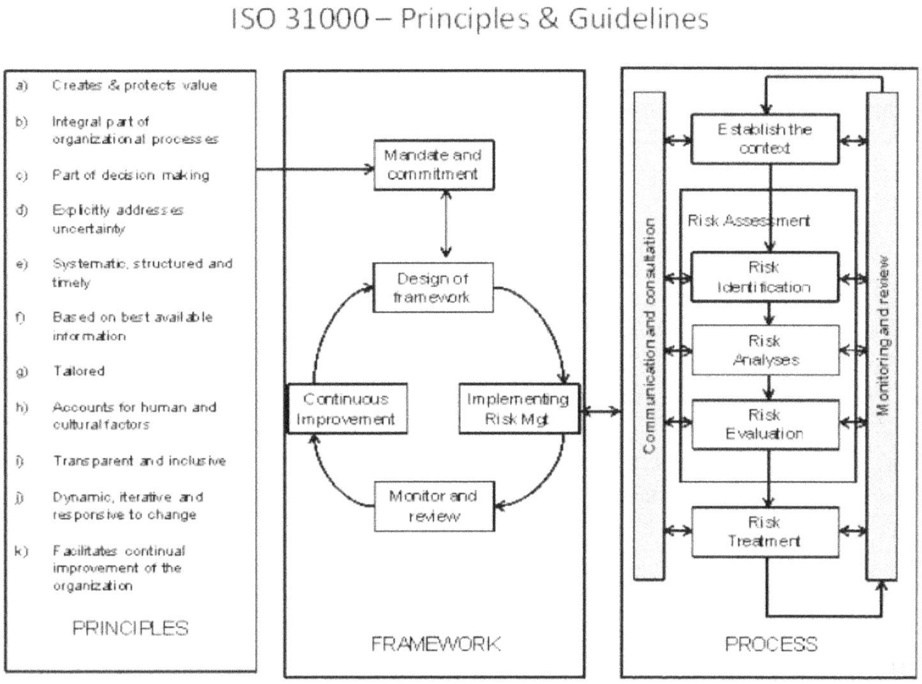

ISO 31000 – Principles & Guidelines
a) Creates & protects value
b) Integral part of organizational processes
c) Part of decision making
d) Explicitly addresses uncertainty
e) Systematic, structured and timely
f) Based on best available information
g) Tailored
h) Accounts for human and cultural factors
i) Transparent and inclusive
j) Dynamic, iterative and responsive to change
k) Facilitates continual improvement of the organization
PRINCIPLES
Mandate and commitment
Design of framework
Continuous Improvement
Implementing Risk Mgt
Monitor and review
FRAMEWORK
Establish the context
Risk Assessment
Risk Identification
Risk Analyses
Risk Evaluation
Risk Treatment
Communication and consultation
Monitoring and review
PROCESS

Figure A-7 ISO Framework

A.2.8 Department of Homeland Security, Risk Management Fundamentals

In response to The Policy of Integrated Risk Management, established by the Secretary of Homeland Security, the Office of Risk Management and Analysis at DHS published a guide, "Risk Management Fundamentals" (DHS, 2011), to promote a common understanding of and approach to risk management. Figure A-8 shows the DHS risk management process.

Define the Context
Identify Potential Risk
Assess and Analyze Risk
Develop Alternatives
Decide and Implement
Evaluate and Monitor

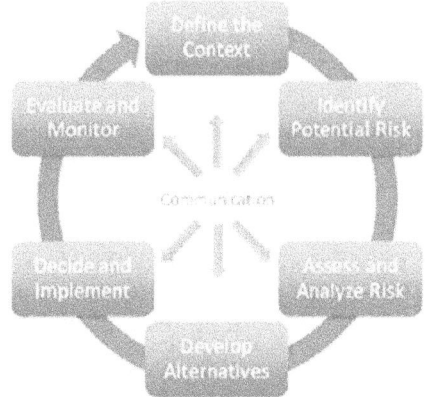

Figure A-8 DHS Risk Management Process

The DHS process is comprised of the following:

- Defining and framing the context of decisions and related goals and objectives.

- Identifying the risks associated with the goals and objectives.

- Analyzing and assessing the identified risks.

- Developing alternative actions for managing the risks and creating opportunities, and analyzing the costs and benefits of those alternatives.

- Making a decision among alternatives and implementing that decision.

- Monitoring the implemented decision and comparing observed and expected effects to help influence subsequent risk management alternatives and decisions.

A discussion of DHS's implementation of its Integrated Risk Management (IRM) Framework is provided below.

A.2.8.1 Implementation of Risk-Informed Security at the Department of Homeland Security

The Department of Homeland Security has begun the implementation of a risk management process similar to that recommended by the RMTF and described in the body of this report. The DHS process is guided by the concept of IRM, which has the following goals:

- Unify efforts among all homeland security partners to ensure that strategies and actions are informed by a common understanding of homeland security risk.

- Ensure that information and analysis about homeland security risks are incorporated into strategic and operational decisionmaking processes.

- Build a common understanding of risk management through development of a risk lexicon, risk-informed planning process, training, and standards of practice.

- Provide mechanisms to share risk data, risk assessments, and risk management decision support and analysis tools across the homeland security enterprise.

In alignment with the IRM goals, DHS has performed risk assessments at the strategic and operational level to make better use of risk information in departmental decisionmaking. The DHS approach to each of the components of the risk triplet (Kaplan and Garrick, 1981) is summarized below based on the more detailed descriptions in "Risk Management Fundamentals" (DHS, 2011). Each summary is followed by specific examples of the concepts applied in current DHS risk assessments.

What can go wrong?

In determining which risks are to be considered for a particular risk assessment, the DHS Risk Management Fundamentals suggest focusing only on the risks that are relevant to the underlying decision. These can include strategic, operational, and institutional risks. They should generally be divided into scenarios that can be analyzed individually, covering the full scope of the decision context while avoiding overlaps.

National Risk Profile Development

As defined in the National Infrastructure Protection Plan (NIPP), the goal of infrastructure protection at DHS is to "build a safer, more secure, and more resilient America by preventing, deterring, neutralizing, or mitigating the effects of deliberate efforts by terrorists to destroy, incapacitate, or exploit elements of our Nation's Critical Infrastructure and to strengthen national preparedness, timely response, and rapid recovery of critical infrastructure in the event of an attack, natural disaster, or other emergency (DHS, 2009)."

As part of the efforts to meet this goal, the DHS National Protection and Programs Directorate's Office of Infrastructure Protection works with the critical infrastructure community to determine the risk to the Nation's infrastructure from natural hazards and terrorism. The results of this discussion are used to create the National Risk Profile ("the Profile"), which "every year identifies the highest relative risks to critical infrastructure and those critical infrastructure sectors (18 total) at a higher risk from the greatest number of hazards. The Profile also identifies other risk management concerns, such as high-likelihood risks and low-likelihood/high-consequence infrastructure protection priorities (DHS, 2009)."

How likely is it?

The "Risk Management Fundamentals" (DHS, 2011) notes that common sources for frequencies and probabilities include historical records, models, simulations, and elicitations of subject-matter experts. These frequencies and probabilities may be qualitative or quantitative with the choice to quantify risk information based on the needs of the decisionmaker.

Integrated Chemical, Biological, Radiological, and Nuclear Terrorism Risk Assessment Threat Estimation

The Science and Technology Directorate's Integrated Chemical, Biological, Radiological, and Nuclear (CBRN) Terrorism Risk Assessment (ITRA) is conducted by DHS as directed under Homeland Security Presidential Directive 18 (HSPD-18) to provide a risk-based decision support tool to agencies across the Federal Government responsible for reducing and mitigating CBRN terrorism risk. To determine the frequencies associated with rare, highly uncertain terrorist attacks, the ITRA relies on the elicitation of subject-matter experts from the intelligence community (DHS, 2012).

Risk Assessment Process for Informed Decisionmaking (RAPID) Vulnerability Estimation

RAPID is a multihazard probabilistic risk assessment created to inform the DHS budget cycle. The assessment includes the calculation of scenario-dependent vulnerabilities to terrorist attacks given the current U.S. security alignment. RAPID uses structured surveys to determine how the different elements of the homeland security enterprise work together to prevent terrorist attacks. The data are collected for each DHS budget program and input into fault trees, allowing for the calculation of the risk reduced by each program. This information can identify potential gaps in the security structure and compare programmatic risk reduction to help guide the Department's strategic direction and influence the allocation of resources.

What are the consequences?

As with frequency and probability, consequences may also be qualitative or quantitative. Depending on the analysis, they include loss of life, injuries, economic impacts, psychological consequences, environmental degradation, and inability to execute essential missions. To determine the relevant consequences for an individual assessment, "Risk Management Fundamentals" (DHS, 2011) recommends using structured techniques, such as value focused thinking.

Strategic National Risk Assessment Consequences

The Strategic National Risk Assessment (SNRA), which was developed and implemented by DHS in support of Presidential Policy Directive 8 for National Preparedness, calls for national preparedness to be based on core capabilities that support "strengthening the security and resilience of the United States through systematic preparation for the threats that pose the greatest risk to the security of the Nation, including acts of terrorism, cyber attacks, pandemics, and catastrophic natural disasters."

The assessment examined the consequences associated with six categories of harm: loss of lives, injuries and illnesses, direct economic costs, and, for the first time at DHS, social displacement, psychological distress, and environmental impact. The SNRA drew upon input from the Federal interagency, including data and information from Government models and assessments, historical records, structured analysis, and judgments of experts from different disciplines (DHS, 2011a).

A.2.8.2   Risk in Regulation at DHS

Executive Order 12866, "Regulatory Planning and Review," published in the *Federal Register* on October 4, 1993 (58 FR 51735), stipulates that Federal agencies must make a reasoned determination that the benefits of an intended regulation justify its costs. In many DHS analyses, the primary benefit of a regulation is the reduction in the probability or consequence of terrorist attacks. In these regulations, the calculation of benefits can be determined by qualitative or quantitative risk reduction assessments.

U.S. Customs and Border Protection Break-even Analyses

U.S. Customs and Border Protection (CBP) has proposed several regulations involving the transit of people and goods across U.S. borders. The costs of the regulations are determined by examining the measures that travelers and businesses will take as a direct result of the proposed rules, estimating the costs of those activities, and adding the costs for the government to implement the regulation. In a typical cost-and-benefit analysis, the economic benefits would then be compared with the costs. For several CBP regulations, the regulatory benefits include the risk reduction associated with preventing certain types of terrorist attacks. With the difficulty associated with determining the probability of major terrorist attacks occurring, CBP had to forgo the calculation of benefits from the expected probability reduction and instead calculated the reduction in probability of an attack necessary to balance the costs. This technique is called break-even analysis and has been used in several of CBP's regulatory decisions, (IEI, 2008 and 2008a).

While the context of the DHS risk management process has fundamental differences with that of the NRC's, the continuing experience of integrating risk management at DHS can provide important insights into the implementation of an agencywide risk management effort and specifically into the potential structure for risk-informing security threats.

A.2.9 Center for Strategic and International Studies

The above sections provide short descriptions of risk management systems developed by organizations and other Federal agencies. There are countless other examples of such programs, and the use of risk management in both Government and private sector activities. A study entitled, "Risk Management in Non-DoD U.S. Government Agencies and the International Community: Best Practices and Lessons Learned" (CSIS, 2011), provides insights and case studies of risk management practices at various agencies (including the NRC) and within several countries.

A.3   Recommended NRC Risk Management Framework

Drawing upon the insights from the various risk management frameworks described above, the RMTF adopted a very similar process for NRC decisionmaking. The steps of the process are similar to those defined by each of the above frameworks and are shown in Figure A-9.

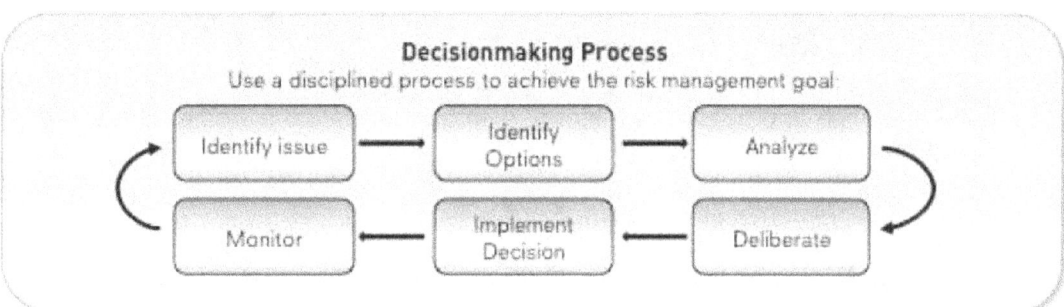

Figure A-9 The Regulatory Decisionmaking Process

The steps of this proposed risk management framework for the NRC include the following:

(1)   Identification and framing (characterization) of an issue, proposal, or problem:

- o   Define the nature of the hazard.

- o   Identify the potential worker or public exposure.

- o   Determine the source of the issue or proposal (e.g., event, application).

(2)   Identification of options and alternatives:

      o   Identify possible alternatives, including no action.

      o   Consider performance-based approaches.

      o   Prepare defense-in-depth proposals to be assessed against tolerability or acceptability criteria and other factors.

(3)   Analysis:

      o   Select risk evaluation technique(s) given hazard, issue, and decisionmaking criteria (can include deterministic, risk assessments, judgment, etc.).

      o   Evaluate options related to appropriate barriers and controls given:

          •   What can go wrong?"

          •   "How likely is it?" and

          •   "What are the consequences?"

(4)   Deliberation (integrated decisionmaking):

      o   Evaluate options, considering uncertainties and acceptability criteria.

      o   Consider external factors.

      o   Determine appropriate risk-informed and performance-based defense-in-depth protection.

(5)   Actions and implementation of selected option:

      o   Follow NRC processes.

          •   rulemaking, licensing, environmental reviews

          •   oversight

      o   Apply regulatory activities (Life Cycle)

          •   design, operations, decommissioning

(6)   Monitoring and feedback:

      o   Gather information from oversight activities.

      o   Gather information from events, stakeholders, and other sources.

      o   Assess problems and identify corrective actions.

(7)   Communicate with stakeholders throughout the process:

- ○   Gather information.

- ○   Deliberate.

- ○   Implement and solicit feedback.

Additional discussions related to the deliberative process are provided in Chapter 2 and Appendix B, "Risk Management – Analysis and Deliberation."

A.4   Applicability

The risk management goal and related risk management process are intended to be used universally for NRC decisions regarding the regulation and oversight of licensed activities, including reactors, materials, waste, fuel cycle, and transportation. In many cases, the current NRC decisionmaking processes (e.g., integrated decisionmaking for nuclear reactors and risk-informed decisionmaking for materials licensees) are similar to the risk management approach described in this report. Major actions, such as licensing of reactor or fuel cycle facilities, agency level initiatives, and rulemakings, include project plans and processes that are often explained in terms of the six steps of the risk management process (or in equivalent terms and steps). The handling of some routine activities may at first glance appear simpler than the risk management process, but this is likely because some steps, such as selecting risk evaluation techniques and decisionmaking criteria, are already incorporated into procedures or guidance documents. Provided the existing guidance and processes ensure that appropriate barriers are established and risks are maintained at an acceptably low level, the move to a risk management framework is unlikely to require significant changes to existing practices. It is not expected that the risk management process will complicate matters such that staff or managers should avoid its adoption. Instead, it is expected that a comprehensive and holistic approach across NRC programs will provide long-term efficiencies, consistency, and other benefits.

Although beyond the immediate scope of the RMTF, the general concepts of risk management and methodical decisionmaking are also applicable to other agency activities. For example, a consistent risk management approach across NRC programs could support changes in management practices such that program managers would have increased flexibility to allocate resources to address risk or safety concerns. Insights from inspections, operating experience, scientific studies, or risk assessments could inform timely changes to programs and focus areas as part of the implementation of budgeted resources. Another example in which the NRC uses risk management processes is in the area of providing security to the agency's information systems.

A.5   References

(Commission, 1997)   The Presidential/Congressional Commission on Risk Assessment and Risk Management, "Framework for Environmental Health Risk Management," Vols. 1 and 2, 1997.

(CSIS, 2011)   Center for Strategic and International Studies, "Risk Management in Non-DoD U.S. Government Agencies and the International Community: Best Practices and Lessons Learned," A Report of the CSIS Defense and National Security Group, March 2011.

(DHS, 2009)   U.S. Department of Homeland Security, "National Infrastructure Protection Plan," 2009.

(DHS, 2010)   U.S. Department of Homeland Security, "DHS Risk Lexicon," September 2010.

(DHS, 2011)   U.S. Department of Homeland Security, "Risk Management Fundamentals," November 2011.

(DHS, 2011a)   U.S. Department of Homeland Security, "The Strategic National Risk Assessment in Support of PP 8: A Comprehensive Risk-Based Approach toward a Secure and Resilient Nation," 2011.

(DHS, 2012)   U.S. Department of Homeland Security, "Companion to the Integrated CBRN Terrorism Risk Assessment (ITRA)," 2012.

(Exec. Order, 2011)   Executive Order 13563, "Improving Regulation and Regulatory Review," January 18, 2011, published in the *Federal Register* on January 21, 2011 (76 FR 821).

(GAO, 2005)   U.S. Government Accountability Office, "Risk Management: Further Refinements Needed to Assess Risks and Prioritize Protective Measures at Ports and Other Critical Infrastructure," December 2005.

(GAO, 2008)   U.S. Government Accountability Office, "Strengthening the Use of Risk Management Principles in Homeland Security," June 2008.

(IEI, 2008)   Industrial Economics, Inc. "The Western Hemisphere Travel Initiative Implemented in the Land Environment," 2008.

(IEI, 2008a)   Industrial Economics, Inc., "Importer Security Filing and Additional Carrier Requirements," 2008.

(IRGC, 2005)   International Risk Governance Council, "White Paper on Risk Governance – Towards an Integrative Approach," September 2005.

(ISO, 2009) — International Organization for Standardization, "Risk Management – Principles and Guidelines," ISO-31000, 2009.

(Kaplan and Garrick, 1981) — S. Kaplan and B.J. Garrick, "On the Quantitative Definition of Risk," *Risk Analysis*, 1:11–27.

(National Research Council, 1996) — National Resource Council, "Understanding Risk: Informing Decisions in a Democratic Society," National Academy Press, 1996.

(NASA, 2010) — U.S. National Aeronautics and Space Administration, "Risk-Informed Decisionmaking Handbook," NASA/SP2010-576, Version 1.0, April 2010.

(NRC, 1999) — U.S. Nuclear Regulatory Commission, Staff Requirements Memorandum Regarding SECY-98-144, "White Paper on Risk-informed and Performance-Based Regulation," March 1, 1999, Agencywide Documents Access and Management System (ADAMS) Accession No. ML003753601.

(OECD, 2010) — Organization for Economic Cooperation and Development, "Risk and Regulatory Policy – Improving the Governance of Risk," 2010.

(USCG, 1999) — U.S. Coast Guard, "Operational Risk Management," Commandant Instruction 3500.3, November 1999.

APPENDIX B

RISK MANAGEMENT – ANALYSES AND DELIBERATION

B.1   Introduction

As discussed in other sections of this report, the risk management framework shown in Figure B-1 is primarily an approach to provide structure and logic to the decisionmaking process. Major elements of the process are the technical analyses and the deliberations to determine what risk-informed, performance-based, defense-in-depth protections are appropriate for a given radiological hazard, relevant scenarios, and the associated uncertainties.

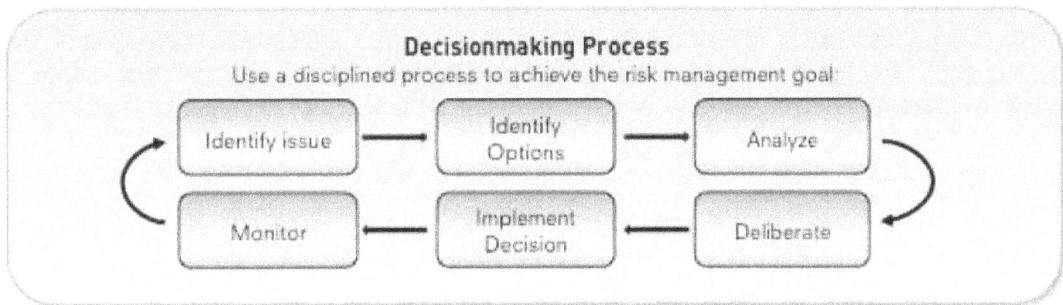

Figure B-1 The Regulatory Decisionmaking Process

B.2   Analysis

A key input into the deliberations within a risk management program is the technical analysis or actual evaluation of risks associated with the subject system or activity. A challenge for introducing risk management at the NRC will be the terminology associated with the "analyze" step in the process and how risk assessments and the "traditional engineering analyses" have been used in the licensing of most materials, devices, and facilities. The efforts to improve the use of risk insights in NRC processes were sometimes interpreted as being separate and distinct from previous engineering and licensing activities. An important part of developing a risk management framework at the NRC will be communicating that the traditional engineering analyses and related acceptance criteria are actually part of risk evaluation and management of risk through establishing certain conventions and standards. NRC regulatory programs for all of the various materials, devices, and facilities have been based, either explicitly or implicitly, on identifying radiological hazards and asking:

* What can go wrong?

* How likely is it?

* What are the consequences?

These questions are commonly referred to as the "risk triplet" (Kaplan and Garrick, 1981). In some traditional approaches, the "what can go wrong" question was addressed by identifying certain events or conditions (e.g., maximum credible accidents, design-basis events) and evaluating structures or systems using acceptance criteria contained in NRC regulations, guidance, and industry codes and standards. Establishing the scenarios and acceptance criteria considered at least a subjective estimation of the "how likely is it" question. The "what are the consequences" question also considered the likelihood of an event or condition in establishing the acceptance criteria, whether it was an actual dose limit or a limit on a system, structure, or a component that helped to mitigate an accident or contain radioactive material (see Appendix F for reactor-related examples). So, while the Risk Management Task Force (RMTF) does not see a move to a risk management system as necessarily at odds with existing NRC regulatory programs, it foresees communication issues with both internal and external stakeholders regarding some aspects of the proposed regulatory framework and the related terminology.

As presented in the main body of this report, the RMTF has found that a risk management framework is a logical way to fulfill the NRC's mission. Figure 2-2 (repeated below) shows the transition or flow from the mission to ensure adequate protection of public health and safety to a framework to manage risks to workers and the public from the use of radioactive materials.

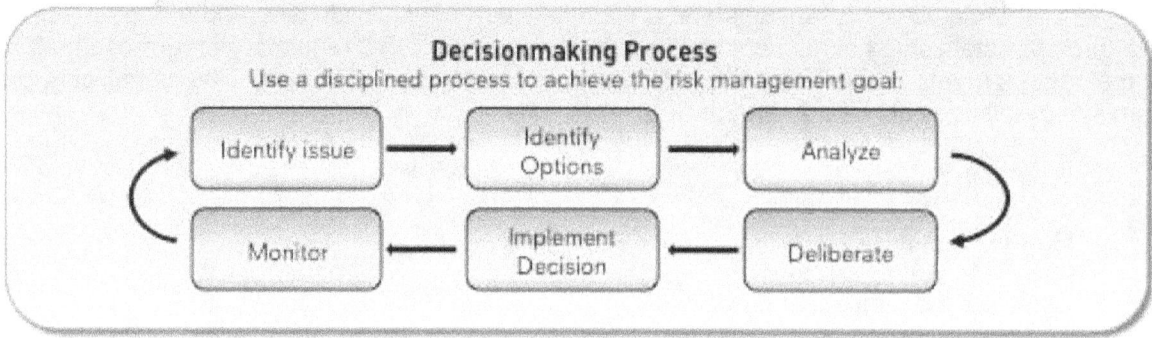

Figure B-2 A Proposed Risk Management Regulatory Framework

Within this framework, a risk-informed and performance-based approach is not an alternative to defense-in-depth concepts, but instead is simply a means to determine, in concert with other techniques, an appropriate level of protection to prevent, contain, and mitigate possible releases of radioactive material from NRC-licensed activities. The selection of a technique or combination of techniques to evaluate risks (i.e., analyze a problem or proposal) considers the radiological hazard and the relative strengths and limitations of specific techniques to address the material, device, or facility during relevant offnormal scenarios. It is also likely that different techniques may better support specific portions of the life cycle or specific concerns in terms of internal and external events, design and subsequent configuration management, and the importance of the man and machine interface.

Given the importance of terminology in the development and implementation of a risk management approach, the RMTF offers the following definitions:[1]

Technical Analysis

An evaluation to support NRC decisionmaking that includes the identification of potential radiological hazards and addresses: (1) what can go wrong, (2) how likely is it, and (3) what are the consequences. The evaluation process uses one or more techniques to address these questions. The possible techniques include expert judgment, reliance on industry or international standards, traditional engineering analyses, hazards or scenario analyses, and risk assessments.

Risk Assessment

An analysis technique that uses estimates of frequencies and consequences to systematically calculate risk and present the results (usually in numerical form). In this context, risk assessments include probabilistic risk assessments (PRAs), probabilistic safety assessments (PSAs), and risk analyses in the reactor programs and with risk-informed decisions in the materials programs.

A key concept in the development of a risk management process at the NRC is that the evaluation of risks can be done any number of ways—including traditional engineering analyses, risk assessments, and other techniques selected to support specific decisions related to particular issues and hazards. This point is reinforced by discussions in International Standards Organization (ISO) 31010, "Risk Management – Risk Assessment [technical analysis] Techniques" (ISO, 2009), which describe a variety of techniques to analyze a problem and inform the decisionmaking process.

1　The RMTF terminology differs somewhat from generic risk management terminology, which generally refers to risk assessment as any technique used to support deliberations (i.e., the RMTF equivalent of technical analysis). The above terminology was chosen because of the longstanding practice at the NRC and within the broader nuclear community to differentiate between probabilistic risk assessments and traditional deterministic type analyses.

These techniques include the following:

- discussions or brainstorming

- expert elicitation

- hazard analyses

- scenario analyses

- failure mode and effects analyses

- fault tree*

- event tree*

- decision trees

- Monte Carlo simulations

- cost-benefit analyses

- frequency-consequence curves*

    * Usually associated with "risk assessment" type technique

Within the proposed Risk Management Regulatory Framework, technical analyses will be used to support decisions on appropriate regulatory controls and oversight, which are described in terms of risk-informed, performance-based defensein depth and related measures to prevent, contain, and mitigate conditions and accidents that could result in radiation exposures to workers or the public. The processes currently in place for operating nuclear power reactors include a balancing of traditional engineering approaches (sometimes referred to as deterministic or mechanistic analyses) and risk assessments. This balancing of technical analysis techniques to support decisions regarding appropriate defense-in-depth measures is shown in Figure B-3.

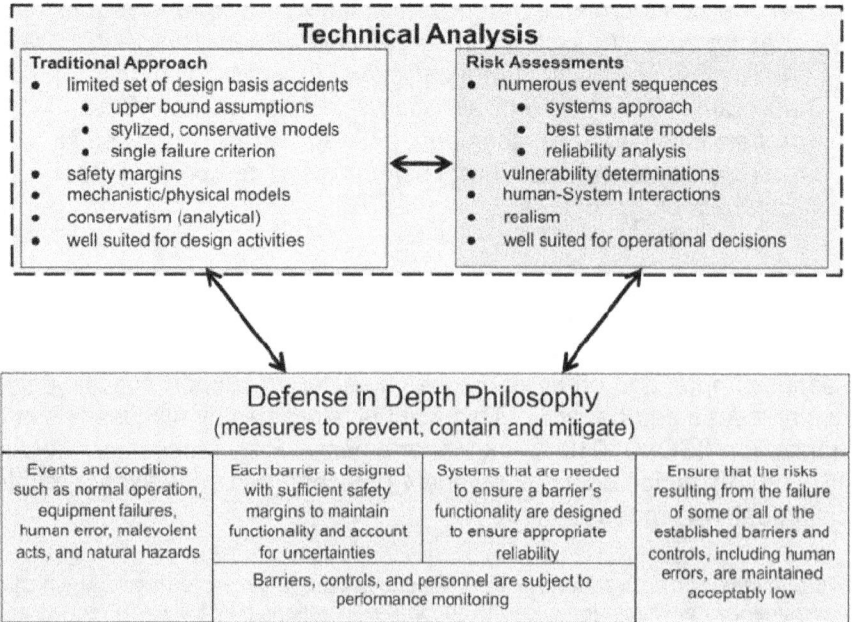

Figure B-3 Balancing Risk Assessments and Deterministic Techniques

From the previous discussions, the purpose of the technical analysis is to support subsequent deliberations on what are appropriate barriers and controls to prevent, contain, and mitigate possible releases of radioactive material and ensure that the risks from events that degrade or challenge the barriers are maintained acceptably low. The appropriate barriers range from simple containers for some radioactive sources to complex structures for nuclear power plants. Likewise, the systems and actions taken to maintain barriers can range from labels and administrative controls to complex mitigation systems. Controls established for materials licensees can also address circumstances where the use of a device requires the temporary bypass of physical barriers (e.g., radiography and irradiation devices). The possible need for emergency preparedness requirements is based on the risk that barriers might be compromised such that the public could be exposed to radioactive materials, and protective actions, such as sheltering or evacuation, might be warranted.

The selection of the technical analysis technique considers the nature of the hazard, the possible challenges to barriers, and the complexity of barriers and supporting systems. In general, decisions regarding simpler devices and frequent events (e.g., activities with many materials licensees) can be supported by technical analyses based on traditional engineering approaches, operating experience, and qualitative risk assessments. Decisions related to more complex facilities and infrequent events (e.g., nuclear power plants) can benefit from analyses that include more robust risk assessments. This relationship between various technical analysis techniques and the nature of the possible radiological hazard is shown in Figure B-4:

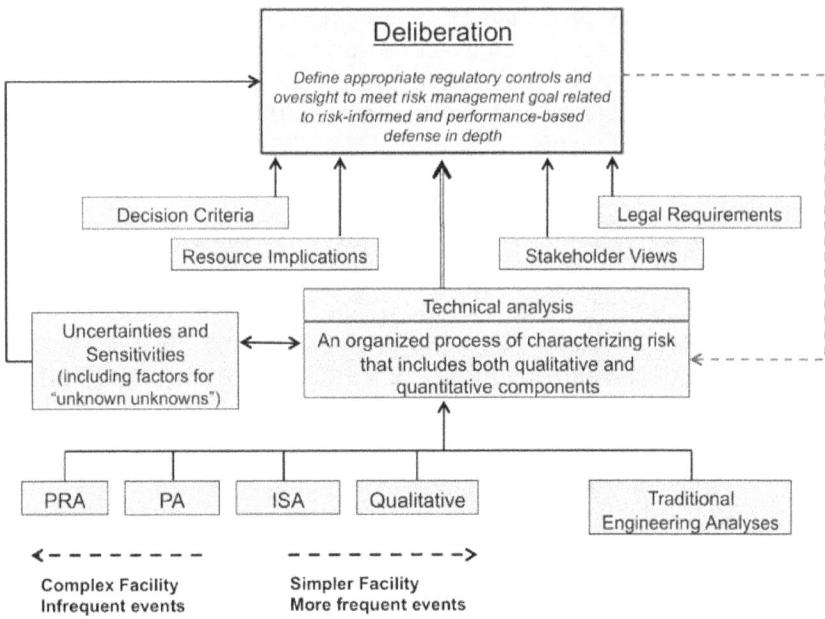

Figure B-4 Technical Analysis Techniques and Deliberation

The discussions of traditional approaches and risk assessments shown in Figures B-3 and B-4 are mostly defined for nuclear power reactors but have some relevance to other regulatory program areas. As used here, the traditional approach evaluates risk by defining certain events that could challenge barriers and performing engineering analyses on systems and structures to ensure that defined acceptance criteria are satisfied. An example is to define a design-basis weather event (e.g., wind or precipitation) and evaluate the loads on structures using established industry codes and standards. Another example is to assume a mechanical failure, such as a specific pipe break, and perform analyses on the capabilities of safety systems to remove heat from the reactor core and maintain temperatures below regulatory requirements. These types of engineering analyses tend to define design-basis events (sometimes referred to as maximum credible or maximum hypothetical events) based on historical data and engineering judgment; include upper bound assumptions; use stylized, conservative analytical models and assumptions (e.g., failure of mitigation systems); and incorporate safety margins. These kinds of analyses are well suited for design activities where engineers are making decisions on structures, materials, cooling capabilities, and other factors affecting the selection and construction of plant systems, structures, and components. Limitations of traditional approaches can include the (1) "over-design" and resultant increase in cost of some plant features resulting from conservative assumptions and (2) "underdesign" and resultant compromise of barriers if the design-basis event is exceeded or a vulnerability is not identified.

The attributes of more detailed risk assessments (e.g., PRAs for power reactors) include the analysis of numerous event sequences, the use of best estimate analytical models, the identification of system importance and establishment of reliability goals, the identification of vulnerabilities, and the interactions and dependencies between systems and operators. The risk assessments have been better suited to support decisions related to configuration management and accident management for events exceeding the traditional design-basis events. Limitations of the risk assessment techniques include (1) completeness of scenario lists, and (2) the difficulties in defining typical design specifications, including safety margins, using existing PRA methodologies. It should be noted that risk assessments have been introduced into the NRC's regulatory system in areas such as the requirements for an integrated safety assessment for fuel cycle facilities, the performance of PRAs for new reactors, and the optional use of PRAs for nuclear reactors to justify changes to special treatment requirements for systems, structures, and components (Title 10 *Code of Federal Regulations* (10 CFR) 50.69, "Risk-informed Categorization and Treatment of Structures, Systems, and Components for Nuclear Power Reactors").

The key to the selection of a technical analysis technique or combination of techniques is to ensure that the analysis will support the associated deliberation or decisionmaking process in an effective and efficient manner. This, in turn, means that there needs to be some acceptance criteria or other rationale for decisionmaking that are established and understood by the decisionmaker and agency stakeholders.

B.3   Deliberate

The technical analyses are a key input to the process but can support a successful outcome only if there is an equally disciplined approach to the actual decisionmaking. The NRC's decisions related to its core mission can be described in terms of ensuring that its licensees provide and maintain the appropriate risk-informed and performance-based defense-in-depth protections to achieve the following:

*   Ensure appropriate barriers, controls, and qualified personnel to prevent, contain, and mitigate possible inadvertent exposure to radioactive material according to the hazard present, the relevant scenarios, and the associated uncertainties.

*   Ensure the risks resulting from the failure of some or all of the established barriers and controls are maintained acceptably low.

The technical analysis is an important input into the deliberation process, but it is not the only factor that influences the final decision. The decisionmaker needs to consider the uncertainty and sensitivity analyses associated with the technical analysis, as well as the analytical results and how they compare to decision criteria established for mechanistic approaches, risk assessment approaches, or a combination thereof. The uncertainties to be considered include the degree of understanding of hazards, scenarios, phenomena, safety margins, and other tangible and intangible factors. There have been numerous cases in which emergent technical issues have been resolved, at least in part, by safety margins or other actions previously taken to address uncertainties, including conservatisms, to cover the so-called "unknown unknowns." The deliberative process also includes consideration of resources and schedules for the agency and its licensees and the input received from various internal and external stakeholders. Other factors include legal requirements and the desire to maintain consistency with various guidance documents, treaties, or standards.

A representation of the deliberative process, including consideration of the various factors, is shown in Figure B-5.

Figure B-5 Deliberations

In a risk management regime, the decisionmaking process involves discussions of what risks are tolerable or acceptable to the NRC, key stakeholders, and the public at large. Various approaches have been developed to represent tolerable or acceptable risks from nuclear plants and other modern industrial facilities. It should be noted that the NRC has established and, in some cases, codified acceptance criteria (deterministic and risk-informed) for regulatory decisions related to license applications, amendments to licenses, and other actions for specific types of licensees.

Examples include the following:

- Reactors

    o   NUREG-0800, "Standard Review Plan for the Review of Safety Analysis Reports for Nuclear Power Plants: LWR Edition" (NRC, 2007)

    o   Regulatory Guide 1.174, "An Approach for Using Probabilistic Risk Assessment in Risk-informed Decisions on Plant-specific Changes to the Licensing-basis," Revision 2 (NRC, 2011)

    o   Regulatory Guide 1.187, "Guidance for Implementation of 10 CFR 50.59, "Changes, Tests, and Experiments" (NRC, 2000)

    o   Regulatory Guide 1.201, "Guidelines for Categorizing Structures, Systems, and Components in Nuclear Power Plants According to Their Safety Significance" (NRC, 2006)

    o   Regulatory Guide 1.205, "Risk-informed, Performance-based Fire Protection for Existing Light-Water Nuclear Power Plants," Revision 1 (NRC, 2009)

- Materials

    o   NUREG-1556 (Volumes 1 – 21), "Consolidated Guidance About Materials Licensees" (NRC, 1998)

    o   "Risk-Informed Decisionmaking for Nuclear Material and Waste Applications," Revision 1 (NRC, 2008)

    o   NUREG/CR-6642, "Risk Analysis and Evaluation of Regulatory Options for Nuclear Byproduct Material Systems" (NRC, 2000a)

- Fuel Cycle Facilities

    o   NUREG-1520, "Standard Review Plan for the Review of a License Application for a Fuel Cycle Facility," Revision 1 (NRC, 2010)

    o   NUREG-1513, "Integrated Safety Analysis Guidance Document" (NRC, 2001)

- Storage and Transportation Systems

    o   NUREG-1536, "Standard Review Plan for Dry Cask Storage Systems," Initial Report, (NRC, 1997)

    o   NUREG-1927, "Standard Review Plan for Renewal of Spent Fuel Dry Cask Storage System Licenses and Certificates of Compliance" (NRC, 2011a)

    o   NUREG-1609, " Standard Review Plan for Transportation Packages for Radioactive Material," Initial Report (NRC, 1999) o   NUREG-1617, "Standard Review Plan for Transportation Packages for Spent Nuclear Fuel," Initial Report (NRC, 2000b)

These guidance documents often outline a decisionmaking process that is similar to the one described in this report. For example, the report entitled, "Risk-informed Decisionmaking for Nuclear Material and Waste Applications," Revision 1, (NRC, 2008), includes steps to define the regulatory issue, determine the appropriate techniques to perform evaluations, and apply the risk-informed decisionmaking process.

In some cases, the technical analysis described in the NRC guidance emphasizes the traditional engineering approaches used in the initial licensing of most NRC facilities. While traditional engineering approaches, sometimes referred to as deterministic analyses, often omit specific discussion of event frequencies, probabilities, and consequences, they are nevertheless an evaluation technique that supports risk management decisions. Defining acceptance criteria in terms of limits on specific barriers (with added safety margins) is a common approach. The traditional regulations and guidance may also define specific assumptions regarding initiating events, such as earthquakes, floods, and other challenges to one or more of the defense-in-depth barriers. A common approach supporting the NRC deliberative processes includes the development and use of consensus codes and standards. This practice is consistent with the National Technology Transfer and Advancement Act of 1995 (Public Law 104113), which requires government agencies to use consensus standards where possible. In the reactor area, the NRC has incorporated specific sections of the ASME Code and some Institute of Electrical and Electronics Engineers (IEEE) Standards by reference into its regulations. Other consensus codes and standards are referenced in material or facility licenses. Increasingly, the incorporation of risk assessments and performance-based approaches has been introduced to various codes and standards. The transition from the current guidance to the decisionmaking process proposed by the RMTF is described in Chapter 4 for the various NRC regulatory programs.

As previously mentioned, a concern with the traditional or deterministic approach is that the establishment of "maximum credible accidents" as the sole basis for defining defense-in-depth protections may not account for unidentified vulnerabilities. Examples of such insights from risk assessments related to nuclear reactors include the identification of concerns regarding intersystem loss-of-coolant accidents, anticipated transients without scram, and station blackout events. Attempts to introduce risk assessments as a way to supplement the traditional approaches gained support following the publication of WASH-1400, "The Reactor Safety Study," and the accident at Three Mile Island (TMI) in 1979. Specific milestones include the issuance of the following NRC policy statements:

- "Policy Statement on Severe Reactor Accidents Regarding Future Designs and Existing Plants," (NRC, 1985)

- "Safety Goals for the Operations of Nuclear Power Plants," (NRC, 1986)

- "Policy Statement on the Regulation of Advanced Reactors," (NRC, 2008a)

- "Use of Probabilistic Risk Assessment Methods in Nuclear Regulatory Activities; Final Policy Statement" (NRC 1995)

Each of these policy statements addressed operating reactors and future reactors and maintained that specific activities for the operating reactors (e.g., TMI action items) were a viable alternative to incorporating risk assessments more formally into the decisionmaking process. NRC actions were informed by various activities and risk assessments, including the "Individual Plant Examination Program: Perspectives on Reactor Safety and Plant Performance" (NRC, 1997a), individual plant examinations for external events (NRC, 2001a), and NUREG-1150, "Severe Accident Risks: An Assessment for Five U.S. Nuclear Power Plants—Final Summary Report" (NRC,1990). The policy statements supported improved use of risk assessment techniques in the design and licensing of future reactors. The PRA policy statement of 1995 encouraged, but did not require, the increased use of risk assessment techniques in all NRC regulatory programs. The absence of decisionmaking criteria complicated the implementation of these policy statements. A step forward in the integration of risk assessments into the decisionmaking process was provided by Regulatory Guide 1.174 in 1998. This Regulatory Guide clarified how insights from risk assessments should be used within an integrated decisionmaking process to support applications for and NRC review of proposed changes to the licensing-basis for operating reactors. Similar guidance was provided for materials, transportation, and waste programs in the risk-informed decisionmaking process developed by FSME and NMSS. The integrated decisionmaking process from Regulatory Guide 1.174 and associated guidance to support decisionmaking are shown below:

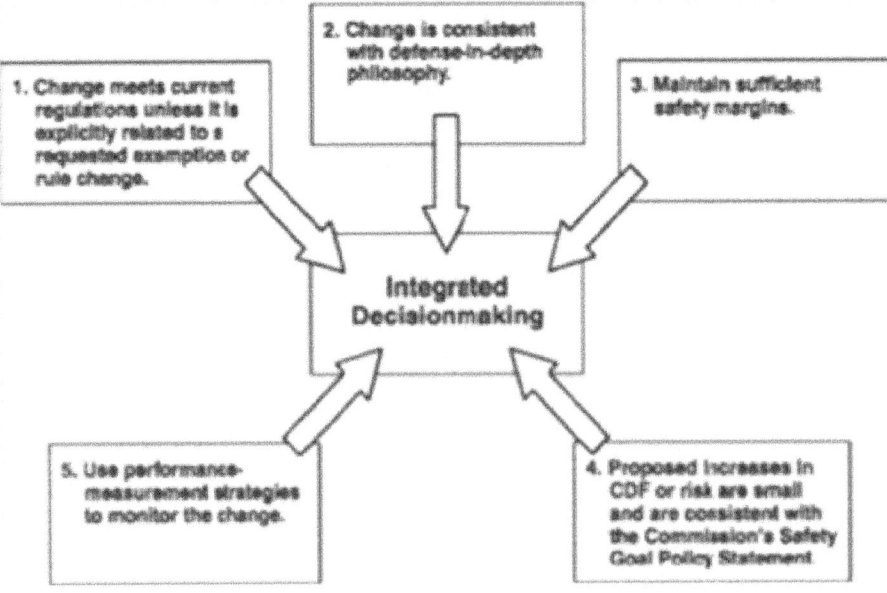

Figure B-6 RG 1.174 Integrated Decisionmaking Process

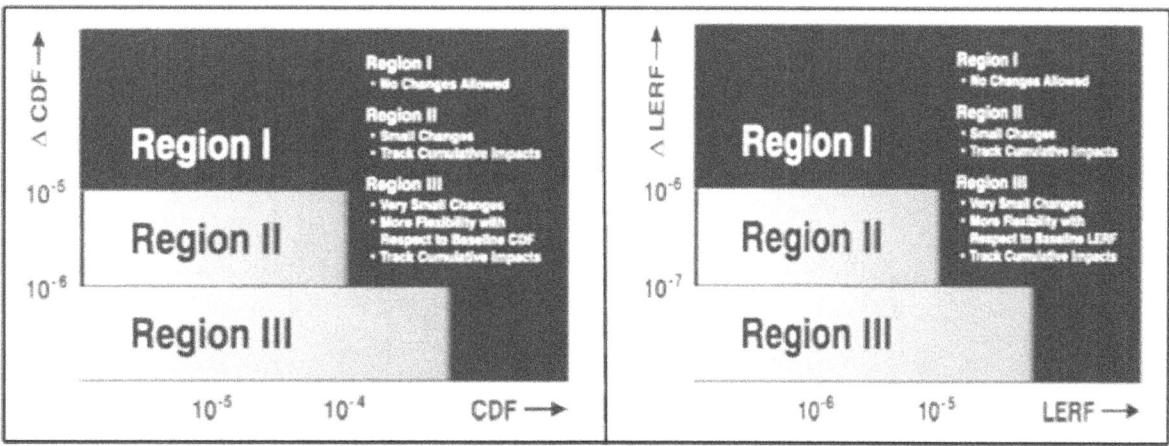

Figure B-7 RG 1.174 Process and Decision Criteria

A risk-informed approach has also been developed for the NRC's oversight of commercial nuclear power plants (NRC, 2006a). The oversight process uses a significance determination process (SDP) to help assess licensees' performance and determine appropriate agency responses. The guidance related to the SDP is shown in Figure B-8.

ΔCDF	Reactor Oversight Process – Significance Determination Process	ΔLERF
10^{-4}	**Red** (high safety or security significance) is quantitatively greater than 10^{-4}ΔCDF or 10^{-5} ΔLERF. Qualitatively, a Red significance indicates a decline in licensee performance that is associated with an unacceptable loss of safety margin. Sufficient safety margin still exists to prevent undue risk to public health and safety.	10^{-5}
10^{-5}	**Yellow** (substantial safety or security significance) is quantitatively greater than 10^{-5} and less than or equal to 10^{-4} ΔCDF or greater than 10^{-6} and less than or equal to 10^{-5} ΔLERF. Qualitatively, a Yellow significance indicates a decline in licensee performance that is still acceptable with cornerstone objectives met, but with significant reduction in safety margin.	10^{-6}
10^{-6}	**White** (low to moderate safety or security significance) is quantitatively greater than 10^{-6} and less than or equal to 10^{-5}ΔCDF or greater than 10^{-7} and less than or equal to 10^{-6} ΔLERF. Qualitatively, a White significance indicates an acceptable level of performance by the licensee, but outside the nominal risk range. Cornerstone objectives are met with minimal reduction in safety margin.	10^{-7}
	Green (very low safety or security significance) is quantitatively less than or equal to 10^{-6} ΔCDF or 10^{-7} ΔLERF. Qualitatively, a Green significance indicates that licensee performance is acceptable and cornerstone objectives are fully met with nominal risk and deviation.	

Figure B-8 ROP Significance Determination Guidance

The guidance documents include a variety of acceptance criteria for different regulatory programs, specific facilities, and for different types of events or operating conditions. There have been several notable attempts to develop acceptance criteria that could support consistent decisionmaking for different types of nuclear facilities—and in some cases for any type of industrial facility. Such criteria are often presented in terms of some measure of consequence (e.g., dose, fatalities, health effects, economic costs) versus probability or estimated frequency of an event or condition. The NRC introduced the possible use of a frequency consequence (F-C) curve in NUREG-1860, "Feasibility Study for a Risk-Informed and Performance-Based Regulatory Structure for Future Plant Licensing" (NRC, 2007a). The F-C curves proposed in NUREG-1860 and in proposals related to the next generation nuclear plant (NGNP, 2010) are provided in Figure B-9.

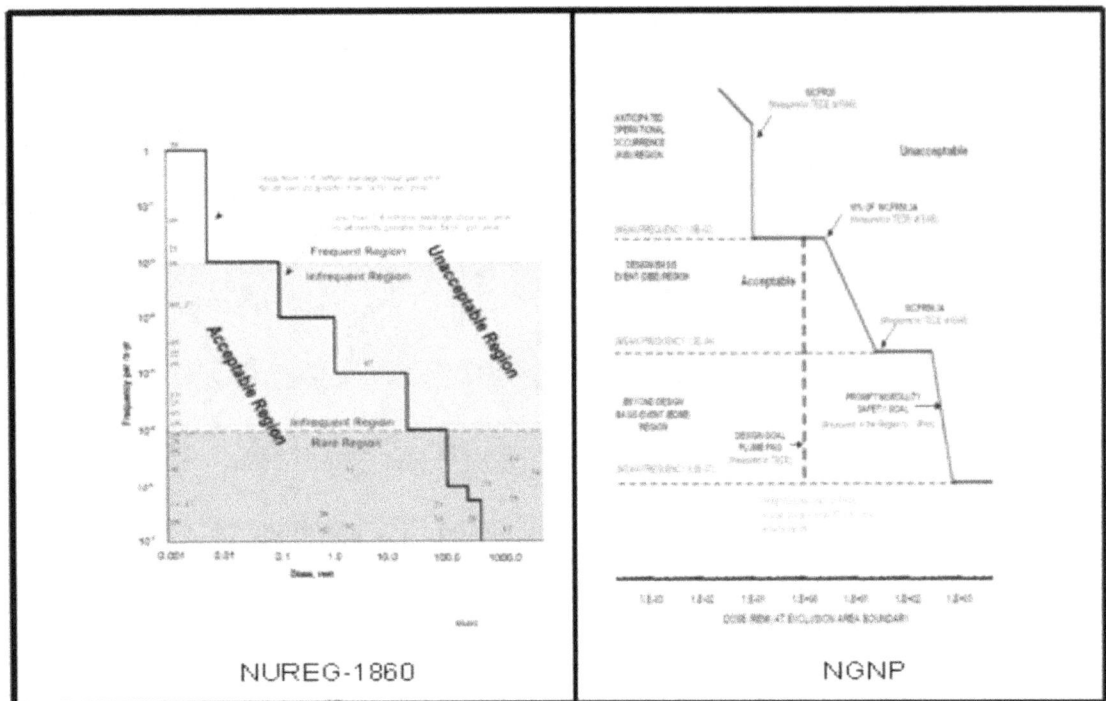

Figure B-9 F-C Curves from NUREG-1860 and NGNP

Some F-C diagrams nclude multiple regions to differentiate between risks generally viewed as tolerable, risks viewed as unacceptable, and risks that should be reduced as practical or cost-effective to do so (e.g., the "as low as is reasonably achievable" (ALARA) region in Figure B-10). A useful discussion of various approaches to decisionmaking using F-C concepts is provided in the report, "Societal Risks," prepared for the Health and Safety Executive (HSE) in the United Kingdom (Ball, 1998).

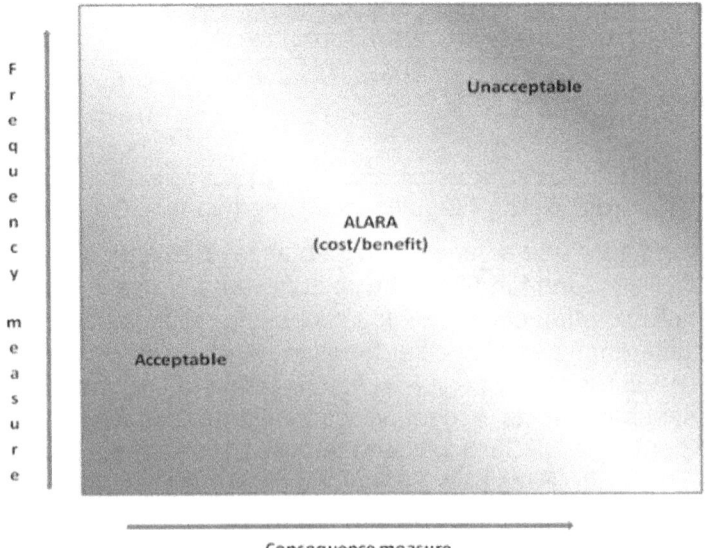

Figure B-10 Generic F-C Curve, with ALARA region

Although not proposing to incorporate a specific F-C curve into the risk management regulatory framework, the RMTF encourages the inclusion of broader topics, such as the HSE reports, in the NRC's training programs. The development and use of the F-C curve similar to Figure B-10 is perceived by some as a dramatic departure from past practices, but many NRC programs already incorporate aspects of the approach by differentiating between high frequency–low consequence activities and the potential for low frequency–high consequence accidents. In the analyses to support licensing of nuclear power reactors, events have traditionally been defined within the categories of (1) normal operation, (2) anticipated operational occurrences, (3) design-basis accidents, and (4) beyond-design-basis accidents. The primary criteria for placing scenarios within the above categories are related to event frequencies. The allowable consequences (defined in terms of degree of fuel damage) are defined for the categories, and generally more damage is acceptable for scenarios with lower frequencies (see Appendix F). The general concept is also evident in the graded treatment of different licensees, with those presenting lesser risks requiring fewer barriers and controls while facilities such as nuclear power plants are required to have multiple barriers, mitigating systems and procedures, and emergency plans. The NRC's safety goals for nuclear power reactors are actually defined in terms of a measure of consequence (prompt fatalities and cancer fatalities) and frequency (as a percentage of other societal risks). The introduction section to the policy statement documented the NRC's view that radiation protection requirements, such as 10 CFR Part 20, "Standards for Protection Against Radiation," ensured that the consequences from higher frequency events (e.g., normal operation) did not warrant additional measures to reduce the frequency of routine releases.

The use of different regions or categories to delineate importance or regulatory treatment has also been used in various NRC programs. The ALARA concept has long been a major part of the radiation protection programs that the NRC requires of its licensees. The backfit provisions of the NRC's reactor regulations and the guidance on regulatory analyses for rulemakings have likewise recognized that there are those barriers and controls that the NRC requires to ensure adequate protection of public health and safety, those that provide a substantial increase in public protection such that the costs of implementation are justified, and those in which the safety benefit does not justify the associated costs (NRC, 2004). Similar categories (regions) using changes in core damage frequency (CDF) and large early release frequency (LERF) were defined in the above figures from Regulatory Guide 1.174 for defining proposed licensing actions that would be considered independent of the total CDF/LERF, considered in relation to total CDF/LERF, or not considered. A similar construct shown in Figure B-11 was proposed in the risk-informed decisionmaking approach for licensing of byproduct material applications.

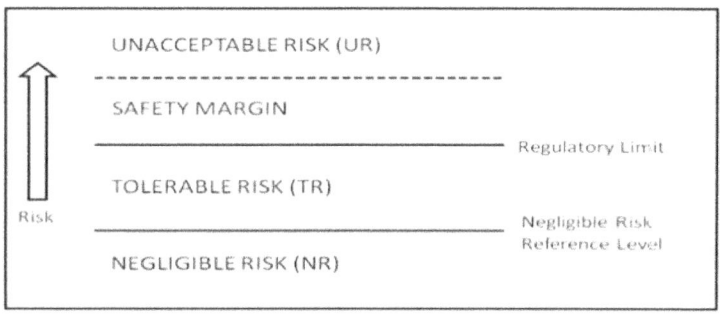

"Risk Informed Decisionmaking for Nuclear Material and Waste Applications," Revision 1

Figure B-11 Risk-Informed Decisionmaking

The NRC goal to move to more performance-based approaches to regulation would be supported by expanding the ALARA principle from radiation protection to the broader area of risk management (as shown in Figure B-10). As discussed in Appendix D, "Performance-Based Regulation," deterministic and prescriptive approaches can limit the flexibility of both the regulated industries and the NRC to respond to lessons learned from operating experience and support the adoption of improved designs or processes. An ALARA approach within the risk management regulatory framework could define performance criteria in terms of risk-informed, performance-based defense-in-depth, while providing licensees with the flexibility to determine the most cost-effective means to provide and maintain appropriate protections to prevent, mitigate, and contain the possible release of radioactive materials. Previous activities—such as the maintenance rule (10 CFR 50.65, "Requirements for Monitoring the Effectiveness of Maintenance at Nuclear Power Plants") for nuclear power reactors—have been successful in establishing objective and measurable criteria for risk-informed, performance-based regulation and oversight of NRC licensees. The adoption of the ALARA principle would reinforce the stated objective in the Advanced Reactor Policy Statement (NRC, 2008a) for new designs to provide enhanced margins of safety and use simplified, inherent, passive, or other innovative means to accomplish their safety functions. SECY-10-0121, "Modifying the Risk-informed Regulatory Guidance for New Reactors" (NRC, 2010a) discusses several issues related to a possible discrepancy between the Advanced Reactor Policy Statement goals for enhanced safety for new plants and the actual regulation and oversight of new reactors based on the same rules and guidance applied to operating reactors.

In addition to comparing the results from the technical analyses to the appropriate decision or acceptance criteria, the deliberative process needs to consider other factors, such as those shown in Figure B-5. A key factor included in the definition of risk-informed, performance-based defense-in-depth is the consideration of uncertainties. Uncertainties include gaps in knowledge regarding the radiological materials and their behavior during scenarios that could result in the exposure of workers or the public. There may also be uncertainties associated with the abilities of barriers and controls to prevent or mitigate accident scenarios or to contain the radioactive materials following such scenarios. Computational models are used to simulate devices and complex systems for infrequent scenarios to help determine appropriate design features and controls, but these models introduce another group of uncertainties to consider during the decisionmaking process. Such uncertainties should be addressed, as practical, within the technical analyses and should be identified and accounted for as part of deliberations on a particular issue or problem. Sensitivity studies can be a useful tool to evaluate uncertainties and otherwise inform the decisionmaking process.

B.4   Implementation Factors

Decisions resulting from deliberations on specific issues or proposals address whether the defense-in-depth protections associated with the subject radiological hazard are adequate, can be relaxed, or need to be strengthened. It is important to recognize that the NRC has a limited and relatively fixed set of regulatory processes by which it can define requirements for its licensees. The implementation processes include preparing regulations and guidance; reviewing proposed licensing actions; performing environmental reviews; and executing oversight programs.

It is likewise necessary to understand that the actual defense-in-depth protections are implemented and maintained by licensees who process and control the radioactive materials or related facilities. However, the use of radioactive materials introduces particular risks and public concerns, and so the NRC's regulatory regime, adopted pursuant to the Atomic Energy Act of 1954, as amended, and other statutes, provides the legal basis for ensuring that NRC licensees provide reasonable assurance of adequate protection of the health and safety of their workers and the public. Licensees within the NRC's various regulatory programs provide the appropriate defense-in-depth protections for workers and the public through: 1) design features incorporated into devices or facilities, 2) operating practices and procedures, 3) maintenance of barriers and supporting systems, 4) inspections and surveillance programs, 5) radiation monitoring, 6) reporting events and other information to their organizations, the NRC, and other agencies, and, finally 7) the decommissioning of facilities.

As discussed throughout this report, appropriate evaluations and actions will vary depending on the type and amount of radioactive material, as well as on the complexity of the device or facility. For some materials and devices, concerns regarding an inadvertent release is small and few controls beyond routine radiation protection programs (time, distance, shielding) are necessary. On the other hand, facilities such as commercial nuclear power plants contain a large amount of radioactive materials that warrants multiple levels of protection to prevent, contain, and mitigate possible releases. These same recognitions are reflected in how the NRC implements actions (e.g., regulations vs. guidance) and how licensees provide appropriate protection (e.g., engineered design features or administrative controls).

A representation of a risk management framework combining NRC processes and associated licensee controls is shown below:

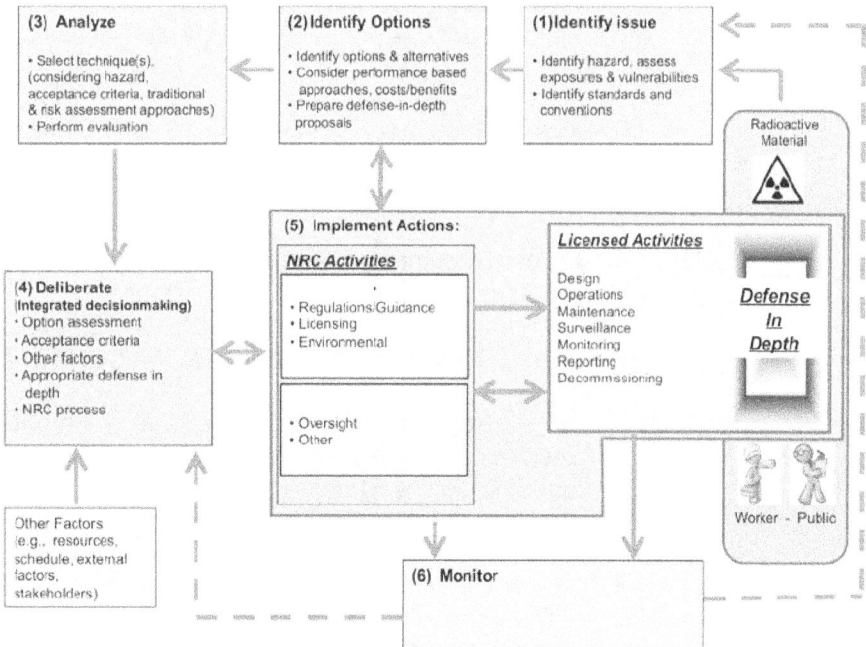

Figure B-12 Implementation of Decisions

The means by which the NRC might define requirements or expectations and the means available for licensees to provide appropriate defense-in-depth protections are actually part of the deliberative process, as well as being the vehicles for implementing decisions. An example is the NRC goal to define performance-based approaches as an alternative to traditional or prescriptive requirements in which a rule or license condition would define specified engineered design features or specific maintenance and surveillance practices. Consideration of a performance-based option could influence the analysis and deliberation steps of the risk management framework.

B.5   References

(Ball, 1998)   D. J. Ball and P. J. Floyd, "Societal Risks," Final Report for the Health and Safety Executive, United Kingdom, 1998

(ISO, 2009)   International Organization for Standardization, "Risk Management – Risk Assessment Techniques," ISO-31010, 2009.

(Kaplan and Garrick, 1981)   S. Kaplan and B.J. Garrick, "On the Quantitative Definition of Risk, *Risk Analysis*," 1:11–27.

(NGNP, 2010)   Idaho National Laboratory, "Next Generation Nuclear Plant Licensing-basis Event Selection White Paper," INL/EXT-1019521, September 2010, Agencywide Documents Access Management System (ADAMS) Accession No. ML102630246.

(NRC, 1985)   U.S. Nuclear Regulatory Commission, "Policy Statement on Severe Reactor Accidents Regarding Future Designs and Existing Plants," August 1985, ADAMS Accession No. ML003711521.

(NRC, 1986)   U.S. Nuclear Regulatory Commission, "Safety Goals for the Operations of Nuclear Power Plants; Policy Statement," August 1986, ADAMS Accession No. ML011210381.

(NRC, 1990)   U.S. Nuclear Regulatory Commission, "Severe Accident Risks: An Assessment for Five U.S. Nuclear Power Plants — Final Summary Report," NUREG-1150, Volume 1, December 1990, ADAMS Accession No. ML040140729.

(NRC, 1995)   U.S. Nuclear Regulatory Commission, "Use of Probabilistic Risk Assessment Methods in Nuclear Regulatory Activities; Final Policy Statement," August 1995, ADAMS Accession No. ML021980535.

(NRC, 1997)   U.S. Nuclear Regulatory Commission, "Standard Review Plan for Dry Cask Storage Systems," Initial Report, NUREG-1536, January 1997, ADAMS Accession No. ML010040237

(NRC, 1997a)   U.S. Nuclear Regulatory Commission, "Individual Plant Examination Program: Perspectives on Reactor Safety and Plant Performance," NUREG-1560, December 1997, ADAMS Accession No. ML063550244.

(NRC, 1998)	U.S. Nuclear Regulatory Commission, "Consolidated Guidance About Materials Licenses," NUREG-1556, 1998 and later, multiple volumes.
(NRC, 1999)	U.S. Nuclear Regulatory Commission, "Standard Review Plan for Transportation Packages for Radioactive Material," Initial Report, NUREG-1609, March 1999.
(NRC, 2000)	U.S. Nuclear Regulatory Commission, "Guidance for Implementation of 10 CFR 50.59, Changes, Tests, and Experiments," Regulatory Guide 1.187, November 2000, ADAMS Accession No. ML003759710.
(NRC, 2000a)	U.S. Nuclear Regulatory Commission, "Risk Analysis and Evaluation of Regulatory Options for Nuclear Byproduct Material Systems," NUREG/CR-6642, February 2000, ADAMS Accession No. ML003684878 (nonpublic).
(NRC, 2000b)	U.S. Nuclear Regulatory Commission, "Standard Review Plan for Transportation Packages for Spent Nuclear Fuel," Initial Report, NUREG-1617, March 2000, ADAMS Accession No. ML003696262.
(NRC, 2001)	U.S. Nuclear Regulatory Commission, "Integrated Safety Analysis Guidance Document," NUREG-1513, May 2001, ADAMS Accession No. ML011440260.
(NRC, 2001a)	U.S. Nuclear Regulatory Commission, "Perspectives Gained From the Individual Plant Examination of External Events (IPEEE) Program," NUREG-1742, September 2001, ADAMS Accession No. Ml021270132.
(NRC, 2004)	U.S. Nuclear Regulatory Commission, "Regulatory Analysis Guidelines of the U.S. Nuclear Regulatory Commission," NUREG/BR-0058, Revision 4, September 2004, ADAMS Accession No. ML042820192.
(NRC, 2006)	U.S. Nuclear Regulatory Commission, "Guidelines for Categorizing Structures, Systems, and Components in Nuclear Power Plants According to Their Safety Significance," Regulatory Guide 1.201, May 2006, ADAMS Accession No. ML061090627.
(NRC, 2006a)	U.S. Nuclear Regulatory Commission, "Reactor Oversight Process," NUREG-1649, Revision 4, December 2006, ADAMS Accession No. ML070890365.
(NRC, 2007)	U.S. Nuclear Regulatory Commission, "Standard Review Plan for the Review of Safety Analysis Reports for Nuclear Power Plants: LWR Edition, NUREG-0800, March 2007.

(NRC, 2007a)　　U.S. Nuclear Regulatory Commission, "Feasibility Study for a Risk-informed and Performance-Based Regulatory Structure for Future Plant Licensing," NUREG-1860, Volume 1, December 2007, ADAMS Accession No. ML080440170.

(NRC, 2008)　　U.S. Nuclear Regulatory Commission, "Risk-informed Decisionmaking for Nuclear Material and Waste Applications," Revision 1, February 2008, ADAMS Accession No. ML080720238.

(NRC, 2008a)　　U.S. Nuclear Regulatory Commission, "Policy Statement on the Regulation of Advanced Reactors," October 2008, ADAMS Accession No. ML082750370.

(NRC, 2009)　　U.S. Nuclear Regulatory Commission, "Risk-informed, Performance-Based Fire Protection for Existing Light-Water Nuclear Power Plants," Regulatory Guide 1.205, December 2009, ADAMS Accession No. ML092730314.

(NRC, 2010)　　U.S. Nuclear Regulatory Commission, "Standard Review Plan for the Review of a License Application for a Fuel Cycle Facility," NUREG-1520, Revision 1, May 2010, ADAMS Accession No. ML101390110.

(NRC, 2010a)　　U.S. Nuclear Regulatory Commission, "Modifying the Risk-Informed Regulatory Guidance for New Reactors," SECY-10-0121, September 14, 2010, ADAMS Accession No. ML102230076.

(NRC, 2011)　　U.S. Nuclear Regulatory Commission, "An Approach for Using Probabilistic Risk Assessment in Risk-Informed Decisions on Plant-Specific Changes to the Licensing Basis," Regulatory Guide 1.174, Revision 2, May 2011, ADAMS Accession No. ML100910006.

(NRC, 2011a)　　U.S. Nuclear Regulatory Commission, "Standard Review Plan for Renewal of Spent Fuel Dry Cask Storage System Licenses and Certificates of Compliance," NUREG-1927, March 2011, ADAMS Accession No. ML111020115.

APPENDIX C

DEFENSE-IN-DEPTH

C.1   Introduction

The term "defense-in-depth" has been used since the 1960s in the context of ensuring nuclear reactor safety. The concept was developed and applied to compensate for the recognized lack of knowledge of nuclear reactor operations and the consequences of potential accidents. While both experience with reactor operations and the knowledge base of potential (and actual) consequences have grown considerably since that time, the concept continues to be highly relevant today.

The Risk Management Task Force (RMTF) has reviewed a number of documents that historically have helped to shape the characterization of defense-in-depth. Since the characterizations provided in these documents are not completely consistent and are focused on operating power reactors, the RMTF concluded that clarifying what the U.S. Nuclear Regulatory Commission (NRC) means by defense-in-depth is a necessary part of the development of a holistic strategic vision. This appendix describes the proposed RMTF characterization, summarizes other descriptions of defense-in-depth (which have focused, as noted above, on power reactors), and compares them with the RMTF characterization. In one case (power reactors), the RMTF has extended this characterization to provide additional perspective.

C.2   Task Force Characterization of Defense-in-depth

The RMTF characterizes defense-in-depth as follows:

> Risk-informed and performance-based defense-in-depth protections:
>
> •   Ensure appropriate barriers, controls, and personnel to prevent, contain, and mitigate exposure to radioactive material according to the hazard present, the relevant scenarios, and the associated uncertainties; and
>
> •   Ensure that the risks resulting from the failure of some or all of the established barriers and controls, including human errors, are maintained acceptably low.

Key terms in this characterization include:

•   Barriers: Barriers can take a number of forms, such as a container, a wall, or a restricted area of land. In some cases, engineered systems such as cooling systems may exist to help ensure barrier functionality. The number and types of barriers needed should be commensurate with and proportional to the extent of the hazard and reflect the current understanding of important associated uncertainties.

- Controls: Administrative actions are established to ensure the performance of a barrier.

- Personnel: Programs are established to ensure that people are properly trained and qualified for the jobs they have and that periodic training is provided to keep those people focused on safety. When performance problems arise, specific lessons-learned training is provided to ensure, in the best way possible, that problems are not repeated.

- Hazard: A hazard is an assemblage of radioactive material licensed by the NRC (e.g., a nuclear reactor core, a well-logging device containing radioactive material, a uranium mine). The extent of the hazard is measured in terms of the amount of radioactive material present and the expected length of time it would exist.

- Associated uncertainties: Uncertainties exist in a number of aspects of facility design and operation, from the properties of materials used to the likelihood of events that could challenge that facility.

- Risks: Risk assessment methods provide an effective means for measuring the adequacy of defense-in-depth. That is, risk assessment measures the likelihood of individual barriers being compromised, the probabilities that additional barriers would be compromised if the previous barrier were to be compromised, and how poorly understood phenomena or new information would affect these likelihoods and probabilities. This information is useful in ensuring an appropriate balance among barrier capabilities, in ensuring the low likelihood that multiple barriers would be compromised, and ensuring that the overall result is an adequately low risk to the potentially harmed individual or group.

C.3   Historical Characterizations

NRC white paper, as modified in 10 CFR 50.69, Statements of Consideration

The NRC's "White Paper on Risk-Informed and Performance-Based Regulation" was completed in 1999 and contained descriptions of many terms, including defense-in-depth (NRC,1999). This description was modified somewhat and was issued in 2004 in the Statements of Consideration for the NRC's final rule, Title 10 of the *Code of Federal Regulations* (10 CFR) 50.69, "Risk-Informed Categorization and Treatment of Structures, Systems, and Components for Nuclear Power Reactors" (NRC, 2004), as follows:

> *Defense-in-depth is an element of the NRC's safety philosophy that employs successive measures to prevent accidents or mitigate damage if a malfunction, accident, or naturally caused event occurs at a nuclear facility. Defense-in-depth is a philosophy used by the NRC to provide redundancy as well as the philosophy of a multiple-barrier approach against fission product releases. The defense-in-depth philosophy ensures that safety will not be wholly dependent on any single element of the design, construction, maintenance, or operation of a nuclear facility. The net effect of incorporating defense-in-depth into design, construction, maintenance, and operation is that the facility or system in question tends to be more tolerant of failures and external challenges.*

The RMTF characterization of defense-in-depth is intended to achieve the same intent of the 2004 characterization and extends the 2004 characterization in two ways. First, while it seems to be implicit in the 2004 characterization, the RMTF has made more explicit that the number and types of barriers are dependent on the radioactive hazard present. Second, the RMTF characterization explicitly uses risk assessment to measure the outcome of defense-in-depth, which the 2004 characterization describes more generally as "more tolerant."

International Atomic Energy Agency

In INSAG-10 (IAEA, 1996), the International Atomic Energy Agency (IAEA) provided a description of defense-in-depth for operating reactors as follows:

> *Defence in depth is generally structured in five levels. Should one level fail, the subsequent level comes into play. The objective of the first level of protection is the prevention of abnormal operation and system failures. If the first level fails, abnormal operation is controlled or failures are detected by the second level of protection. Should the second level fail, the third level ensures that safety functions are further performed by activating specific safety systems and other safety features. Should the third level fail, the fourth level limits accident progression through accident management, so as to prevent or mitigate severe accident conditions with external releases of radioactive materials. The last objective (fifth level of protection) is the mitigation of the radiological consequences of significant external releases through the off-site emergency response.*

The RMTF characterization is intended to apply to a broader set of regulated activities, some of which would not need to have the five levels discussed in INSAG-10. However, the RMTF characterization should achieve the same outcomes as the INSAG-10 characterization when applied to operating reactors.

Next Generation Nuclear Plant (NGNP)

In 2009, Idaho National Laboratory, under contract to the U.S. Department of Energy, published a paper, "Next Generation Nuclear Plant Defense-in-Depth Approach" (NGNP, 2009), that reviewed the existing literature on defense-in-depth and outlined a definition of defense in depth that would be used in the design of new power reactors. The definition was summarized as follows:

> • *Plant Capability Defense-in-Depth reflects the decisions made by the designer in the selection of functions, structures, systems, and components for the design that ensure defense-in-depth in the physical plant.*

> • *Programmatic Defense-in-Depth reflects the decisions made regarding the processes of manufacturing, constructing, operating, maintaining, testing, and inspecting the plant and the processes undertaken that ensure plant safety throughout the lifetime of the plant.*

- *Risk-Informed Evaluation of defense-in-depth reflects the development and evaluation of strategies that manage the risks of accidents, including the strategies of accident prevention and mitigation. This aspect of defense-in-depth also provides the framework for performing deterministic and probabilistic safety evaluations, which help determine how well various Plant Capability Defense-in-Depth and Programmatic Defense-in-Depth strategies have been implemented.*

As noted previously, the RMTF characterization is intended to apply to a broader set of regulated activities. The RMTF characterization, extended for operating power reactors as discussed in Section C.4 of this appendix, has important similarities to the NGNP characterization. More specifically, both the NGNP and RMTF characterizations include the concept of using risk assessment methods as a measure of effectiveness.

Davis Besse petition— Director's decision

The report of NRC's Near-Term Task Force (NTTF) (NRC, 2011) on the Fukushima accident cites and discusses a 2003 NRC description of defense-in-depth. The NTTF discussion is as follows:

> *An instructive discussion of the defense-in-depth philosophy also appears in director's decisions relating to a petition on Davis-Besse (FirstEnergy Nuclear Operating Company (Davis-Besse Nuclear Power Station, Unit 1), DD033, 58 NRC 151, 163 (2003)).*
>
> *The decision described defense-in-depth as encompassing the following requirements:*
>
> *(1) require the application of conservative codes and standards to establish substantial safety margins in the design of nuclear plants;*
>
> *(2) require high quality in the design, construction, and operation of nuclear plants to reduce the likelihood of malfunctions, and promote the use of automatic safety system actuation features;*
>
> *(3) recognize that equipment can fail and operators can make mistakes and, therefore, require redundancy in safety systems and components to reduce the chance that malfunctions or mistakes will lead to accidents that release fission products from the fuel;*
>
> *(4) recognize that, in spite of these precautions, serious fuel-damage accidents may not be completely prevented and, therefore, require containment structures and safety features to prevent the release of fission products; and*
>
> *(5) further require that comprehensive emergency plans be prepared and periodically exercised to ensure that actions can and will be taken to notify and protect citizens in the vicinity of a nuclear facility.*

The [Near-Term] Task Force has found that the defense-in-depth philosophy is a useful and broadly applied concept. It is not, however, susceptible to a rigid definition because it is a philosophy. For the purposes of its review, the [Near-Term] Task Force focused on the following application of the defense-in-depth concept:

- *protection from external events that could lead to fuel damage*

- *mitigation of the consequences of such accidents should they occur, with a focus on preventing core and spent fuel damage and uncontrolled releases of radioactive material to the environment*

- *emergency preparedness (EP) to mitigate the effects of radiological releases to the public and the environment, should they occur*

These levels of defense-in-depth are appropriate for significant external challenges to a facility. In applying these defense-in-depth features, the [Near-Term] Task Force sought to ensure that the Commission's regulatory requirements, processes, and programs effectively address each layer of protection while maintaining appropriate balance among them. The [Near-Term] Task Force notes that this approach is also consistent with Levels of Defense 3, 4, and 5 in IAEA Draft Safety Standard DS 414, "Safety of Nuclear Power Plants: Design," dated January 2010.

The 2003 characterization was focused on operating reactors, and the NTTF discussion of it has a similar focus. The RMTF characterization, extended for operating power reactors as discussed later in this appendix, encompasses the elements of the 2003 characterization. As part of this extension, it uses reliability assessment to address more quantitatively concepts such as redundancy and the use of automatic safety system actuations, and risk assessment, as means to measure the effectiveness of the set of barriers.

NUREG-1860

With the advent of new reactor licensing activities and the potential development of reactor designs quite different from current operating reactors, the NRC decided to assess the feasibility of implementing a new regulatory structure as an alternative to 10 CFR Part 50. This new structure would take advantage of the extensive experience with 10 CFR Part 50 and would align better with alternative reactor designs. NUREG-1860, "Feasibility Study for a Risk-Informed and Performance-Based Regulatory Structure for Future Plant Licensing" (NRC, 2007), documents this feasibility assessment and provides a framework "that provides an approach, scope and criteria that could be used to develop a set of requirements that would serve as an alternative to 10 CFR [Part] 50 for licensing future" nuclear power plants.

One key part of the NUREG-1860 framework is defense-in-depth, characterized as having the following principles:

- Measures against intentional acts as well as inadvertent events are provided.

- The design provides accident prevention and mitigation capability.

- Accomplishment of key safety functions is not dependent upon a single element of design, construction, maintenance, or operation.

- Uncertainties in SSCs and human performance are accounted for in the safety analyses.

- The design has the capability to prevent an unacceptable release of radioactive material.

- Plants are sited at locations that facilitate the protection of public health and safety.

Like the other characterizations discussed above, NUREG-1860 focused on power reactors. As noted previously, the RMTF characterization is intended to apply to a broader set of regulated activities. The RMTF characterization, as extended for power reactors as discussed in Section C.4, encompasses the NUREG-1860 characterization.

C.4 Extension of RMTF Characterization for Power Reactors

The RMTF concluded that its characterization of defense of depth could be extended for power reactors to be more specific and to reflect the availability of quantitative methods (probabilistic risk assessments). This extension (with additions in italics) is:

Provide risk-informed and performance-based defense-in-depth protections to:

- Ensure appropriate barriers, controls, and personnel to prevent, contain, and mitigate exposure to radioactive material according to the hazard present, the relevant scenarios, and the associated uncertainties.

    - Each barrier is designed with sufficient safety margins to maintain its functionality for relevant scenarios and account for uncertainties.

    - Systems that are needed to ensure a barrier's functionality are designed to ensure appropriate reliability for relevant scenarios.

    - Barriers and systems are subject to performance monitoring.

and

- Ensure that the risks resulting from the failure of some or all of the established barriers and controls, including human errors, are maintained acceptably low.

In addition to those discussed in Section C.2, key terms in this characterization for power reactors include:

- Safety margins: The traditional engineering approach of including margins in equipment design to account for uncertainties.

- Appropriate reliability: The likelihood of successful operation needed by a system such that, considering the frequency of systems challenges, the risk is acceptably low.

- Performance monitoring: Periodically assessing the performance of a barrier or system to ensure its continued functionality. Barrier monitoring can be more qualitative in nature while system monitoring can be accomplished by programs such as the mitigating system performance indicators used in the NRC's Reactor Oversight Process.

C.5   References

(IAEA, 1996)   International Atomic Energy Agency, "Defence in Depth in Nuclear Safety," INSAG-10, A Report by the International Nuclear Safety Advisory Group, 1996.

(NGNP, 2009)   Idaho National Laboratory, "Next Generation Nuclear Plant Defense-in-Depth Approach," INL/EXT-0917139, December 2009.

(NRC,1999)   U.S. Nuclear Regulatory Commission, Staff Requirements Memorandum Regarding SECY-98-44, "White Paper on Risk-Informed and Performance-Based Regulation," March 1, 1999, Agencywide Documents Access and Management System (ADAMS) Accession No. ML003753601.

(NRC, 2004)   U.S. Nuclear Regulatory Commission, "Risk-Informed Categorization and Treatment of Structures, Systems, and Components for Nuclear Power Reactors," 10 CFR 50.69, Published in the *Federal Register* on November 22, 2004 (69 FR 68008).

(NRC, 2007)   U.S. Nuclear Regulatory Commission, "Feasibility Study for a Risk-Informed and Performance-Based Regulatory Structure for Future Plant Licensing," NUREG-1860, Volume 1, December 2007, ADAMS Accession No. ML080440170.

(NRC, 2011)   U.S. Nuclear Regulatory Commission, "Recommendations for Enhancing Reactor Safety in the 21st Century; The Near-Term Task Force Review of Insights from the Fukushima Dai-Ichi Accident," July 2011, ADAMS Accession No. ML112510271.

APPENDIX D

PERFORMANCE-BASED REGULATION

D.1   Background

The U.S. Nuclear Regulatory Commission (NRC) has a longstanding goal to move toward more risk-informed and performance-based approaches in its regulatory programs. The Risk Management Task Force (RMTF) has found that such efforts have been useful, and it has incorporated "risk-informed and performance-based defense-in-depth" as a major part of the proposed Risk Management Regulatory Framework. The Commission has previously directed the NRC staff to solicit input from industry and other stakeholders on performance-based initiatives, including areas that are not amenable to risk-informed approaches, to supplement the agency's traditional deterministic system of licensing and oversight. It should be noted that deterministic[1] and prescriptive[2] regulatory requirements were based mostly on experience, testing programs, and expert judgment, considering factors such as engineering margins and the principle of defense-in-depth. These requirements are viewed as being successful in establishing and maintaining adequate safety margins for NRC-licensed activities. The NRC has recognized, however, that deterministic and prescriptive approaches can limit the flexibility of both the regulated industries and the NRC to respond to lessons learned from operating experience and to adopt improved designs or processes.

The agency has as one of its primary safety goal strategies the use of sound science and state-of-the-art methods to establish, where appropriate, risk-informed and performance-based regulations. The NRC issued a white paper on risk-informed and performance-based regulation to define the terminology and expectations for evaluating and implementing the initiatives related to risk-informed, performance-based approaches (NRC, 1999). The white paper defines a performance-based approach as follows:

A performance-based regulatory approach is one that establishes performance and results as the primary basis for regulatory decisionmaking, and incorporates the following attributes:

> 1.   measurable (or calculable) parameters (i.e., direct measurement of the physical parameter of interest or of related parameters that can be used to calculate the parameter of interest) exist to monitor system, including facility and licensee, performance,

---

1   A deterministic approach to regulation establishes requirements for engineering margin and for quality assurance in design, manufacture, and construction. In addition, it assumes that adverse conditions can exist and establishes a specific set of design-basis events and related acceptance criteria for specific systems, structures, and components based on historical information, engineering judgment, and desired safety margins. An example is a defined load on a structure (e.g., from wind, seismic events, or pipe rupture) and an engineering analysis to show that the structure maintains its integrity.

2   A prescriptive requirement specifies particular features, actions, or programmatic elements to be included in the design or process, as the means for achieving a desired objective. An example is a requirement for specific equipment (e.g., pumps, valves, heat exchangers) needed to accomplish a particular function (e.g., remove a defined heat load).

2. objective criteria to assess performance are established based on risk insights, deterministic analyses, and/or performance history,

3. licensees have flexibility to determine how to meet the established performance criteria in ways that will encourage and reward improved outcomes, and

4. a framework exists in which the failure to meet a performance criterion, while undesirable, will not in and of itself constitute or result in an immediate safety concern.[3]

Performance-based approaches can be pursued either independently or in combination with risk-informed approaches. The white paper mentioned above was a followup to the NRC's assessment of Direction Setting Issue (DSI) 12, "Risk-Informed Performance-Based Regulation," in the mid-1990s (NRC, 1996). The NRC staff and Commission continued to make progress on developing policies and guidance related to performance-based approaches and subsequently issued documents such as SECY-00-0191, "High Level Guidelines for Performance-based Activities" (NRC, 2000), and NUREG/BR-0303, "Guidance for Performance-based Regulation" (NRC, 2002).

D.2   Nuclear Power Reactors

Within the nuclear power reactor arena, performance-based approaches have been proposed and implemented for regulatory programs related to designrelated activities and operational activities. As with the risk-informed initiatives, there are meaningful distinctions between performance-based activities that primarily deal with requirements for plant design and those that deal with plant operation.

D.2.1   Operation

The NRC has traditionally considered the performance of equipment and personnel in its development of regulatory and oversight programs. Examples of NRC activities that incorporated some aspects of performance-based approaches before defining a more structured methodology include the incremental changes to surveillance test intervals (e.g., Generic Letter 93-05, "Line Item Technical Specifications Improvements to Reduce Surveillance Requirements for Testing During Power Operation" (NRC, 1993)), implementation of the leakbeforebreak methodology for nuclear power reactors (NRC, 2007), and the "as low as is reasonably achievable" (ALARA) provisions in Title 10 of the Code of Federal Regulations (10 CFR) Part 20, "Standards for Protection Against Radiation."

Although initiated prior to the development of the formal performance-based approach, the maintenance rule (10 CFR 50.65, "Requirements for Monitoring the Effectiveness of Maintenance at Nuclear Power Plants") is often cited as a good example of a risk-informed, performance-based approach. The maintenance rule requires monitoring of the overall continuing effectiveness of a licensee's maintenance programs to ensure that (1) safety-related and certain non-safety-related structures, systems, and components (SSCs) are capable of performing their intended functions, and (2) for non-safety-related equipment, failures will not

---

3   Using the previous example (footnote 2), a performance-based approach might provide additional flexibility to a licensee on plant equipment and configurations used to accomplish a safety function (e.g., removing a heat load), but the performance criteria could not be the actual loss of a safety function that would result in the release of radioactive materials.

occur that prevent the fulfillment of safety-related functions, and failures resulting in scrams and unnecessary actuations of safety-related systems are minimized. The rule requires licensees to monitor the performance of the defined SSCs against licensee-established goals, which are established based on safety significance and operating experience, and take corrective action when the goals are not met. In addition, the rule requires that licensees assess and manage the increase in risk that might result from proposed maintenance activities (e.g., taking safety significant systems out of service). The scope of the maintenance rule therefore brought more SSCs into the NRC's regulatory programs but did so while providing licensees with flexibility to define methods of monitoring, performance criteria, and reliability goals for specific SSCs.

In 1995, the NRC amended 10 CFR Part 50, "Domestic Licensing of Production and Utilization Facilities," Appendix J, "Primary Reactor Containment Leakage Testing for WaterCooled Power Reactors" (NRC, 1995), to provide a performance-based Option B for the containment leakage testing requirements. Option B requires that test intervals for various leakage tests be determined by using a performance-based approach. Under this approach, performance-based test intervals are based on consideration of the operating history of the component (e.g., an isolation valve or the containment structure) and resulting risk from its failure. Option B, in concert with technical reports and guidance documents, has been used to allow licensees with a satisfactory integrated leak rate testing (ILRT) performance history (i.e., two consecutive, successful Type A tests) to reduce the test frequency for the Type A containment ILRT from three tests in 10 years to one test in 15 years. This relaxation was based on assessments performed by the nuclear industry and the NRC that showed that the risk increase associated with extending the ILRT surveillance interval was very small.

A recent initiative that adopts a risk-informed and performance-based approach is the incorporation of the National Fire Protection Association (NFPA) standard NFPA 805, "Performance-based Standard for Fire Protection for Light-Water Reactor Electric Generating Plants," into NRC's regulations (NRC, 2009). NFPA 805 provides deterministic requirements that are very similar to those in the NRC's traditional fire protection regulations, but also includes performance-based methods for evaluating plant configurations that would not meet the conservative deterministic requirements. The performance-based methods allow engineering analyses to demonstrate that the changes in overall plant risk that result from these plant configurations is acceptably small and that fire protection defense-in-depth is maintained.[4] Defense-in-depth, as applied to fire protection, means that an appropriate balance is maintained between (1) preventing fires from starting, (2) timely detection and extinguishing of fires that might occur, and (3) protection of SSCs important to safety from a fire that is not promptly extinguished. The adoption of NFPA 805 provides a licensee with flexibility on how to implement its fire protection program while maintaining an acceptable level of fire safety.

---

[4] Building upon the guidance in Regulatory Guide 1.174, "An Approach for Using Probabilistic Risk Assessment in Risk-informed Decisions on Plant-specific Changes to the Licensing-basis" (NRC, 2011), Regulatory Guide 1.205, "Risk-informed, Performance-based Fire Protection for Existing Light-Water Nuclear Power Plants" (NRC, 2009), states:

> Prior NRC review and approval is not required for individual changes that result in a risk increase less than 1×10^{-7}/year (yr) for CDF [core damage frequency] and less than 1×10^{-8}/ yr for LERF [large early release frequency]. The proposed change must also be consistent with the defense-in-depth philosophy and must maintain sufficient safety margins. The change may be implemented following completion of the plant change evaluation.

D.2.2 Design

The concept of performance-based approaches in the design of nuclear plant SSCs is harder to grasp than is its incorporation into operational requirements. Much of the focus in this area relates to improvements in the consideration of seismic hazards. The NRC issued Regulatory Guide 1.208, "A Performance-based Approach To Define the Site-specific Earthquake Ground Motion" (NRC, 2007a) to provide an alternative to the traditional approach for addressing the requirements in 10 CFR 100.23, "Geologic and Seismic Siting Criteria." The regulatory guide provides guidance on the development of site-specific ground motion response spectrum (GMRS). The goal of the site-specific, performance-based GMRS is to achieve approximately consistent performance for SSCs, across a range of seismic environments, annual probabilities, and structural failure frequencies. The approach is termed performance-based because it defines a process built on establishing design criteria for SSCs that are derived from performance goals related to withstanding site-specific seismic hazards (e.g., less than about 1 percent probability of unacceptable performance for the site-specific response spectrum ground motion). While this effort establishes alternate design criteria for use with traditional or risk-informed engineering analyses, other design efforts could consider performance-based approaches by coupling the design efforts with operational information, such as inspections, performancegoals, and riskinsights. The use of inspections, tests, analyses, and acceptance criteria within 10 CFR Part 52, "Licenses, Certifications, and Approvals for Nuclear Power Plants," as the means of confirming nuclear plant construction conforms to licensing-basis documents can be considered such a performance-based approach.

D.3 Nonreactor Activities

In the late 1990s to early 2000s, the staff undertook a number of programmatic activities to better include performance considerations into the nuclear materials program. In addition to DSI-12 mentioned above, these activities were documented in SECY-99-062, "Nuclear Byproduct Material Risk Review" (NRC, 1999a), SECY-99-100, "Framework for Risk-informed Regulation in the Office of Nuclear Material Safety and Safeguards" (NRC, 1999b), NUREG/ CR-6642, "Risk Analysis and Evaluation of Regulatory Options for Nuclear Byproduct Material Systems" (NRC, 2000a), and the "Phase II – Byproduct Materials Review" (NRC, 2001). As a result of this work, significant changes were made to the licensing and inspection program to better incorporate performance considerations.

The NUREG-1556 series, Volumes 1-21, "Consolidated Guidance About Materials Licensees," was developed in the late 1990s to pull together into one place the various guidance documents written over the years for the wide variety of materials licensees (NRC, 1998). These documents allow license applicants to find the applicable regulations, guidance, and acceptance criteria used in granting a materials license. Operational experience (performance) and risk insights guided the development of these documents—higher risk activities with significant performance challenges have more prescriptive regulations and guidance than do lower risk activities. Over time, the guidance in NUREG-1556 has been revised to further incorporate performance considerations, and a new revision to the series is under development to address security issues.

The materials inspection program was fundamentally revised in 2001—both in terms of approach and frequency—based on performance. The inspection approach was modified to emphasize licensee knowledge and performance of NRC-licensed activities over document review. Inspectors now review a licensee's program against focus areas that reflect those attributes that are considered (based on insights from NUREG/CR-6642) to be most risk-significant. If a licensee's performance against a given focus element during the inspection is considered to be acceptable, the inspector moves on to the next focus element. Performance concerns or questions lead an inspector to go deeper into that area. Inspection frequencies were also revised to take into account the risk of a given operation and the performance of licensees over time. At the present time, the Office of Federal and State Materials and Environmental Management Programs has tasked a working group with conducting a reevaluation of the inspection program to determine whether further adjustments to the program are warranted to make it more efficient and effective. A key part of that reevaluation is consideration of operational performance, based on event reporting, enforcement history, and stakeholder interviews.

The materials program routinely—at least annually—assesses licensee performance-based on events reported by NRC and Agreement State licensees to the Nuclear Materials Events Database (NMED). NMED provides a tool for capturing, assessing, and trending licensee performance in terms of reportable events. This trending function is an important part of the feedback loop for changes to the materials regulatory program: rulemaking, licensing, and inspection.

D.4 NRC Activities

The NRC has implemented some performance-based elements into areas other than establishing regulatory requirements, such as its inspection program and reactor oversight process. Inspection procedures have been changed to increase the focus on performance, reliability, and availability, and comparison of performance to requirements and goals. The reactor oversight process complements the NRC's inspection program by using performance indicators related to the performance of SSCs and licensee personnel. Although not involving revisions to NRC regulations, these activities have not only improved the effectiveness of the NRC, but also focused licensee activities on improving performance and overall plant safety.

D.5 Challenges and Recommendations

Efforts to increase the consideration of performance-based approaches for NRC-licensed facilities and activities face many of the same obstacles as those identified for risk-informed approaches (see Appendix E) and, as a general matter, any proposal to change wellestablished ways of doing business. This is true within the NRC and also within the regulated communities, which have sometimes expressed their preference for well defined, prescriptive requirements versus performance-based proposals. A specific challenge for performance-based approaches relates to terminology and perception. The phrase "risk-informed, performance-based" may have already become an overused buzzword as all parties seek to meet expectations that their work activities or proposals support the NRC's policy statement. In addition, the phrase has generally become synonymous with the risk-informed aspects of the agency's goal since riskrelated initiatives have received much more attention and resources. It should be noted, however, that performance-based approaches often complement and support risk-informed approaches, in terms of maintaining validity of assumptions (e.g., availability and reliability), and there have been significant successes in using them together. A question going forward is how much attention should be given to advancing performance-based approaches separate from possible risk-informed initiatives with related performance elements. The proposed Risk Management Regulatory Framework being recommended by the RMTF may provide an opportunity to help resolve the issues related to terminology and implementation.

Another potential issue facing performance-based approaches relates to the perception that problems or negative trends are to be avoided and that reaching a threshold for corrective actions is indicative of failure. This "zero-failure" expectation is sometimes applied independent of realistic safety or health considerations. A common example is the public perception regarding the release of or exposure to very low amounts of radioactive material. This may be an unsolvable communication problem, which also applies to risk-informed initiatives, but it should be recognized as a challenge when developing proposals for performance-based approaches. This issue was raised in DSI-12, which included the question of "What should be NRC's strategy and philosophy with respect to changing NRC's responsibilities and authority in areas of little public risk?" Specific examples of such difficulties include the NRC's efforts to address the release of materials contaminated at very low levels through the Below Regulatory Concern Policy Statement (NRC, 1993a) or the later Clearance rulemaking effort (NRC, 2005). The risk management approach being proposed by the RMTF may facilitate a graded approach for the regulation of NRC-licensed materials, devices, and facilities (e.g., introduction of the "design-enhancement" category for nuclear power reactors). This could help in matters where the NRC and its stakeholders perceived choices as either "full regulation" or "no regulation," and therefore had difficulty in crafting compromises amenable to all parties.

The development and use of performance-based approaches have introduced some other policy and legal issues. These were evident in the implementation of the maintenance rule, as well as in the rollout of the NRC's reactor oversight process. Each of these efforts required numerous meetings, preparation of guidance, revisions to guidance documents, use of the NRC Web site for frequently asked questions and other communications, and extensive training of the NRC staff and industry personnel. NRC evaluations included in activities such as the Risk-Informed Environment initiative (Clark, 2008) and Office of Inspector General audits (NRC, 2006) have confirmed the need for additional guidance and training in this area. Specific implementation challenges would exist for the various types of facilities and activities regulated by the NRC. The challenges facing the proposed Risk Management Regulatory Framework would likely be similar to those identified and overcome during the development and implementation of these previous initiatives.

D.6   References

(Clark, 2008)   Theresa V. Clark, Mark Caruso, Garreth Parry, and Lynn Mrowca, "Fostering a Risk-informed Environment in Nuclear Reactor Regulation," American Nuclear Society, International Topical Meeting on Probabilistic Safety Assessment and Analysis (PSA 200), Knoxville, TN, September 2008.

(NRC, 1993)   U.S. Nuclear Regulatory Commission, "Line Item Technical Specifications Improvements to Reduce Surveillance Requirements for Testing During Power Operation," Generic Letter 93-05, September 27, 1993, .Agencywide Documents Access and Management System (ADAMS) Accession No. ML031070342.

(NRC, 1993a)   U.S. Nuclear Regulatory Commission, "NRC Withdraws Below Regulatory Concern Policy Statements," Press Release, August 18, 1993.

(NRC, 1995)   U.S. Nuclear Regulatory Commission, "Primary Reactor Containment Leakage Testing for WaterCooled Power Reactors," 10 CFR Part 50, Appendix J, Final Rule, published in the *Federal Register* on September 26, 1995 (60 FR 49495).

(NRC, 1996)   U.S. Nuclear Regulatory Commission, "Staff Requirements – COMSECY-96-061– Risk-informed, Performance-based Regulation (DSI 12)," Staff Requirements Memorandum, April 15, 1997, ADAMS Accession No. ML003671740.

(NRC, 1998)   U.S. Nuclear Regulatory Commission, "Consolidated Guidance About Materials Licensees," NUREG-1556, Multiple Volumes, 1998 and later.

(NRC, 1999)   U.S. Nuclear Regulatory Commission, Staff Requirements Memorandum Regarding SECY-98-144, "White Paper on Risk-informed and Performance-based Regulation," March 1, 1999, (ADAMS) Accession No. ML003753601.

(NRC, 1999a)   U.S. Nuclear Regulatory Commission, "Nuclear Byproduct Material Risk Review," SECY-99-062, Commission Paper, March 1, 1999, ADAMS Accession No. ML003671237.

(NRC, 1999b)   U.S. Nuclear Regulatory Commission, "Framework for Risk-informed Regulation in the Office of Nuclear Material Safety and Safeguards," SECY-99-100, Commission Paper, March 31, 1999, ADAMS Accession No. ML003671288.

(NRC, 2000)   U.S. Nuclear Regulatory Commission, "High Level Guidelines for Performance-based Activities," SECY-00-191. Commission Paper, September 1, 2000, ADAMS Accession No. ML003742883.

(NRC, 2000a)   U.S. Nuclear Regulatory Commission, "Risk Analysis and Evaluation of Regulatory Options for Nuclear Byproduct Materials Systems," NUREG/CR-6642, February 2000, ADAMS Accession No. ML003693052. (Not publically available)

(NRC, 2001)   U.S. Nuclear Regulatory Commission, "Phase II – Byproduct Materials Review," NRC Report, August 2001, ADAMS Accession No. ML012270095.

(NRC, 2002)   U.S. Nuclear Regulatory Commission, "Guidance for Performance-Based Regulation," NUREG/BR-0303, December 2002, ADAMS Accession No. ML023470659.

(NRC, 2005)	U.S. Nuclear Regulatory Commission, "Staff Requirements – SECY-05-0054, "Proposed Rule: Radiological Criteria for Controlling the Disposition of Solid Materials (RIN 3150-AH18)"," Staff Requirements Memorandum, June 1, 2005, ADAMS Accession No. ML051520185.

(NRC, 2006)	U.S. Nuclear Regulatory Commission, "Evaluation of NRC's Use of Probabilistic Risk Assessment (PRA) in Regulating the Commercial Nuclear Power Industry," Office of the Inspector General, Audit Report OIG-06-A-24, September 29, 2006, . ADAMS Accession No. ML062720275.

(NRC, 2007)	U.S. Nuclear Regulatory Commission, "LeakBeforeBreak Evaluation Procedures, NUREG-0800, Standard Review Plan for the Review of Safety Analysis Reports for Nuclear Power Plants: LWR Edition, Section 3.6.3, March 2007, ADAMS Accession No. ML063600396

(NRC, 2007a)	U.S. Nuclear Regulatory Commission, "A Performance-based Approach To Define the Site-specific Earthquake Ground Motion," Regulatory Guide 1.208, March 2007, ADAMS Accession No. ML070310619.

(NRC, 2009)	U.S. Nuclear Regulatory Commission, "Risk-informed, Performance-based Fire Protection for Existing Light-Water Nuclear Power Plants," Regulatory Guide 1.205, December 2009, ADAMS Accession No. ML092730314.

(NRC, 2011)	U.S. Nuclear Regulatory Commission, "An Approach for Using Probabilistic Risk Assessment in Risk-informed Decisions on Plant-specific Changes to the Licensing-basis," Regulatory Guide 1.174, Revision 2, May 2011, ADAMS Accession No. ML100910006.

APPENDIX E

EXPERIENCE WITH RISK-INFORMED REGULATION

E.1   Background

The Energy Research, Inc., report ERI 02-001, "Implementation of Risk-informed Regulation at U.S. Nuclear Power Plants" (Dube, 2002), included a survey of a representative sample of probabilistic risk assessment (PRA) managers (corresponding to 25 percent of the nuclear units) regarding their views on risk-informed initiatives. The report noted that substantial progress had been made regarding the implementation of risk-informed regulation. However, the report also noted that this progress was rather uneven across programs and across licensees. A few licensees, such as the South Texas Project (STP), had made significant strides, while a number of plants, particularly single-unit sites, continued to linger. A universal theme expressed by PRA managers in the 2002 timeframe was that maintaining the PRA models to the necessarily high standards was resource-intensive. Additionally, a sizable fraction of in-house PRA staff supported day-to-day plant operations and maintenance, consistent with Title 10 of the Code of Federal Regulations (10 CFR) 50.65(a)(4), the "maintenance rule." The result was an already too thin staff spread even thinner by the demands of the plant and existing regulations.

Figure E-1 shows the results for several risk-informed applications from the 2002 survey.

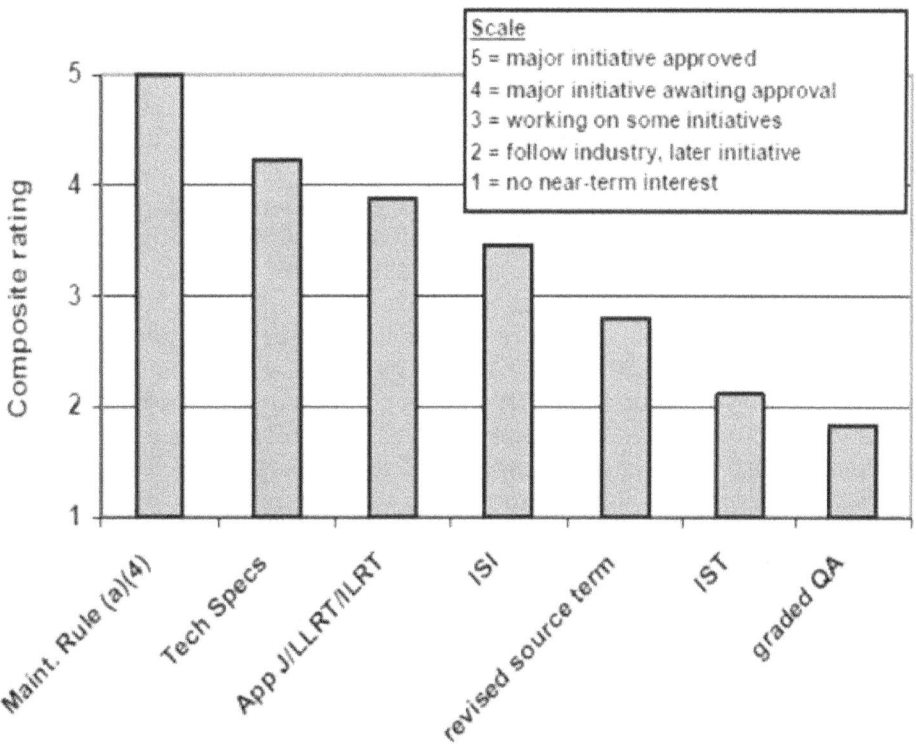

Figure E-1 Status of Risk-Informed Activities (2002)

E.2   Update to 2011

In response to the Commission's staff requirements memorandum on SECY-10-0121, "Modifying the Risk-informed Regulatory Guidance for New Reactors" (NRC, 2010; NRC 2011), the NRC staff, with the participation of stakeholders, conducted tabletop exercises regarding the implementation of a number of key risk-informed initiatives at operating power reactors. The adequacy of the current guidance, if and when applied to new reactor designs (with generally lower risk profiles), was the primary focus of the assessment. Although it was not the intent of these exercises to perform a formal survey of licensee attitudes toward risk-informed initiatives, the degree of interest in risk-informed applications for operating power reactors, as well as new reactors, was evident during the discussions.

Maintenance Rule (a)(4):

The views from 2002 have not diminished over the past decade. Discussions with every PRA manager and lead engineer at the time reinforced the notion that this requirement has added measurably to nuclear safety. It has resulted in an unprecedented increase in the prestige (and workload) of PRA organizations, which heretofore may have been just obscure analysts in a remote part of the overall company. Nuclear stations have created a dedicated panel of plant personnel, called integrated decisionmaking panels (IDPs), to assess the safety significance of performing maintenance during all phases of plant operation. The IDP brings together PRA staff, reactor operators, maintenance staff, and planners whose goal is to exercise the PRA models (and judgment) to their fullest. This has fostered an awareness of the safety impact of working online on balance-of-plant equipment, such as circulating water pumps, which could measurably increase reactor trip frequency.

During the conduct of the recent tabletop exercises discussed above, staff observed that many, if not most, current licensees used presolved risk configuration tables or online risk monitors to assess risk before and during plant maintenance activities. The "blended" approach, whereby the PRA is combined with inputs on the degree of defense-in-depth and planttransient assessment was highlighted. Stakeholders demonstrated how factors other than PRA were often more limiting in terms of the risk management action level. While the staff noted during the exercise that there have been 116 violations of the (a)(4) provision of the maintenance rule since May 1, 2001, they all have been green findings, with 115 of those being noncited violations (NCV) and 1 cited as failure to correct a previous NCV. Implementation of 10 CFR 50.65(a)(4), by almost any measure, appears to be a successful risk-informed regulation. New reactors will be required by regulation to have and maintain an essentially fullscope PRA (all operating modes and external hazards for which staff has endorsed consensus PRA standards), so the online risk management tools are expected to be even more robust.

Risk-Informed Technical Specifications:

The scope of risk-informed technical specification initiatives (RITS) span the entire spectrum— from wholesale changes that have been proposed to highly strategic isolated changes in allowed outage times (AOT). Most licensees have participated in owners group efforts to increase surveillance test intervals (STI), such as has been done on instrumentation and reactor trip channels by the now superseded Westinghouse Owners Group (WOG). Others, such as the previous Combustion Engineering Owners Group, coordinated proposed AOT changes to lowpressure safety injection pumps, safety injection tanks, containment spray, and emergency diesel generators (EDG). Most plants that have applied for extensions in AOT for EDGs have been approved for as much as 14 days. Since 2002, the importance of this application has not diminished, but it is fair to say that over the past decade much of the "low-hanging fruit" in terms of greatest gain in safety and operational flexibility for the lowest implementation cost has been gathered. Nonetheless, nearly half of the reactor units have received NRC approval to implement RITS 5b (surveillance frequency control program). The more comprehensive RITS 4b (completion times) has to-date received limited interest with only STP currently applying the technology and several other plants expressing near-term interest. For new reactor designs, the USAPWR design center has developed a 4b and 5b option in the technical specifications, with Comanche Peak Units 3 and 4 pursuing the technology. Other combined license (COL) applicants have taken a "wait-and-see" approach, not willing to risk holding up their COL approval until the groundwork has been established by pilot applications. Many applicants have stated informally that once they obtain their COLs, they intend to pursue a wide range of risk-informed initiatives, including 4b and 5b.

Appendix J, Integrated Leak Rate Testing (ILRT):

Another application, where nearly all licensees who have tried have had some degree of success, is risk-informed containment leak rate testing in its various forms. Requirements are found in Appendix J of 10 CFR Part 50, "Domestic Licensing of Production and Utilization Facilities. Type A testing relates to the overall measurement of containment air mass and leakage rate by pressurizing the containment during an outage (i.e., ILRT). Type B testing is a series of pneumatic tests to detect and measure local leak rates across specified pressure retaining boundaries, such as the personnel air lock door seal. Type C testing is a set of pneumatic tests to measure containment leakage rates through containment isolation valves. Current requirements call for Type A ILRT once in 10 years. Based in part on successful past performance on these tests, as well as risk assessments, many licensees have requested extensions of the test interval to once every 15 years. Both the WOG and Nuclear Energy Institute have had major efforts in this area. Again, depending on the particulars of a refueling outage, the ILRT may often be a critical path item because of the necessity to isolate and pressurize the containment. While Appendix J was not one of the topical areas that was exercised during the aforementioned tabletop workshops in 2011, discussions with industry representatives and staff indicate a large fraction of the nuclear power plant units have used this initiative.

In-Service Inspection (ISI) of piping:

Operating nuclear power plants must inspect piping in accordance with the ASME Boiler and Pressure Vessel Code, Section XI. The original inspection programs are based on deterministic analyses using design stress reports. These analyses are extremely conservative. Service experience has shown that historical pipe failures occur caused by corrosion and fatigue and in locations not addressed by the existing inspection programs. American Society of Mechanical Engineers (ASME) Code Cases have been providing much of the technical basis for proceeding with these revisions. About half the nuclear plant PRA managers surveyed in 2002 had a major submittal for relief, either awaiting approval or already approved, and that percentage has increased over the past decade to about 90 percent of the U.S. nuclear units. Those who have had success with risk-informed ISI (RI-ISI) identify this activity as "low hanging fruit," in that the risk benefits are realized while reducing burden and worker radiation exposures. During the 2011 tabletop exercises, the NRC staff concluded that implementation of RI-ISI was "riskneutral" for the current reactor fleet, and likely to be so for the new reactor designs. The Electric Power Research Institute (EPRI) has a research program underway to implement the

EPRI approach to RI-ISI at many of the new reactor design centers. EPRI has also proposed the implementation of risk-informed preservice inspection, but the staff has not approved this proposed technology at this time.

Alternative Source Term:

In 1995, more than 30 years after the Technical Information Document TID-14844 (AEC, 1962), the revised or "alternative" source term was published as NUREG-1465 (NRC, 1995). This alternative source term assumes the timedependent rather than instantaneous releases of fission products, and it envelops all light-water reactor plants with the representative accident source terms. A licensee may choose to implement all aspects of the alternative source term, including composition and magnitude, timing of release, and chemical and physical form of the radiological source term; or the licensee may choose to implement only selected aspects of the alternate source term. Guidance for licensees of operating nuclear power reactors in applying the alternative source term was provided in Regulatory Guide (RG) 1.183, "Alternative Radiological Source Terms for Evaluating Design Basis Accidents at Nuclear Power Reactors," (NRC, 2000) and Standard Review Plan, Section 15.0.1, Revision 0, "Radiological Consequences Analysis Using Alternate Source Terms," (NRC, 2000a).

Alternative source terms have the potential benefit primarily in the relaxing of design requirements for a number of systems, such as control room heating, ventilation, and air conditioning and secondary containment. To meet control room dose requirements, stringent design criteria regarding rate of room pressurization, leakage rates, and other factors often must be made. Design deficiencies or system performance issues can place an operating unit in a condition outside the design basis of the plant, in some cases jeopardizing continued plant operation. Realistic source terms can relax the timing of the actuation of certain systems and pose less of a problem in terms of issues, such as emergency diesel generator loading. At the time of the original survey in 2002, about two-thirds of the PRA managers were planning to have some sort of initiative to credit alternative source terms, but it is not evident that this has come to full fruition. The 2011 tabletop exercises did not assess alternative source terms, but the standard design applications for certification, including the AP1000, Economic Simplified BoilingWater Reactor, USAPWR, and U.S. EPR have availed themselves of the alternative as described in RG 1.183 (the ABWR has not).

In-Service Testing (IST):

Much like ISI, IST is an ASME Code requirement to provide periodic performance testing of check valves, motoroperated valves, airoperated valves, and pumps. RI-IST could potentially improve plant safety by assigning higher priority to key equipment, reducing the duration of plant outage, and allowing for more efficient use of plant resources. NRC guidance is set forth in RG 1.175, "An Approach for Plant-specific Risk-informed Decisionmaking: In-Service Testing," (NRC, 1998). However, the response of the industry in this arena has been less than enthusiastic. In the 2002 survey, PRA managers representing about two-thirds of the nuclear units had no intention of issuing a licensing submittal in the near future, and that attitude has not changed significantly in the past decade. The major argument for not making a submittal is that the licensees do not see a major payback. The high cost of the analysis outweighs what little savings they project to achieve from the revised IST program. Because there was no expressed interest in RI-IST on the part of the new reactor applicants, this topical area was not evaluated during the 2011 tabletop exercises.

Categorization of SSCs (10 CFR 50.69), (Also Referred to as Graded QA):

The implementation of risk-informed categorization and treatment of structures, systems, and components (SSCs) under 10 CFR 50.69, "Risk-informed Categorization and Treatment of Structures, Systems and Components for Nuclear Power Reactors," remains an underachieving initiative for currently operating reactors. PRA managers surveyed in 2002 did not see a clear and immediate benefit, either in reduction of costs nor in terms of increased safety, and that view has not changed over the past decade. Only STP has implemented an earlier form of 10 CFR 50.69 under a waiver of regulations. Since then, only one other licensee has expressed any interest in the program. While the potential payback would not seem to justify the effort and cost that must be put into such a program for many if not most operating reactors, new reactor applicants see large potential savings if the program is implemented before equipment procurement for new build. However, during the 2011 tabletop exercises, none of the early COL applicants expressed interest in applying 10 CFR 50.69 before receiving their COLs. By then, the COL holder would have already issued most of the design specifications for equipment to be procured. Hence, some of the greatest benefits of using 10 CFR 50.69 will not be realized for the first round of COL holders.

Fire Protection (NFPA 805):

One other risk-informed activity, not included in the 2002 survey, deserves mentioning. Changes to fire protection programs (NRC, 2009) through implementation of 10 CFR 50.48(c) (National Fire Protection Association (NFPA) 805) is, at this point, one of the major risk-informed initiatives in the industry, with over 40 units pursuing this alternative approach to fire protection. In the views of the industry, the effort is consuming most of the industry's "spare" PRA resources. The activity can be considered "high cost, high payback." High cost means that a high quality PRA that substantially conforms to RG 1.200 (NRC, 2009a) for both internally initiated events and internal fire is needed. The cost of implementing fire protection enhancements can run into the tens of millions of dollars. High payback—in that the alternative, continuing with the current fire protection measures, or plant shutdown—is even more prohibitive.

Summary

The NRC staff summarized the results of the tabletop exercises in a public meeting held on February 28, 2012 (NRC, 2012). The NRC staff did not identify potential significant decreases in the enhanced safety margins for new reactors arising from the various initiatives and practices discussed above. The tabletop exercises likewise concluded that the current risk thresholds used in the reactor oversight process are appropriate for new reactors.

E.3   References

(AEC, 1962)   U.S. Atomic Energy Commission, "Calculation of Distance Factors for Power and Test Reactor Sites," TID-14844, March 23, 1962.

(Dube, 2002)   D.A. Dube, "Implementation of Risk-informed Regulation at U.S. Nuclear Power Plants," ERI 02-001, Energy Research, Inc., May 2002.

(NRC, 1995)   U.S. Nuclear Regulatory Commission, "Accident Source Terms for Light-Water Nuclear Power Plants," NUREG-1465, February 1995, Agencywide Documents Access and Management System (ADAMS) Accession No. ML041040063.

(NRC, 1998)   U.S. Nuclear Regulatory Commission, "An Approach for Plant-specific, Risk-informed Decisionmaking: In-Service Testing," Regulatory Guide 1.175, August 1998, ADAMS Accession No ML003740149.

(NRC, 2000)   U.S. Nuclear Regulatory Commission, "Alternative Radiological Source Terms for Evaluating Design Basis Accidents at Nuclear Power Reactors," Regulatory Guide 1.183, July 2000, ADAMS Accession No ML003716792.

(NRC, 2000a)   U.S. Nuclear Regulatory Commission, "Radiological Consequence Analysis Using Alternate Source Terms," NUREG-0800, Standard Review Plan, Section 15.0.1, Revision 0, July 2000, ADAMS Accession No ML003734190.

(NRC, 2009)   U.S. Nuclear Regulatory Commission, "Risk-informed, Performance-based Fire Protection for Existing Light-Water Nuclear Power Plants," Regulatory Guide 1.205, December 2009, ADAMS Accession No ML092730314.

(NRC, 2009a)   U.S. Nuclear Regulatory Commission, "An Approach for Determining the Technical Adequacy of Probabilistic Risk Assessment Results for Risk-informed Activities," Regulatory Guide 1.200, Revision 2, March 2009, ADAMS Accession No ML090410014

(NRC, 2010)   U.S. Nuclear Regulatory Commission, "Modifying the Risk-informed Regulatory Guidance for New Reactors," SECY-10-0121, September 14, 2010, ADAMS Accession No ML102230076.

(NRC, 2011)   U.S. Nuclear Regulatory Commission, "Modifying the Risk-Informed Regulatory Guidance for New Reactors"," SECY-10-0121, Staff Requirements Memorandum, March 2, 2011, ADAMS Accession No ML110610166.

(NRC, 2012)   U.S. Nuclear Regulatory Commission, "Risk-informed Regulatory Framework for New Reactors," Public Meeting Presentation, February 28, 2012, ADAMS Accession No ML12062A198

APPENDIX F

NUCLEAR POWER REACTORS – LICENSING-BASIS EVENTS

F.1   Introduction

From the beginning, the licensing and oversight of commercial nuclear reactors has focused on potential hazards and the establishment of requirements to prevent, contain, and mitigate possible releases of radioactive material. The safety analyses for plants have considered both the failure and malfunction of plant equipment (e.g., pumps, piping, and control rods) and external hazards (e.g., earthquakes and floods) that might initiate transients and compromise barriers. The failures and malfunctions of plant equipment have traditionally been described in Chapter 15 of each plant's final safety analysis report (FSAR). The larger-scale events that might affect multiple plant areas or structures are usually discussed in Chapters 2 and 3 of the FSAR, as well as in specific sections on systems, structures, and components (SSCs) and related operating programs.

F.2   Internal Events

In the analyses of the failure or malfunction of specific plant SSCs, many plants use a version of an American Nuclear Society (ANS) standard that includes classification of plant conditions (ANS, 1983). This approach divides plant conditions into four categories, in accordance with anticipated frequency of occurrence and potential radiological consequences to the public. The four categories are as follows:

   Condition I   - Normal operation and operational transients

   Condition II   - Faults of moderate frequency

   Condition III   - Infrequent faults

   Condition IV   - Limiting faults

The basic principle applied in relating design requirements to each of the conditions is that the most probable occurrences should yield the least radiological risk to the public, and those extreme situations having the potential for the greatest risk to the public shall be those least likely to occur. The events are analyzed using detailed computer codes that have been benchmarked to tests and experiments. Reactor trip and engineered safety systems are modeled in the analyses according to established conventions (e.g., the single failure criterion and the crediting of only safety-related SSCs). The safety analysis typically includes acceptance criteria similar to the following for the various events, in keeping with the established relationships between expected event frequency and radiological consequences:

*   Fuel damage (defined as penetration of the fission product barrier, i.e., the fuel rod clad) is not expected during normal operation and operational transients (Condition I) or any transient conditions arising from faults of moderate frequency (Condition II). It is not possible, however, to preclude a very small number of rod failures. These will be within the capability of the plant cleanup system and are consistent with the plant design bases.

- The reactor can be brought to a safe state following a Condition III event with only a small fraction of fuel rods damaged (see above definition), although sufficient fuel damage might occur to preclude resumption of operation without considerable outage time.

- The reactor can be brought to a safe state and the core can be kept subcritical with acceptable heat transfer geometry following transients arising from Condition IV events.

The analysis of SSC failures and malfunctions can be described in the above terms of postulated initiating events (PIEs), analytical models and assumptions, and acceptance criteria. As described in Appendix B, this traditional approach can generally be summarized as follows:

Traditional Approach

- limited set of design-basis accidents
  - upper bound assumptions
  - stylized, conservative models
  - single failure criterion
- safety margins
- mechanistic and physical models
- analytical conservatisms
- well suited to support design activities

Routine plant operations are generally controlled by established radiation protection practices and requirements. For power reactors, radiation controls for the protection of the public are provided in Title 10 of the *Code of Federal Regulations* (10 CFR) Part 20, "Standards for Protection Against Radiation"; 10 CFR Part 50, "Domestic Licenses of Production and Utilization Facilities," Appendix I, "Numerical Guides for Design Objectives and Limiting Conditions for Operation to Meet the Criterion 'As Low is As Reasonably Achievable' for Radioactive Material in Light-WaterCooled Nuclear Power Reactor Effluents"; and effluent controls in the plant license and procedures (e.g., 10 CFR 50.36a and offsite dose calculation manual).

Condition II events are also referred to as anticipated operational occurrences (AOOs) and are evaluated to ensure limited fuel failures and, thereby, minimal releases of radioactive material. The protection against AOOs is provided primarily by the reactor protection system, and analyses are performed to confirm that safety limits are not exceeded. Safety limits are defined for adequate heat removal from the reactor's fuel rods (e.g., specified acceptable fuel design limits (SAFDLs)) and pressure relief to protect the reactor coolant pressure boundary (for transients resulting in increased reactor coolant system (RCS) pressure). Condition II events are selected through evaluation of various events, such as changes (increases or decreases) in heat removal by secondary systems, decreases in reactor coolant system flow, reactivity and power distribution anomalies, and changes (increases or decreases) in reactor coolant inventory. Assuming plant systems perform as expected, safety functions are not challenged by AOOs, and a plant would be able to perform repairs, if needed, and return to operation within a short time.

Infrequent and limiting faults are also referred to as design-basis accidents (DBAs). These events are evaluated using acceptance criteria that may include limited fuel failures but maintain barriers such that core cooling functions are maintained. This, in turn, would demonstrate that the calculated offsite dose would not exceed established criteria related to short-term health effects. In general, the actual occurrence of a DBA-related initiating event would be expected to challenge plant personnel and require actuation of engineered safety features, such as the emergency core cooling system. Although plant conditions may warrant entry into emergency operating procedures and would trigger emergency preparedness activities when emergency action levels were exceeded, the plant response would be expected to stabilize, and it is unlikely to warrant evacuations because of offsite doses challenging the protection action guidelines.

F.3   External Hazards

In addition to the failure or malfunction of plant equipment, the challenges to and protections of nuclear power plants from various external hazards are described in plant licensing documents (e.g., FSAR Chapter 2). External hazards are those events that could initiate a variety of plant transients or accidents while also challenging one or more of the barriers provided to mitigate or contain potential releases of radioactive materials. The failure of a specific SSC may be evaluated using failure modes and effects analysis and mechanistic models for the thermal-hydraulic response of the plant for comparison to established acceptance criteria. However, the large number of SSCs that could be affected by a major external event led to a different approach for the evaluations and protections from such events. Examples of external hazards are:

- nearby industrial, transportation, and military facilities

- meteorological events (e.g., precipitation, wind, storms)

- flooding events (external or internal)

- seismic events

- fires

- malevolent acts

The general approach for external hazards is to incorporate protections or qualification requirements related to the hazard to all safety-related components, or at least to a large set of equipment subject to the event. Seismic qualification requirements are, for example, generally applied to safety-related SSCs. A protection feature added to a subset of safety-related components might be flood protection for those rooms possibly affected by the flooding of a nearby river. As with the evaluation of AOOs and DBAs, an important part of selecting and evaluating the external hazards is the evaluation of expected probability or frequency of challenges from external events. Information and estimates related to various external hazards are provided in NRC regulatory guides and by agencies such as the U.S. Geological Survey and U.S. Army Corps of Engineers. The goal in these evaluations is to provide protections against events that can reasonably be predicted to occur at a site. For some PIEs, the NRC guidance has established acceptance criteria related to ensuring a low probability of exceeding the assumptions (e.g., wind speeds) for external hazards. Site specific information is used to select meteorological events, such as the probable maximum precipitation and the potential for wind speeds from tornadoes and hurricanes exceeding design values. Regional information is also used to determine the maximum probable flood. Plant structures are evaluated to ensure they can withstand and provide the necessary protections against such events. Geological records are evaluated to estimate the potential effects of seismic events on the plant site and its specific SSCs. Several internal events, such as fire and internal flooding scenarios, are addressed in a similar manner. Existing guidance related to selected plant level events is summarized in Table F-1.

Table F-1 Approaches to External Hazards

Plant-level Event	Guidance	Frequency Discussion
Precipitation (rain, snow)	Regulatory Guide (RG) 1.206, "Combined License Applications for Nuclear Power Plants" (NRC, 2007) and "Interim Staff Guidance on Assessment of Normal and Extreme Winter Precipitation Loads on the Roofs of Seismic Category I Structures" (NRC, 2009)	n/a (historical information)
Hurricanes	RG 1.221, "Design Basis Hurricane and Hurricane Missiles for Nuclear Power Plants" (NRC, 2011)	design-basis hurricane wind speeds should correspond to an exceedance frequency of 10^{-7} per year (calculated as a best estimate)
Tornadoes	RG 1.76, "design-basis Tornado and Tornado Missiles for Nuclear Power Plants" (NRC, 2007a)	design-basis tornado wind speeds should correspond to an exceedance frequency of 10^{-7} per year (calculated as a best estimate)
Nearby Industrial	RG 1.206 (NRC, 2007) and RG 1.91, "Evaluation of Explosions Postulated To Occur on Transportation Routes Near Nuclear Power Plants" (NRC, 1978)	Exceedance frequency of going over the 10 CFR Part 100, "Reactor Site Criteria," dose limits on the order of magnitude of 10^{-6} per year
External Flooding	RG 1.59, "Design Basis Floods for Nuclear Power Plants" (NRC, 1977)	n/a (historical information)
Seismic	RG 1.208, "A Performance-based Approach to Define the Site-specific Earthquake Ground Motion" (NRC, 2007b)	The desired performance being expressed as the target value of 10^{-5} for the mean annual probability of exceedance (frequency) of the onset of significant inelastic deformation
Aircraft Impact (intentional)	RG 1.217, "Guidance for the Assessment of beyond-design-basis Aircraft Impacts" (NRC, 2011a)	n/a (defined aircraft impact characteristics)
Security Assaults	(various)	n/a (defined design-basis threat)
Internal Fires	RG 1.205, "Risk-informed, Performance-based Fire Protection for Existing Light-Water Nuclear Power Plants" (NRC, 2009a)	Prior NRC review and approval is not required for individual changes that result in a risk increase less than 10^{-7}/year (yr) for CDF and less than 10^{-8}/yr for LERF
Loss of large areas caused by fires or explosions	Interim Staff Guidance DC/COLISG-016 – Compliance with 10 CFR 50.54(hh)(2) and 10 CFR 52.80(d) (NRC, 2010)	n/a (defined mitigation strategies)

F.4   Deterministic Methods

The terminology related to nuclear plant licensing and relationships between design basis, design-basis events, beyond-design-basis accidents, and licensing-basis is confusing. The complexity of the terminology has increased over the last several decades as new methodologies, such as PRA, were introduced and as the NRC and industry responded to specific issues or concerns (e.g., station blackout (SBO)). As explained in "A Short History of Nuclear Regulation, 1946–2009" (NRC, 2010a), the initial design and licensing of nuclear power plants were approached as follows:

> *Regulators using a deterministic approach simply tried to imagine "credible" mishaps and their consequences at a nuclear facility and then required the defense-in-depth approach—layers of redundant safety features—to guard against them.*

These "credible mishaps" were, in turn, used to define design-basis events, which were, then, used to determine the safety classification of SSCs, the contents of licensing-basis documents, such as FSARs and technical specifications, and supporting documents, such as plant procedures. The licensing efforts for early plants focused therefore on "design-basis events." Regulator and licensee attention were centered on the mitigation of AOOs and DBAs and on ensuring that plant structures and layouts addressed design-basis external hazards such that safety-related equipment was protected and plants could proceed from operations to a safe shutdown condition following a design-basis event.

Safety Margins

The traditional or deterministic approach addressed uncertainties and provided conservatisms through the establishment of design limits, acceptance criteria, and safety margins. There are several descriptions of the term "safety margin," with each recognizing that plants and specific barriers are designed and constructed such that actual failures are not expected until key parameters well exceed the values assumed in the supporting engineering evaluations. However, like many other terms commonly used in licensing processes, "safety margin" is not officially defined. 10 CFR 50.59, "Changes, Tests, and Experiments," once included a criterion related to a reduction in the "margin of safety," but the criterion was replaced specifically because of the differing interpretations of the phrase (NRC, 1999). The supplementary information associated with the rulemaking included the following characterization:

> *…Margins within the plant design and in the established licensing-basis exist on many levels. There are margins from the assumptions of initial conditions, conservatisms such as computer modeling and codes to account for uncertainties, allowances for instrument drift and system response time, redundancy, and independence of components. Margins are built into the facility to account for routine plant fluctuations and transients and response to accident conditions. Margins also exist in the established regulatory acceptance criteria to be met for response to various accidents and transients. As a result, substantial margins are provided by the regulatory envelope within which a plant has demonstrated its ability to respond to a spectrum of design basis accidents…*

A representation of various margins and how they contribute to global plant safety is shown in Figure F-1 (recreated from NEA/CSNI/R(2007)9, "Task Group on Safety Margin Action Plan (SMAP), Safety Margins Action Plan – Final Report" (NEA, 2007).

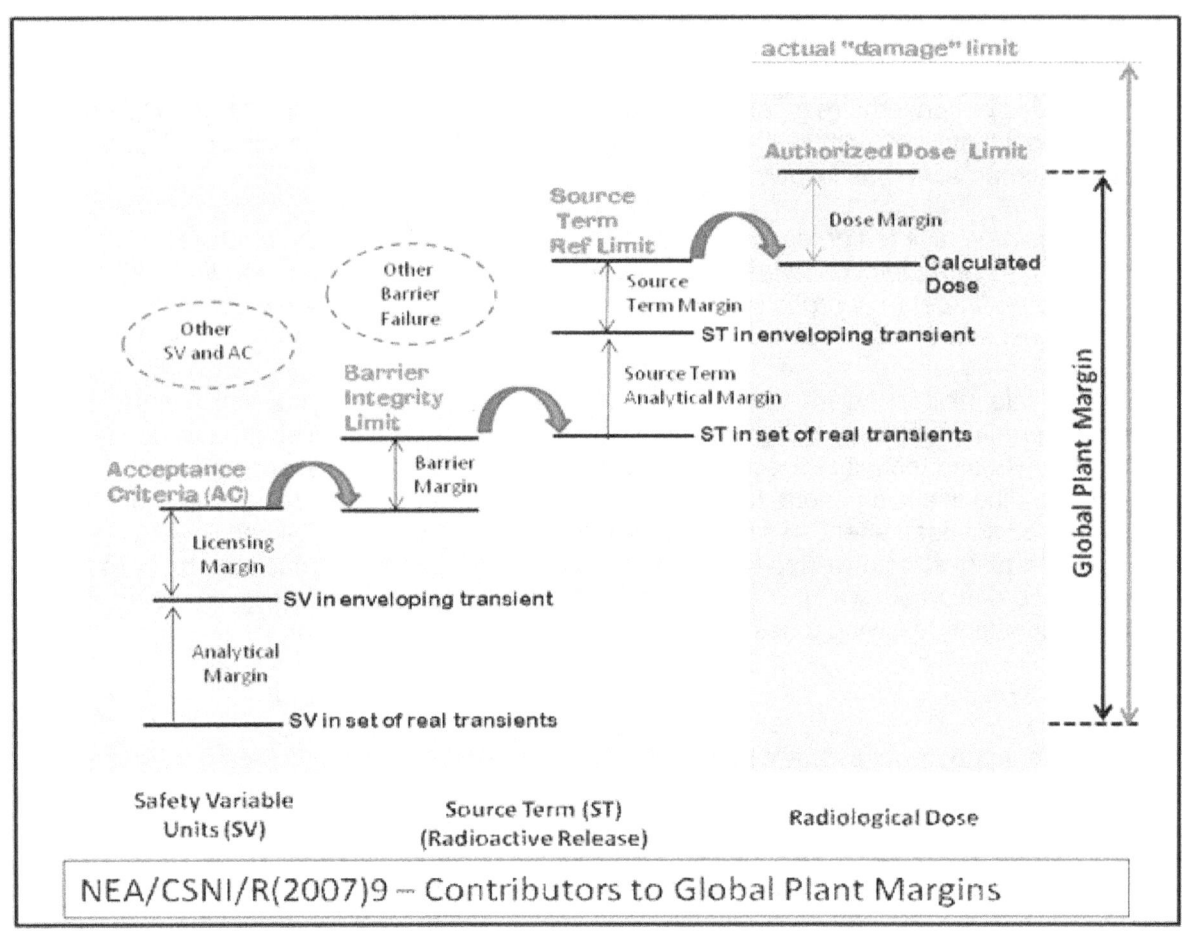

Figure F-1 Nuclear Event Analysis Margins (NEA, 2007)

NUREG-1860 (NRC, 2007c) proposed a definition for safety margin that basically equates to the combination of licensing and barrier margins from the above figure. There are continuing efforts, such as the Idaho National Laboratory's Risk-informed Safety Margin Characterization project, which are attempting to better define and integrate safety margins into risk assessment methodologies to support nuclear plant life extensions or other initiatives (INL, 2009).

F.5   Risk Assessments

Additional insights regarding the risks related to nuclear power plants (what can go wrong, how likely is it, what are the consequences) were developed in the last several decades, and questions arose as to how to address them in the design and licensing processes. The increasing use of risk assessment methodologies resulted in additional events being considered during licensing (e.g., SBO, anticipated transient without scram (ATWS)), which informed areas such as emergency planning, and led to various initiatives to incorporate risk-informed, performance-based activities into the NRC regulation and oversight of nuclear power plants. As discussed in Appendix B, the risk assessment approach has the following characteristics:

Risk Assessment Characteristics

- numerous event sequences
  - systems approach best estimate models
  - uncertainty analysis
- vulnerability determinations
- human and system interactions
- realism
- well suited to support operational decisions

Increasingly, the NRC's regulations began to address and be informed by risk assessments and consideration of beyond-design-basis events, which involve the failure of safety-related functions or PIEs not included in the design-basis events. Although not using the terms "design-enhancement events or conditions," the NRC did establish special regulatory requirements for selected events, such as SBO, ATWS, and aircraft impact. In addition, regulations such as the maintenance rule and programs such as the Reactor Oversight Process used risk insights to require or encourage licensees to manage the overall risks associated with plant operation. Other beyond-design-basis scenarios were evaluated to identify potential vulnerabilities (i.e., events dominating the risk profile) and to support aggregate estimates of plant risk. Similarly, the NRC approach to environmental reviews evolved to consider risk insights and possible severe accident mitigation alternatives that should be included in applicants' environmental reports. The addition of the beyond-design-basis events is shown in the lower right region of Figures F-2 and F-3. As discussed in Chapter 4 and Appendix H, the distinction between "adequate protection" and "safety enhancement" is also an issue when discussing the categorization of events and related events and programs.

As a general matter, the frequency of events shown in Figure F-2 decreases as they go from normal operations to AOOs to DBAs to the design-enhancement and severe-accident events within the beyond-design-basis region. The potential consequences in terms of challenges to barriers (primarily fuel cladding) and ultimate release of radioactive material increases across event categories, from little to no damage for AOOs to significant core damage for some severe accidents. The result is a relationship that is similar to the F-C curve discussed in Appendix B. This similarity is important in the identification and evaluation of possible options to revise the licensing and oversight of nuclear power plants to better incorporate risk insights and performance-based approaches. Figure F-2 addresses the relationship of SSC classification within the licensing-basis event categories, but similar associations could be shown for procedures or other regulatory programs. Figure F-3 provides a simplified version of the figure and includes an estimation of the International Atomic Energy Agency (IAEA) defense-in-depth scale (IAEA, 2010) to show that there are similarities between the U.S. and international approaches. Figure F-4 provides a simplified depiction of various regulatory documents and programs and their relationship to the various event categories.

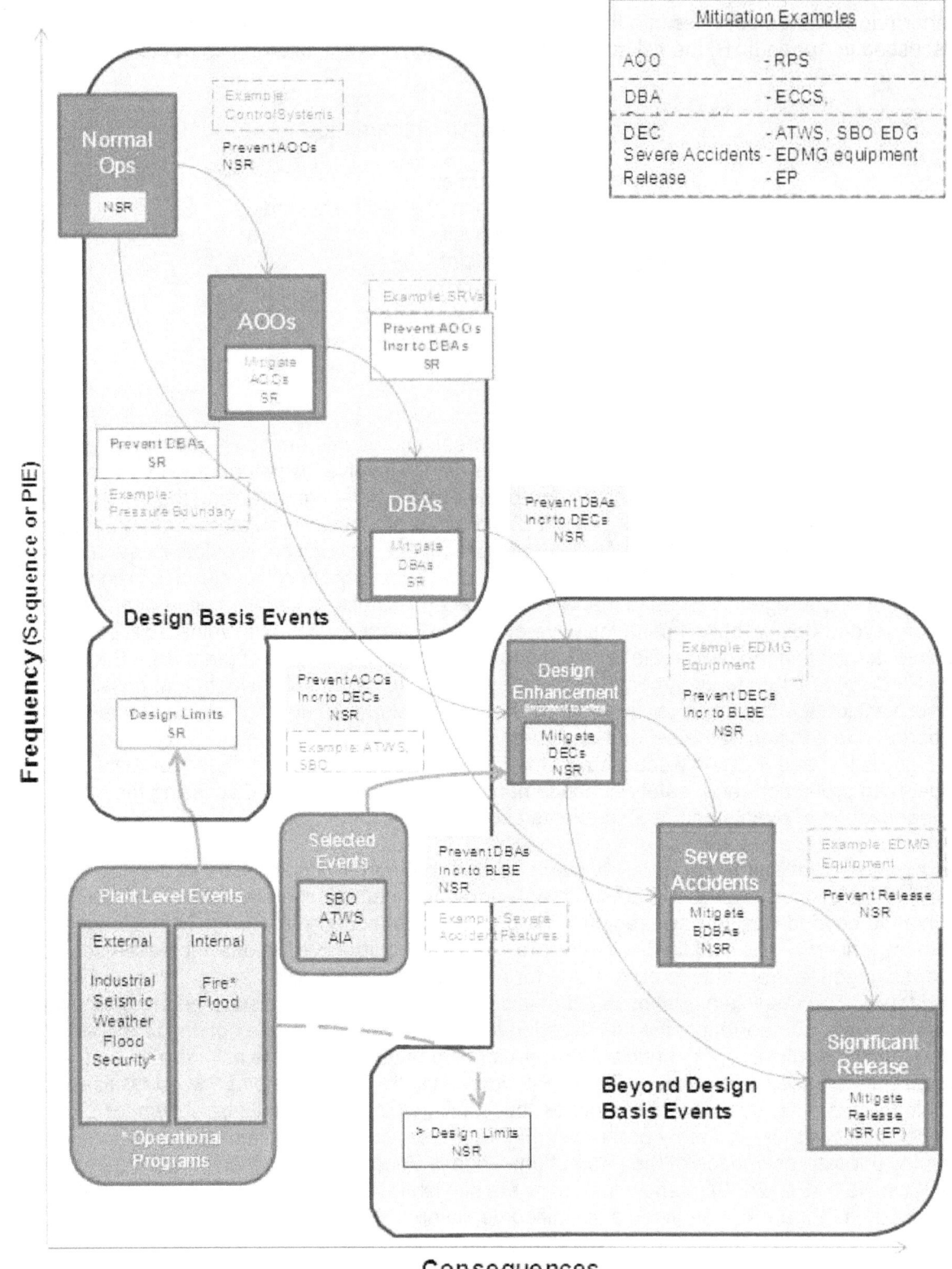

Figure F-2 Event Categories

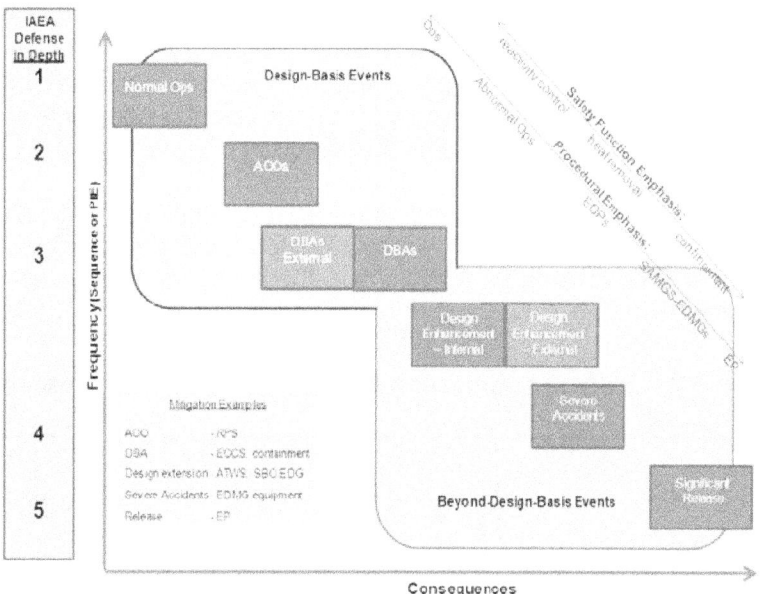

Figure F-3 Event Categories and IAEA Defense-in-depth

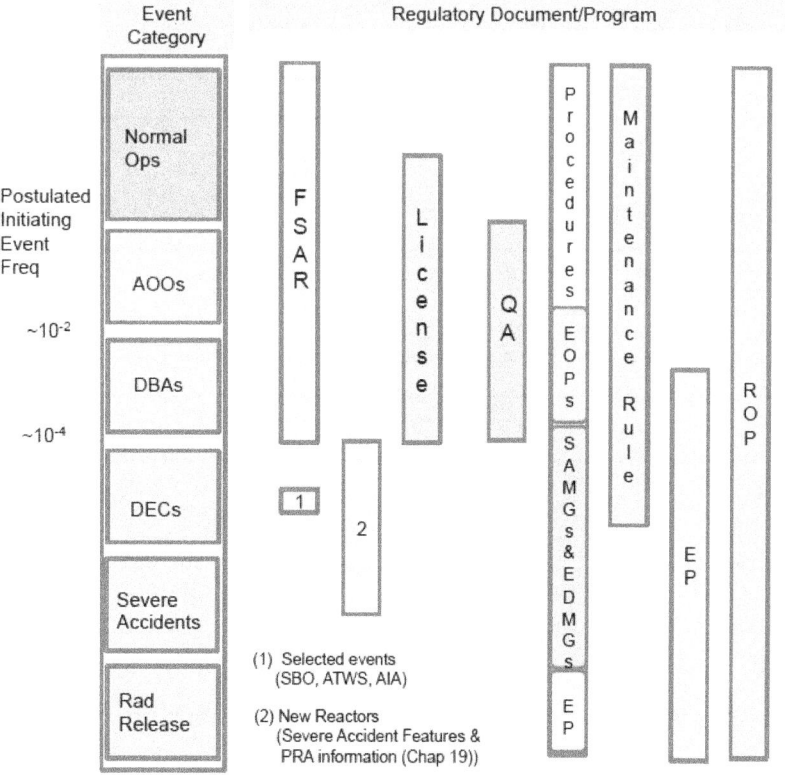

Figure F-4 Licensing-basis Events, Documents and Programs

As discussed throughout this report, the existing construct has generally provided sufficient levels of protection. To better integrate the traditional engineering approaches and risk assessment methodologies, the RMTF has developed the risk-informed and performance-based defense-in-depth definition for all NRC regulatory programs and the additional clarification of the terminology for nuclear power reactors that was presented in Chapter 4.

In attempting to define options and possible implementation transitions, the concept of risk-informed and performance-based defense-in-depth provides general guidance, but to be more useful, it needs to be translated into the actual parts of the licensing process shown in Figure F-2 and the related general construct of the event analyses summarized in Table F-2:

Table F-2 Licensing Approach to Events (current approach)

	PIEs	Key Assumptions		Acceptance Criteria
Normal Ops	n/a	n/a Maintain initial conditions, standby equipment		Part 20, ALARA
AOOs	Functional list (e.g., ANS 51.1/52.1 Cond II) (PIE frequency < ~10^{-2}/yr)	Credit SR SSCs single failure		Maintain safety limits, such as SAFDLs, RCS pressure
DBAs	Functional list (e.g., ANS 51.1/52.1 Cond III/IV) (PIE frequency > ~10^{-4}/yr)	Credit SR SSCs single failure		Maintain coolable core, containment
Design-basis External Events	Maximum credible scenario (e.g., frequency of exceedance > ~10^{-7} for tornado conditions)		Engineering analysis, safety margins	Codes and standards related to barrier design and equipment qualification
Design-enhancement (in concept but term not currently used)	Specific events: ATWS SBO AIA	As defined in each rule – generally best estimate		As defined in each rule (alternate shutdown, coping time, AIA functions)
Residual Risks (in concept but term not currently used)	n/a	n/a		n/a
Beyond-design-basis External Events	n/a	n/a		n/a

A hurdle to developing a more coherent structure for licensing-basis events and the related analyses (deterministic and risk assessments) is the lack of a requirement for licensees to have and maintain plant-specific PRAs.

Operating Reactors

Licensees for operating nuclear power reactors are not required to have a PRA, but they have generally completed assessments of internal and external events in response to Generic Letter 88-20, "Individual Plant Examinations for Severe Accident Vulnerabilities" (NRC, 1988), including Supplement 4, "Individual Plant Examination of External Events (IPEEE) for Severe Accident Vulnerabilities" (NRC, 1991). In addition, most licensees have performed and maintain some form of a plant-specific PRA to support assessments required by the maintenance rule (10 CFR 50.65), environmental reviews for license renewal, and the significance determination process within the reactor oversight program. These assessments might have led to plant modifications or other actions to address identified risk concerns, but the assessments themselves were not incorporated into NRC's licensing-related activities. Many operating reactor licensees have performed PRAs that comply with NRC-endorsed codes and standards (NRC, 2009b) to support risk-informed licensing actions or voluntary adoption of risk-informed regulations related to fire protection (10 CFR 50.48) or treatment of SSCs (10 CFR 50.69). These activities may include NRC reviews or audits of the licensee's PRAs and establishment of regulatory programs for maintaining PRAs, but they do not usually result in the incorporation of the PRA or its results into the licensing-basis documents. Most of the risk assessments performed for operating reactors have been Level 1 PRAs, with some licensees performing Level 2 PRAs.[1]

The NRC has developed standardized plant analysis risk (SPAR) models for all operating plants in the United States, with most of the models limited to internal events, although some include external hazards, a few including shutdown conditions, and three progress to Level 2 PRAs. The NRC has also performed two studies that have included Level 3 PRAs for selected operating plants (NRC, 1975; NRC, 1990).

New Reactors

New reactors are licensed under the provisions of 10 CFR Part 52 and are subject to the requirements in 10 CFR 50.71, "Maintenance of Records, Making of Reports," which states:

> (h)(1) No later than the scheduled date for initial loading of fuel, each holder of a combined license under subpart C of 10 CFR part 52 shall develop a level 1 and a level 2 probabilistic risk assessment (PRA). The PRA must cover those initiating events and modes for which NRC-endorsed consensus standards on PRA exist one year prior to the scheduled date for initial loading of fuel.

---

1   For the type of nuclear plant currently operating in the United States, a PRA can estimate three levels of risk:

- A Level 1 PRA, which estimates the frequency of accidents that cause damage to the nuclear reactor core. This is commonly called core damage frequency.

- A Level 2 PRA, which starts with the Level 1 core damage accidents and estimates the frequency of accidents that release radioactivity from the containment.

- A Level 3 PRA, which starts with the Level 2 radioactivity release accidents and estimates the consequences in terms of injury to the public and damage to the environment.

(2) Each holder of a combined license shall maintain and upgrade the PRA required by paragraph (h)(1) of this section. The upgraded PRA must cover initiating events and modes of operation contained in NRC-endorsed consensus standards on PRA in effect one year prior to each required upgrade. The PRA must be upgraded every four years until the permanent cessation of operations under § 52.110(a) of this chapter.

(3) Each holder of a combined license shall, no later than the date on which the licensee submits an application for a renewed license, upgrade the PRA required by paragraph (h)(1) of this section to cover all modes and all initiating events.

Summaries of PRAs are submitted by reactor designers seeking certification of their design, as well as applicants and holders of combined licenses. The NRC performs reviews and audits, and the PRA summaries are included in Chapter 19 of final safety analysis reports (FSARs). The PRAs support decisions related to the regulatory treatment of non-safety systems (RTNSS), availability programs, severe accident mitigation alternatives, and various risk-informed proposals included in applications (e.g., risk-informed technical specifications). The NRC has developed SPAR models for new reactor designs certified by the NRC.

Generation IV Reactor Designs

As discussed in Chapter 4 and Appendix H, Generation IV reactors introduce technologies such as liquidmetal and helium as coolants, different fuel forms, and other significant differences from light-water reactors. In addition, international activities and goals for Generation IV reactors include the proposed use of inherent safety features and a reduced reliance on offsite emergency preparedness. The technology differences and deployment goals have led the designers and proponents of Generation IV reactors to include Level 3 PRAs as part of their design and licensing plans. An example is the set of activities related to the next generation nuclear plant (NGNP, 2011).

F.5   Possible Changes to or Additions of Event Categories

The proposed Risk Management Regulatory Framework is intended to improve the NRC's regulation and oversight of nuclear power reactors by providing a logical way to determine the appropriate barriers and controls needed to provide risk-informed and performance-based defense-in-depth. To accomplish this, the concept of risk-informed and performance-based defense-in-depth needs to be integrated into the overall assessment of events and hazards that is shown in Figures F-2 and F-3. As discussed in Chapter 4 and Appendix H, the most notable change in the RMTF proposals is the formalization of the design-enhancement category within the traditional beyond-design-basis regime. Although it might be beneficial to simply add the design-enhancement category, the best approach would use this opportunity to ensure an overall construct of events, analyses, and resultant plant safety features and programs within categories and logical transitions between categories.

A longstanding issue related to the regulation of nuclear power plants relates to the distinction between actions needed to ensure "adequate protection" and those pursued as "safety enhancements." The RMTF has included recommendations in Chapter 4 and Appendix H to clarify this issue by relating the two tiers (adequate protection and safety enhancements) to design-basis events and the design-enhancement category. This is represented in Figure F-5:

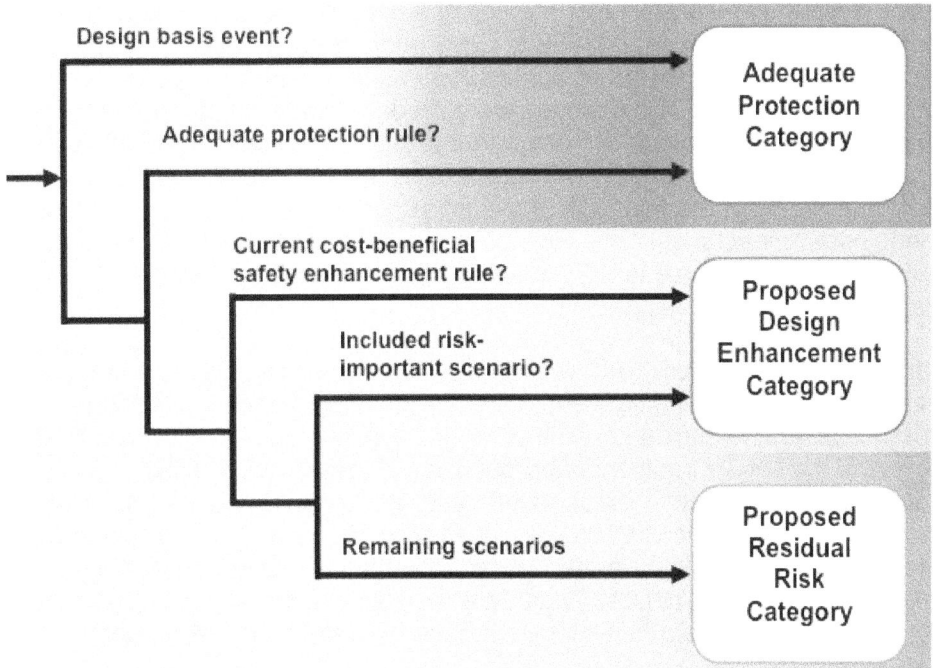

Figure F-5 Regulatory Framework for Nuclear Power Reactors

Unfortunately, it is very difficult to consistently align defense-in-depth, core safety functions, licensing-basis events, safety classification, and other topics with the tiers of adequate protection and safety enhancements. Other organizations (WENRA, 2008; IAEA, 2010) have developed a more coherent set of standards that, although similar in many respects to the NRC's processes, would be difficult to adopt without a significant change in the overall regulatory framework (i.e., Option C in Chapter 4). Several alternatives for constructing the design-enhancement category with limited changes to the existing regulatory framework (i.e., Option B in Chapter 4) are provided in Appendix H.

It is possible that the creation of the design-enhancement category could help efforts related to improving the identification, analysis, and protections against design-basis events. Although there has been a general trend of improving computer models and addressing some overconservatisms in analyzing the design-basis events (e.g., statistical treatment of uncertainties), the assumptions and analyses in Chapter 15 of the FSAR for a specific plant or reactor design have been basically unchanged for decades. Revised analyses to support power uprates or major plant modifications have taken advantage of limited changes in computer codes but are otherwise similar to analyses performed in the 1970s. Previous attempts to revise the handling of design-basis events have encountered significant resistance, partly because it was difficult to reach agreement on the regulatory treatment of barriers and controls related to the revised scenarios. The scenarios in this case are comprised of specified initiating events, behavior of mitigating systems, inclusion of coincident failures and single failure criterion, and the limits established on fission product barriers, such as the fuel cladding, reactor coolant pressure boundary, and containment structure. Several of the previous initiatives are briefly described below:

- "Best-Estimate" or Realistic Loss-of-coolant Accident Analysis: The NRC revised its regulations (10 CFR 50.46) in 1988 to allow methods other than those prescribed in Appendix K to 10 CFR Part 50 for the evaluation of emergency core cooling systems (NRC, 1989). The changes were made to credit research that showed the Appendix K methods resulted in estimates of system performance that were significantly more conservative than originally thought. The revised rule allowed, but did not require, use of more realistic models, with sufficient supporting justifications and treatment of uncertainties. Reactor vendors have developed related topical reports, and a significant number of licensees have taken advantage of the revised analyses to support power uprates or other plant changes.

- Changes to Single Failure Criterion: The NRC adopted the single failure criterion as a means to promote redundancy and reliability of safety systems needed to respond to plant transients and accidents. In this regard, the single failure criterion is an important element of the NRC's traditional approach to defense-in-depth (NRC, 1977a). The need to identify the most limiting single failure also resulted in improved analyses of failure modes and effects, which in turn influenced plant designs. Some have argued, however, that more modern analytical methods can better support plant designs by identifying risk-significant failure combinations and that alternatives to the single failure criterion may do better at ensuring system reliability than does the established approach (NRC, 2005).

- Development of Risk-informed Safety Analysis Approach and Pilot Application: A response to the RMTF solicitation of public comments on a possible Risk Management Regulatory Framework identified a previous industry proposal related to taking a more risk-informed approach to transient and accident analyses. The Westinghouse Owners Group (WOG) submitted WCAP-16084, "Development of Risk-informed Safety Analysis Approach and Pilot Application" (WOG, 2003), to address the classification of transients and accidents into realistic categories by considering the frequency of occurrence of the event combination (i.e., the initiating event in combination with coincident occurrences and a single failure). The WOG subsequently withdrew the topical report based on feedback from the NRC staff that the scope of the report was too broad for effective review and approval.

F.6   References

(ANS, 1983)   American National Standards Institute/American Nuclear Society, ANSI /ANS 51.1-1983, "Nuclear Safety Criteria for the Design of Stationary Pressurized Water Reactor Plants," ANS, LaGrange Park, IL.

(IAEA, 2010)   International Atomic Energy Agency, "Draft – Safety of Nuclear Power Plants: Design," Draft Safety Guide DS-414, September 2010.

(INL, 2009)   Idaho National Laboratory (Hess, et. al.), "Risk-Informed Safety Margin Characterization," INL/CON-09-15549, July 2009.

(NEA, 2007)   Nuclear Energy Agency, "Task Group on Safety Margin Action Plan (SMAP), Safety Margins Action Plan – Final Report," NEA/CSNI/R (2007)9, 2007.

(NGNP, 2011)   Idaho National Laboratory, "Next Generation Nuclear Plant-Probabilistic Risk Assessment White Paper," INL/EXT-11-21270, September 2011.

(NRC, 1975)   U.S. Nuclear Regulatory Commission, "Reactor Safety Study: An Assessment of Accident Risks in U.S. Commercial Nuclear Power Plants," NUREG-75-014 (WASH-1400), October 1975.

(NRC, 1977)   U.S. Nuclear Regulatory Commission, "Design Basis Floods for Nuclear Power Plants," Regulatory Guide 1.59, August 1977, Agencywide Documents Access and Management System (ADAMS) Accession No. ML003740388.

(NRC, 1977a)   U.S. Nuclear Regulatory Commission, "Single Failure Criterion," SECY-77-439, August 17, 1977, ADAMS Accession No ML060260236.

(NRC, 1978)   U.S. Nuclear Regulatory Commission, "Evaluations of Explosions Postulated To Occur on Transportation Routes Near Nuclear Power Plants," Regulatory Guide 1.91, February 1978, ADAMS Accession No ML003740286.

(NRC, 1988)   U.S. Nuclear Regulatory Commission, "Individual Plant Examination for Severe Accident Vulnerabilities," Generic Letter 88-20, November 23, 1988, ADAMS Accession No ML031470299.

(NRC, 1989)   U.S. Nuclear Regulatory Commission, "Best Estimate Calculations of Emergency Core Cooling System Performance," Regulatory Guide 1.157, May 1989, ADAMS Accession No ML003739584.

(NRC, 1990)   U.S. Nuclear Regulatory Commission, Severe Accident Risks: An Assessment for Five U.S. Nuclear Power Plants - Final Summary Report," NUREG-1150, Volume 1, December 1990, ADAMS Accession No. ML040140729.

(NRC, 1991)   U.S. Nuclear Regulatory Commission, "Individual Plant Examination of External Events (IPEEE) for Severe Accident Vulnerabilities," Generic Letter 88-20, Supplement 4, June 28, 1991, ADAMS Accession No ML031150485.

(NRC, 1999)   U.S. Nuclear Regulatory Commission, "Changes, Tests, and Experiments (10 CFR. 50.59)," Final Rule, published in the *Federal Register* on October 4, 1999 (64 FR 53582)

(NRC, 2005)   U.S. Nuclear Regulatory Commission, "Risk-Informed and Performance-Based Alternatives to the Single Failure Criterion," SECY-05-0138, August 2, 2005, ADAMS Accession No. ML051950619.

(NRC, 2007)   U.S. Nuclear Regulatory Commission, "Combined License Applications for Nuclear Power Plants," Regulatory Guide 1.206, June 2007, ADAMS Accession No. ML070720184.

(NRC, 2007a)   U.S. Nuclear Regulatory Commission, "Design-basis Tornado and Tornado Missiles for Nuclear Power Plants," Regulatory Guide 1.76, March 2007, ADAMS Accession No. ML070360253.

(NRC, 2007b)   U.S. Nuclear Regulatory Commission, "A Performance-based Approach To Define the Site-specific Earthquake Ground Motion," Regulatory Guide 1.208, March 2007, ADAMS Accession No. ML070310619.

(NRC, 2007c)   U.S. Nuclear Regulatory Commission, "Feasibility Study for a Risk-informed and Performance-based Regulatory Structure for Future Plant Licensing," NUREG-1860, Volume 1, December 2007, ADAMS Accession No. ML080440170.

(NRC, 2009)   U.S. Nuclear Regulatory Commission, "Interim Staff Guidance on Assessment of Normal and Extreme Winter Precipitation Loads on the Roofs of Seismic Category I Structures," DC/COL-ISG-7, July 1, 2009, ADAMS Accession No. ML091490556.

(NRC, 2009a)   U.S. Nuclear Regulatory Commission, "Risk-informed, Performance-based Fire Protection for Existing Light-Water Nuclear Power Plants," Regulatory Guide 1.205, December 2009, ADAMS Accession No. ML092730314.

(NRC, 2009b)   U.S. Nuclear Regulatory Commission, "An Approach for Determining the Technical Adequacy of Probabilistic Risk Assessment Results for Risk-Informed Activities," Regulatory Guide 1.200, Revision 2, March 2009, ADAMS Accession No. ML090410014.

(NRC, 2010)   U.S. Nuclear Regulatory Commission, "Interim Staff Guidance DC/COL-ISG-016 – Compliance with 10 CFR 50.54(hh)(2) and 10 CFR 52.80(d)," June 2010 (Not publically available)

(NRC, 2010a)   U.S. Nuclear Regulatory Commission, "A Short History of Nuclear Regulation, 1946 – 2009," NUREG/BR-0175, Revision 2, October 2010, ADAMS Accession No. ML102980443.

(NRC, 2011) U.S. Nuclear Regulatory Commission, "Design-basis Hurricane and Hurricane Missiles for Nuclear Power Plants," Regulatory Guide 1.221, October 2011, ADAMS Accession No. ML110940300.

(NRC, 2011a) U.S. Nuclear Regulatory Commission, "Guidance for the Assessment of Beyond-design-basis Aircraft Impacts," Regulatory Guide 1.217, August 2011, ADAMS Accession No. ML092900004.

(WENRA, 2008) Western European Nuclear Regulators' Association, "WENRA Reactor Safety Reference Levels," Reactor Harmonization Working Group, January 2008.

(WOG, 2003) Westinghouse Owners Group, "Development of Risk-informed Safety Analysis Approach and Pilot Applications," WCAP-16084NP, Revision 0, September 2003.

APPENDIX G

SAFETY CLASSIFICATION OF SYSTEMS, STRUCTURES, AND COMPONENTS

G.1 Background

Nuclear power plants are large industrial facilities that include systems, structures, and components (SSCs) to generate electricity and to ensure that the nuclear reactor does not pose an undue risk to public health and safety. It is in the economic interests of the plant operator for the power generation equipment to be manufactured and maintained such that it supports reliable plant operation. The need to define requirements on SSCs that help ensure plant safety has also been recognized from the early development of nuclear power. A logic or system to classify equipment according to its role in ensuring plant safety is a natural part of the process to define requirements for design, operation, and other features of SSCs with safety functions.

The term "safety-related" has been used to define requirements for the protection of SSCs from safe shutdown earthquakes (Title 10 of the *Code of Federal Regulations* (10 CFR) Part 100, "Reactor Cite Criteria") and is more widely used to distinguish those SSCs warranting special treatment in terms of quality assurance, environmental qualification, and applicability of various industry codes and standards. The terminology of "safety-related," "important to safety," and related phrases have also been intertwined with the broader debates about the role of the NRC and how it regulates commercial nuclear power plants.

Safety-related SSCs are defined in 10 CFR 50.2, "Definitions," as follows:

> *Safety-related structures, systems, and components means those structures, systems, and components that are relied upon to remain functional during and following design-basis events to assure:*
>
> (1)   the integrity of the reactor coolant pressure boundary
>
> (2)   the capability to shut down the reactor and maintain it in a safe shutdown condition; or
>
> (3)   the capability to prevent or mitigate the consequences of accidents that could result in potential offsite exposures comparable to the applicable guideline exposures set forth in 10 CFR 50.34(a)(1) or 10 CFR 100.11 ["Determination of Exclusion Area, Low Population Zone, and Population Center Distance,"] of this chapter, as applicable.

Design-basis events are defined in 10 CFR 50.49, "Environmental Qualification of Electric Equipment Important to Safety for Nuclear Power Plants," as follows:

> Design-basis events are defined as conditions of normal operation, including anticipated operational occurrences, design-basis accidents, external events, and natural phenomena for which the plant must be designed to ensure functions (b)(1)(i) (A) through (C) [see above items 1, 2 and 3] of this section.

The terminology associated with SSC classification and related special treatment has been a longstanding source of confusion and discussions between the staff and industry. This is in part because of the frequent use of the terms "safety-related" and "important to safety" in similar discussions and sometimes without distinction. Although attempts to clarify the terminology have been ongoing for several decades, there continue to be some communication issues regarding the classification of SSCs. As will be discussed later, the introduction of Risk-Informed approaches and creation of additional categories and classifications have created additional complexity to the discussions. For the purpose of this report, the distinction between "safety-related" and "important to safety" is taken from Generic Letter 8401, "NRC Use of the Terms 'Important to Safety' and 'Safety Related'" (NRC,1984), and a related memorandum entitled "Standard Definitions for Commonly-Used Safety Classification Terms" (NRC,1981). The clarification of these terms was also related to policy issues on whether the NRC should limit its regulations and oversight to safety-related equipment or the path taken, which was to broaden its scope to other matters that contributed to the risks posed by nuclear power plants to the public health and safety.

The term "important to safety" was defined as follows:

> *Definition (from 10 CFR Part 50, "Domestic Licensing of Production and Utilization Facilities, Appendix A, "General Design Criteria for Nuclear Power Plants")* Important to Safety
>
> "Those structures, systems, and components that provide reasonable assurance that the facility can be operated without undue risk to the health and safety of the public."

Additional insights related to "important to safety" include:

- Encompasses the broad class of plant features, covered (not necessarily explicitly) in the General Design Criteria, that contribute in important ways to safe operation and protection of the public in all phases and aspects of facility operation (i.e., normal operation and transient control as well as accident mitigation)

- Includes Safety-Grade (or Safety-Related) as a subset

- Note that "important to safety" in 10 CFR 50.49 includes safety-related, non-safety-related whose failure could prevent accomplishment of safety functions defined in definitions of safety-related, and certain post-accident monitoring equipment.

The discussions of safety-related reinforced the definition from NRC regulations (subsequently incorporated into 10 CFR 50.2, "Definitions," and made consistent across various regulations) and clarified that "safety-grade" was synonymous with "safety-related." The class of "safety-related" SSCs was described as a subset of "important to safety" SSCs, and reference was made to Regulatory Guide (RG) 1.29, "Seismic Design Classification" (NRC, 2007), for a typical listing of "safety-related" SSCs for light-water reactors.

Various regulatory guides, industry codes and standards, and other guidance documents include discussions of safety-related SSCs and have created subcategories within the broader class of safety-related SSCs. For example, the Standards Committee within the American Nuclear Society (ANS) has identified safety classes (ANS, 1983) that are summarized as follows:

Table G-1 ANS/ASME Safety Classes

Safety Class 1	Major reactor coolant pressure boundary functions (reactor pressure vessel, connected piping, reactor coolant loops (including primary side of steam generators), pressurizers, and recirculation loops)
Safety Class 2	More important pressure boundary functions (primary containment, containment penetrations, sprays and auxiliary systems, emergency core cooling systems, residual heat removal systems, secondary side of steam generators, standby liquid control, and reactor core isolation cooling)
Safety Class 3	All other nuclear safety-related functions (nuclear safety-related cooling water, residual heat removal secondary side, nuclear safety-related ventilation, chemical and volume control system, fuel pool, nuclear safety-related electrical equipment, secondary containments)
Non-safety Related Class(es)	Specified nonnuclear safety-related functions (radwaste, fire protection, fuel pool cooling and cleanup, effluent monitoring, fuel storage and handling, reactor water cleanup, turbinegenerator, condenser)

These or similar classifications are used to define appropriate requirements related to design, manufacture, and quality assurance for different types of SSCs or by different standard-developing organizations. For example, the American Society of Mechanical Engineers (ASME) Boiler and Pressure Vessel Code defines requirements for various quality group classifications using Quality Groups A through D that generally align to the above ANS safety classes (see RG 1.26, "Quality Group Classifications and Standards for Water, Steam, and Radioactive-Waste-Containing Components of Nuclear Power Plants" (NRC, 2007a)). The safety-related designation is likewise used directly or by reference in standards issued by the Institute of Electrical and Electronics Engineers, American Concrete Institute, and the American Society for Testing and Materials. Much of the specific guidance provided in the various regulatory guides and industry standards is related to the special requirements for an SSC given its safety-related designation and possible service conditions associated with nuclear power plants.

As implied by the wide application of the concept and terminology of safety-related SSCs, the classification of SSCs follows from the definition in the NRC regulations and has evolved, for the currently operating reactors, to a point of general consensus and stability. The related concept of design-basis events has likewise become fairly well established, and discussions usually involve details about assumptions or analyses. However, the Fukushima accident in Japan has resulted in some renewed discussions regarding the identification and maintenance of design-basis events. The subjects of design-basis and beyond-design-basis events are discussed in more detail in Appendix F.

G.2   Additional Regulatory Requirements ("Design-Enhancement Events")

The evolution of nuclear plant designs, operating practices, and regulatory requirements has involved various approaches and compromises that have complicated the relationships between design-basis events, safetyclass designations, and special treatment requirements, such as the applicability of 10 CFR 50, Appendix B, "Quality Assurance Criteria for Nuclear Power Plants and Fuel Reprocessing Plants." Such changes have resulted from risk insights produced by probabilistic risk assessments (PRAs) and operating experience. They also include the development of 10 CFR 50.62, "Requirements for Reduction of Risk from Anticipated Transients Without Scram (ATWS) Events for Light-Water-Cooled Nuclear Power Plants," and 10 CFR 50.150, "Aircraft Impact Assessment." The use of PRAs has also challenged the traditional designations and treatment of safety-related SSCs and greatly increased discussions regarding beyond-design basis events.

Although not used in the context of NRC regulations, the terms "design extension events" or "design extension conditions" have been introduced by some international organizations and other national regulatory bodies (WENRA, 2008; IAEA, 2011). In the RMTF proposed risk management framework, this category of events has been called the "design enhancement category." This category is used to provide a transition between design-basis events and beyond-design-basis events, provide a more logical approach for the regulatory treatment of some events, and alleviate concerns about uncertainties where risk insights are used as part of the decisionmaking process for regulatory requirements. A reason for introducing the concept here is that the terminology may be helpful in discussing events such as ATWS, station blackout (SBO), and aircraft impact, which have introduced regulatory requirements for non-safety-related SSCs. In some sense, these are simply examples of the NRC determining that SSCs needed for certain events or conditions fall within the broader category of important to safety without needing to be classified as safety-related. The advantage of adopting a term like "design-enhancement events" is that it supports a logical construct of events and related non-safety-related SSCs that are important to safety instead of thinking of ATWS, SBO, aircraft impact, and possible future actions as individual outliers from the traditional approach of design-basis events and safety-related SSCs.

A short discussion of the several events or requirements falling within the potential category of design-enhancement events is provided below.

Anticipated Transients Without Scram (10 CFR 50.62)

The requirements for addressing ATWS were published on June 26, 1984, in the *Federal Register* (49 FR 26036) and were issued as a result of concerns identified in PRAs and operating events. The Commission intended the ATWS rule requirements to provide further assurance that failure of the reactor to scram following an anticipated operational transient would not adversely affect public health and safety. In recognizing that the ATWS scenarios were expected to be infrequent occurrences, the regulation and related guidance acknowledged that the related SSCs need not be categorized as safety-related (see Generic Letter 85-06,

"Quality Assurance Guidance for ATWS Equipment That Is Not Safety Related" (NRC, 1985)). Instead, the Commission required that the equipment be designed to perform its function in a reliable manner and encouraged reliability assurance programs for reactor protection systems to minimize the chances of an ATWS event.

The requirements related to ATWS highlight the difficulty in explaining clearly NRC regulations. For example, the NRC's acknowledgment that ATWS-related SSCs need not be safety-related would seem consistent with the scenario not being a "design-basis event," and many (but not all) references to ATWS refer to it as a "beyond-design-basis accident." However, certain SSCs (e.g., ATWS mitigation system actuation circuitry (AMSAC) or redundant reactivity control system (RRCS)) are required by the regulations to address ATWS events. Therefore, they do perform a design-basis function as that term is defined in Appendix B to Nuclear Energy Institute (NEI) 97-04, "Design Bases Program Guidelines" (NEI, 2000), which is endorsed in RG 1.186, "Guidance and Examples for Identifying 10 CFR 50.2 Design Basis"[1] (NRC, 2000). This apparent contradiction (beyond-design-basis event, yet serving a design-basis function) is applicable to the SBO and aircraft impact examples and might be alleviated by adopting a concept similar to "design-enhancement category."

Station Blackout (10 CFR 50.63)

In 1988, the Commission concluded that additional regulatory requirements were justified to address concerns about SBO events (NRC, 1988). The term "station blackout" refers to the complete loss of alternating-current (ac) electric power to the essential and nonessential switchgear buses in a nuclear power plant. SBO, therefore, involves the loss of offsite power concurrent with turbine trip and failure of the onsite emergency ac power system, but not the loss of available ac power to buses fed by station batteries through inverters or the loss of power from "alternate ac sources." As with the ATWS rule, SBO requirements were established as a result of insights from risk assessments and some operational events.

The rule requires plants to be able to withstand for a specified duration and recover from an SBO. The SBO duration for each plant is determined based on:

- the redundancy of the onsite emergency ac power sources

- the reliability of the onsite emergency ac power sources

- the expected frequency of loss of offsite power

- the probable time needed to restore offsite power

---

1   The definition of design basis in 10 CFR 50.2 is:
     Design bases means that information which identifies the specific functions to be performed by a structure, system, or component of a facility, and the specific values or ranges of values chosen for controlling parameters as reference bounds for design. These values may be (1) restraints derived from generally accepted "state of the art" practices for achieving functional goals, or (2) requirements derived from analysis (based on calculation and/or experiments) of the effects of a postulated accident for which a structure, system, or component must meet its functional goals.specific functions to be performed by a structure, system, or component of a facility, and the specific values or ranges of values chosen for controlling parameters as reference bounds for design. These values may be (1) restraints derived from generally accepted 'state of the art' practices for achieving functional goals, or (2) requirements derived from analysis (based on calculation and/or experiments) of the effects of a postulated accident for which a structure, system, or component must meet its functional goals."

RG 1.155, "Station Blackout" (NRC, 1988a) provides quality assurance guidance for non-safety-related systems and equipment used to meet the requirements of 10 CFR 50.63, "Loss of Alternating Current Power."

Aircraft Impact

As a result of the terrorist events of September 11, 2001, the NRC developed and issued a new regulation, 10 CFR 50.150, "Aircraft Impact Assessment" (NRC, 2009), which requires applicants for new nuclear power plants to perform a designspecific assessment of the effects on the facility of the impact of a large, commercial aircraft. The applicant is required to use realistic analyses to identify and incorporate design features and functional capabilities to show, with reduced use of operator actions, that either the reactor core remains cooled or the containment remains intact and that either spent fuel cooling or spent fuel pool integrity is maintained.

As discussed in the supplementary information for the aircraft impact assessment rulemaking, the Commission determined that the intentional crash of a large, commercial aircraft into a nuclear power plant is a beyond-design-basis event, and the NRC's requirements that apply to the design, construction, testing, operation, and maintenance of design features and functional capabilities for design-basis events will not apply to design features or functional capabilities selected by the applicant solely to meet the requirements of the aircraft impact rule.

Consideration of a rule to require applicants for new nuclear power reactors to perform an aircraft impact assessment and describe design features and functional capabilities addressing such impacts, which are beyond-design-basis scenarios, is similar to the Commission's consideration in the mid-1980s of new rules addressing accidents more severe than design-basis accidents. The Policy Statement on Severe Reactor Accidents (NRC, 1985a) explained the Commission's conclusion that, although it was proposing criteria to show new reactor designs to be acceptable for severe accident concerns, thenexisting plants posed no undue risk to public health and safety, and thus, there was no need for action on operating reactors based on severe accident risks.

While clearly stating that the aircraft impacts are considered beyond-design-basis events, the Commission also defined requirements to evaluate and report changes to the design features and functional capabilities credited with meeting the acceptance criteria in the rule. Such controls were defined for the aircraft impact rulemaking based on similar requirements for severe accident features that are described in the design description documents for new reactors. The controls added for aircraft impact and severe accident features (e.g., change controls and reporting requirements) are an improvement over those defined for the ATWS and SBO rules and provide an example of how the NRC might address updating of information for items considered "important to safety" but not safety-related.

G.3 Regulatory Treatment of Non-safety Systems

In the early 1990s, the NRC developed an approach to address the proposed increased use of passive safety features in advanced reactor designs. Unlike operating reactors, the passive advanced light-water reactor designs (ALWR), such as the AP600 design, proposed extensive use of safety systems that rely on the driving forces of buoyancy, gravity, and stored energy sources. In addition to the active systems used during normal plant operations, the passive ALWR designs proposed non-safety-grade active systems to provide defense-in-depth capabilities for reactor coolant makeup and decay heat removal. These systems would be the first line of defense to reduce challenges to the passive systems in the event of transients or plant upsets. The licensing-related analyses proposed by the industry for the passive designs rely solely on the passive safety systems to demonstrate compliance with the acceptance criteria of various design-basis transients and accidents. To incorporate the defense-in-depth measures into the licensing process, while recognizing the passive safety features' role in responding to design-basis events, the staff and industry developed a process for the regulatory treatment of non-safety systems (RTNSS) (NRC, 1995).

To further illustrate the concept, the criteria used to identify SSCs within the RTNSS regime for the economic simplified boilingwater reactor (ESBWR) design are as follows:

- SSC functions relied upon to meet deterministic NRC performance requirements, such as 10 CFR 50.62 (ATWS) and 10 CFR 50.63 (SBO),

- SSC functions relied upon to ensure long-term safety (beyond 72 hours) and to address seismic events,

- SSC functions relied upon under power operating and shutdown conditions to meet the Commission's safety goal guidelines of a CDF of less than 1×10^{-4} per reactor year and a large release frequency of less than 1×10^{-6} per reactor year;

- SSC functions needed to meet the containment performance goal, including containment bypass, during severe accidents, and

- SSC functions relied upon to prevent significant adverse systems interactions.

This can be viewed as a more formal system to identify and address SSCs determined to be important to safety but not safety-related, including the design-enhancement events such as ATWS and SBO. The remaining criteria address the characteristics of passive plants (e.g., use of non-safety-related systems to replenish passive systems or perform cooling functions after 72 hours), and Risk-Informed insights (for both CDF and containment performance goals). A key feature of the RTNSS regime is the determination of appropriate regulatory oversight for the RTNSS SSCs based on evaluations of risk significance. The controls are defined in the licensing documents such as the reliability assurance program.

G.4   Risk-Informed Categorization (10 CFR 50.69)

The NRC staff prepared SECY-98-300, "Options for Risk-Informed Revisions to 10 CFR Part 50 – "Domestic Licensing of Production and Utilization Facilities" (NRC, 1998) as part of the continuing efforts to increase the use of Risk-Informed approaches in the regulation of nuclear power plants. The staff proposed to change the regulatory scope of SSCs needing special treatment in such areas as quality assurance, environmental qualification, technical specifications, and the ASME Code. The paper described how this could be accomplished, in part, by developing Risk-Informed definitions (i.e., classifications) for safety-related and safety-important SSCs and "grading" special treatment requirements for SSCs based upon their risk importance. This effort included many interactions with the Commission and various stakeholders and ultimately led to the issuance of a new regulation in 10 CFR 50.69, "Risk-Informed Categorization and Treatment of Structures, Systems and Components for Nuclear Power Reactors." The rule defines four classes of SSCs based on the traditional categories of safety-related and non-safety-related and consideration of risk insights, including severe or beyond-design-basis accidents, to define SSCs as being either safety-significant or low-safety-significant. The representation provided below for the four Risk-Informed safety class (RISC) categories in 10 CFR 50.69 is taken from RG 1.201, "Guidelines for Categorizing Structures, Systems, and Components in Nuclear Power Plants According to Their Safety Significance" (NRC, 2006).

Figure G-1   10 CFR 50.69 Risk-Informed Safety Class (RISC) Categories

Regulatory Guide 1.201 and the related industry guidance in NEI 00-04, "10 CFR 50.69 SSC Categorization Guideline" (NEI, 2005), provides detailed guidance on the considerations and process for determining the classification of SSCs into the RISC 1-4 categories. The process includes risk characterizations, defense-in-depth characterizations, risk sensitivity studies, and reviews by an integrated decisionmaking panel. Much of the focus of the guidance documents related to 10 CFR 50.69 is on the categorization of SSCs. Less discussion is provided for the actual treatment programs being implemented under 10 CFR 50.69, which has raised implementation questions, especially for equipment in RISC categories 2 and 3. At this time, no licensee has submitted an application requesting to implement 10 CFR 50.69, although at least one licensee is preparing a pilot application. Following the initial pilot application, lessons learned from the application review will be used to revise the associated industry guidance and RG 1.201.

G.5   SSC Classification for Advanced Reactors (non-LWRs)

The need to revise various regulatory requirements for non-LWR reactor technologies has resulted in some more dramatic proposals for changing the selection of design-basis events and the classification of SSCs. Two proposals are described below:

NUREG-1860

The staff issued NUREG-1860, "Feasibility Study for a Risk-Informed and Performance-Based Regulatory Structure for Future Plant Licensing" (NRC, 2007b), to document a possible framework that (1) could be used to develop an alternative set of technical requirements to 10 CFR Part 50 applicable for future non-LWR nuclear power plants (the framework includes a proposed draft set of technical requirements), and (2) could be used to establish Risk-Informed licensing-basis events and the safety classification of structures, systems, and components. The approaches discussed in NUREG-1860 are sometimes referred to as the technology neutral framework.

NUREG-1860 proposes a safety classification scheme in which all the plant SSCs fall into two categories—safety-significant or nonsafety-significant—distinguished by whether the SSCs need special treatment or not. Consistent with the framework's discussion of selecting licensing-basis events (LBEs), the term "safety-significant" is assigned to those SSCs whose functionality plays a role in meeting the acceptance criteria imposed on the LBEs. The term "special treatment" is used to designate requirements imposed on SSCs that go beyond industry-established requirements for equipment classified as "commercial grade." These requirements provide additional confidence that the equipment is capable of meeting its functional requirements under PRAanalyzed conditions.

The type of special treatment varies depending on the function and importance of the SSC. The treatment helps to ensure that the SSCs will perform reliably (as postulated in the PRA) under the conditions (temperature, pressure, radiation,) assumed to prevail in the event sequences for which the SSC's successful function is credited in the risk analysis. A basic special treatment requirement for all safety-significant SSCs will be the establishment and monitoring of reliability and availability goals. All safety-significant SSCs will have reliability and availability consistent with the values assumed in the PRA. During operation, a process similar to the monitoring of the performance and condition of structures, systems, or components, against licensee-established goals of 10 CFR 50.65, "Requirements for Monitoring the Effectiveness of Maintenance at Nuclear Power Plants" (the maintenance rule), is expected to be an integral part of the monitoring program for this special treatment requirement. Monitoring will consist of periodically gathering, trending, and evaluating information pertinent to the performance and availability of PRA-related SSCs and comparing the result with the established goals and performance criteria to verify that the goals are being met.

Next Generation Nuclear Plant Program

The NRC staff is assessing various white papers submitted as part of the preapplication interactions with the U.S. Department of Energy (DOE) related to the Next Generation Nuclear Power (NGNP) program. The NGNP program is adopting a licensing approach similar to earlier proposals related to high-temperature gas reactors, such as the modular high-temperature gas-cooled reactor and the pebble bed modular reactor. The same process for selecting licensing-basis events and determining equipment classifications has also been incorporated into industry and international standards (e.g., ANS 53.1, "Nuclear Safety Design Standard for Modular Helium-Cooled Reactor Plants" (ANS, 2011)).

The NGNP safety classification approach (NGNP, 2010) results in the classification of SSCs as safety-related, non-safety-related with special treatment, or non-safety-related. Based upon the selected licensing-basis events and considering defense-in-depth attributes, the plant's SSCs are evaluated for the safety significant role they may play in preventing or mitigating the radiological consequence of such events to classify which SSCs are safety related. With its event selections derived, in part, from frequency-and-consequence curves in a process similar to NUREG-1860, the NGNP approach is a mix of the classification approaches from the technology neutral framework and 10 CFR 50.69. The classes are described as follows:

- Safety-Related:

  - SSCs relied on to perform required safety functions to prevent or mitigate the consequences of design-basis events (DBE) to comply with the top level regulatory criteria (TLRC)

  - SSCs relied on to perform required safety functions to prevent the frequency of beyond-design-basis events (BDBE) with consequences greater than the 10 CFR 50.34, "Contents of Applications; Technical Information," dose limits from increasing into the DBE region of the frequency-consequence curve

- Non-Safety-Related with Special Treatment:

  - SSCs relied on to perform functions to mitigate the consequences of anticipated operational occurrences (AOO) to comply with the TLRC

  - SSCs relied on to perform functions to prevent the frequency of DBEs with consequences greater than the 10 CFR 20, "Standards for Protection Against Radiation," offsite dose limits from increasing into the AOO region of the frequency-and-consequence curve

- Non-Safety-Related:

  - All other SSCs (with no special treatment required)

The NRC staff is currently interacting with DOE on the NGNP white paper on SSC classification, and so the process has not yet been finalized or endorsed by the NRC.

G.6   International Approaches

International Atomic Energy Agency

Several International Atomic Energy Agency (IAEA) documents discuss the safety classification of SSCs, including Draft Safety Guide DS-367, "Safety Classification of Structures, Systems, and Components in Nuclear Power Plants" (IAEA, 2011). This draft safety guide updates previous discussions on this topic in documents such as the "IAEA Safety Glossary" (IAEA, 2007), which describes plant equipment in terms of "items important to safety," "safety systems," and "safety-related items."[2] Draft Safety Guide DS-367 describes a more structured evaluation and classification of SSCs, which starts from the higher level requirement defined in Draft Safety Requirements No. SSR 2/1, "Safety of Nuclear Power Plants: Design" (IAEA, 2010).

---

2   In the IAEA Safety Glossary, a "safety system" is a system important to safety, provided to ensure the safe shutdown of the reactor or the residual heat removal from the core, or provided to limit the consequences of anticipated operational occurrences and design-basis accidents. A safety-related item is important to safety but is not part of a safety system. In terms of NRC terminology, the IAEA safety system would be safety-related and the categories of RTNSS or RISC-2 would correspond to safety-related items as discussed in the IAEA Safety Glossary.

The method for classifying the safety significance of items important to safety shall be based primarily on deterministic methodologies complemented where appropriate by probabilistic methods, with account taken of factors such as the following:

- the safety function(s) to be performed by the item

- the consequences of failure to perform the safety function

- the frequency at which the item will be called upon to perform a safety function

- the time following a postulated initiating event at which, or the period for which, it will be called upon to operate.

The draft safety classification guide categorizes safety functions, which can be either preventive or mitigative, based on the severity of the consequences of failure of that specific function. The severity is divided into three levels: high, medium, and low, as follows:

- The severity should be considered "high" if a failure of the safety function could:

  - lead directly to a release of radioactive material that exceeds the limits for design-basis accidents accepted by the regulatory body, or

  - cause the values of key physical parameters to challenge or exceed design limits for design-basis accidents.

- The severity should be considered "medium" if a failure of the safety function could, at worst:

  - lead to a release of radioactive material below the limits for design basis accidents accepted by the regulatory body, or

  - cause the values of key physical parameters to exceed the design limits for anticipated operational occurrences, but remain within the specified design limits for design-basis accidents

- The severity should be considered "low" if a failure of the safety function could, at worst:

  - lead to a release of radioactive material below the limits for the plant conditions for anticipated operational occurrences, or

  - cause the values of key physical parameters to exceed the specified design limits for normal operation, but remain within the specified design limits for anticipated operational occurrences

The IAEA guide calls for the safety significance of all plantspecific safety functions to be established and categorized according to the risk. The approach recommended in the safety guide has three categories (which align with the approaches adopted in many Member States for existing nuclear power plants), supplemented by a fourth category (reflecting the approach followed for some newer reactor designs) to cater to additional features in the design to control consequences in design extension conditions.[3] It should be noted that the safety classification guide discusses design extension events or conditions as follows:

> Safety functions for the mitigation of "design extension conditions" are intended to limit accident progression (e.g. in-vessel mitigation before significant core degradation occurs) and to mitigate the consequences of a severe accident (e.g., ex-vessel mitigation to control the remains of a significantly degraded core).

Design extension conditions are thereby equated to the fourth level of defense-in-depth, as described in INSAG-10, "Defense-in-depth in Nuclear Safety" (IAEA, 1996), which is the control of severe conditions, including the prevention of accident progression and mitigation of the consequences of a severe accident.

The four safety categories, in relation to the safety functions, are summarized in Table G2.

- Level A mitigative safety functions for design-basis accidents should establish a controlled state following a design-basis accident. A controlled state should be reached as soon as possible.

- Level B mitigative safety functions for design-basis accidents should:

  a)   after a controlled state is reached, achieve and maintain a safe shutdown state following a design-basis accident, and

  b)   minimize the challenge to the remaining barriers from the design-basis accident.

Table G-2 IAEA Safety Categories

Type of safety function	Severity of the consequences of the failure of plantspecific safety functions		
	High	Medium	Low
Preventive safety functions	Safety Category 1	Safety Category 2	Safety Category 3
Safety functions for mitigation of anticipated operational occurrences	Safety Category 1	Safety Category 2	Safety Category 3
Safety functions for control or mitigation of design-basis accidents (level A)	Safety Category 1	Safety Category 2	Safety Category 3
Safety functions for control or mitigation of design-basis accidents (level B)	Safety Category 2	Safety Category 3	Safety Category 3
Safety functions for mitigation of consequences in design extension conditions	Safety Category 4	Safety Category 4	**No** Safety Category

---

3   Design extension events or conditions are generally the same as the design enhancement category proposed by the RMTF.

An evaluation is performed for all plant SSCs, and the SSCs are placed in safety functional groups based on the functions they provide (e.g., control of reactivity, removal of heat from the core and spent fuel, and confinement of radioactive material). The assignment of SSCs to safety categories also considers factors such as the availability of other SSCs to fulfill the specific-safety function.

Annex II to the draft IAEA safety classification guide provides examples of engineering design rules (i.e., special treatment requirements) for the various safety categories

Western European Nuclear Regulators' Association

Western European Nuclear Regulators' Association (WENRA) has developed "reference levels," with the objective to attain a common approach to nuclear safety within Europe. Although not as detailed as the IAEA draft safety classification guidance, the WENRA Reactor Harmonization Working Group (WENRA, 2008) has defined the following reference level related to safety classification:

1. **Objective**

    1.1   *All SSCs important to safety shall be identified and classified on the basis of their importance for safety.*

2. **Classification process**

    1.2   *The classification of SSCs shall be primarily based on deterministic methods, complemented where appropriate by probabilistic methods and engineering judgment.*

    1.3   *The classification shall identify for each safety class:*

    –   *The appropriate codes and standards in design, manufacturing, construction and inspection;*

    –   *Need for emergency power supply, qualification to environmental conditions;*

    –   *The availability or unavailability status of systems serving the safety functions to be considered in deterministic safety analysis;*

    –   *The applicable quality requirements.*

3. **Ensuring reliability**

    3.1   *SSCs important to safety shall be designed, constructed and maintained such that their quality and reliability is commensurate with their classification.*

      3.2   The failure of a SSC in one safety class shall not cause the failure of other SSCs in a higher safety class. Auxiliary systems supporting equipment important to safety shall be classified accordingly.

4.   Selection of materials and qualification of equipment

      4.1   The design of SSCs important to safety and the materials used shall consider the effects of operational conditions over the plant lifetime and the effects of design basis accidents on their characteristics and performance.

      4.2   A qualification procedure shall be adopted to confirm that SSCs important to safety meet throughout their design operational lives the demands for performing their function, taking into account environmental conditions over the lifetime of the plant and when required in anticipated operational occurrences and accident conditions.

Similar to the discussions of design extension events in the IAEA guidance, WENRA has summarized design extension events as follows:

• **Design Extension Conditions (DECs):** The plant can be brought into a controlled state and the integrity of the containment is maintained (the containment shall cope with a core melt situation). Significant releases are *practically eliminated*. DECs may be analyzed using the best-estimate approach:

    ○   For those conditions that are not practically eliminated, design provisions shall be made such that only protective measures that are of limited scope in terms of area and time are necessary for the protection of the public, and sufficient time is available to implement these measures

    ○   The possibility of certain conditions occurring is considered to have been practically eliminated if it is physically impossible for the conditions to occur or if the conditions can be considered with a high degree of confidence to be extremely unlikely to arise.

A more detailed discussion in the report of the WENRA Reactor Harmonization Working Group is provided in the following reference level for "Design Extension of Existing Reactors:"

1.   Objective

      1.1   The design extension analysis shall examine the performance of the plant in specified accidents beyond the design basis, including selected severe accidents, in order to minimize as far as reasonably practicable radioactive releases harmful to the public and the environment in cases of events with very low probability of occurrence.

2. **Selection and analysis of Beyond Design Basis Events**

2.1 *Beyond design basis events shall be selected and considered in the safety analysis to determine those sequences for which reasonable practicable preventive or mitigative measures can be identified and implemented.*

2.2 *Realistic assumptions and modified acceptance criteria may be used for the analysis of the beyond design basis events.*

3. **Instrumentation for the management of beyond design basis accident conditions**

3.1 *Adequate instrumentation shall exist which can be used in severe accident environmental conditions in order to manage such accidents according to guidelines/procedures for severe accidents.*

3.2 *Necessary information from instruments shall be relayed to the control room as well as to a separately located supplementary control room/post and be presented in such a way to enable a timely assessment of the plant status and critical safety functions in severe accident conditions.*

Where:

- *Design extension is understood as measures taken to cope with additional events or combination of events, not foreseen in the design of the plant. Such measures need not involve application of conservative engineering practices but could be based on realistic, probabilistic or best estimate assumptions, methods and analytical criteria.*

- *These reference levels aim at providing protection at the level 4 of the defense-in-depth. Such protection could be provided by existing equipment that has been assessed, and if needed modified, to perform the relevant function in a severe accident condition or additional equipment on a best estimate basis*

- *Special attention needs to be given for certain reactor types to the analysis of severe accident conditions with an open containment during certain shutdown states. Should such an accident occur, it should be possible to achieve timely containment isolation or implement equally effective compensatory measures. Therefore consideration has to be given to the time needed for the restoration of containment isolation and effective leaktightness, taking into account factors such as the progression of the accident sequences.*

4. **Protection of the containment against selected beyond design basis accidents**

   4.1   *Isolation of the containment shall be possible in a beyond design basis accident. However, if an event leads to bypass of the containment, consequences shall be mitigated.*

   4.2   *The leaktightness of the containment shall not degrade significantly for a reasonable time after a severe accident.*

   4.3   *Pressure and temperature in the containment shall be managed in a severe accident.*

   4.4   *Combustible gases shall be managed in a severe accident.*

   4.5   *The containment shall be protected from overpressure in a severe accident.*

   4.6   *High pressure core melt scenarios shall be prevented.*

   4.7   *Containment degradation by molten fuel shall be prevented or mitigated as far as reasonably practicable.*

The WENRA description states that the types of events to be analyzed for design extension (if not already analyzed as a design-basis accident) would include:

- ATWS

- SBO

- total loss of feedwater

- loss-of-coolant accident, together with the complete loss of one emergency core cooling system; uncontrolled level drop during midloop operation (PWR) or during refueling

- total loss of the component cooling water system

- loss of core cooling in the residual heat removal mode

- loss of fuel pool cooling

- loss of ultimate heat sink function

- uncontrolled boron dilution (PWR)

- multiple steam generator tube ruptures (PWR)

- loss of required safety systems in the long term after a postulated initiating event

G.7   Discussion

Additional information related to the development of the current NRC structure was provided to the Commission in SECY-11-0156, "Feasibility of Including Risk Information in Categorizing Structures, Systems, and Components as Safety-related or Non-Safety-related" (NRC, 2011). A simplified representation of the various categorization schemes is shown in Figure G-2.

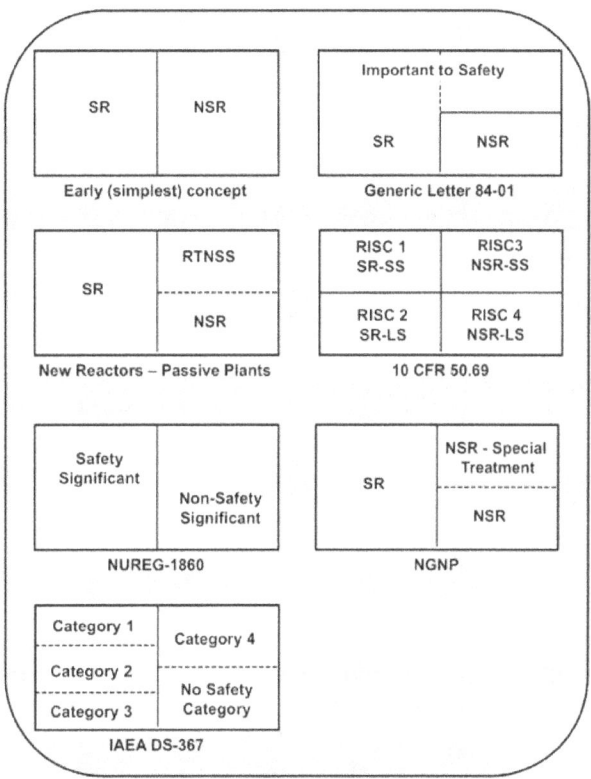

Figure G-2   Representations of Safety
Classification Concepts

The classification of SSCs is also closely related to the introduction of a design-enhancement category and the overall approach to risk-informed and performance-based defense-in-depth, which is discussed in Chapter 4 and Appendix H. The previous discussions in this appendix of various ways to classify SSCs according to their safety role and significance identify several key points that should be considered in developing such schemes for currently operating plants or future plants. These key points include the following:

- Clearly define the reason for the classification system. In general, the safety classification systems are used to differentiate special treatment requirements for SSCs in areas such as quality assurance, environmental qualification, and applicability of various industry codes and standards. Special treatment implies requirements imposed on SSCs that go beyond industry-established requirements for equipment classified as "commercial grade." More recent proposed evaluations of safety classification systems (e.g., NUREG-1860 and IAEA DS-367) include specific discussions on safety functions; safety categories; and the expectations for equipment in terms of capability, dependability, robustness, and reliability. It is important to think of these matters early since the safety classification of SSCs is merely a tool to achieve the desired outcomes. The task of changing the current system should not be underestimated since the existing infrastructure (e.g., NRC regulations, licensee procedures, industry codes and standards) has been developed to define special treatment requirements on the basis of the definition of safety-related SSCs and the premise that the designation "safety-related" helps ensure the desired outcomes.

- The definition of safety-related SSCs includes that equipment "relied upon to remain functional during and following design-basis events" to address the critical safety functions of (1) maintaining the reactor coolant pressure boundary, (2) shutting down the reactor, and (3) preventing or mitigating accidents that could result in significant offsite radiation exposures. These are very similar to the three fundamental safety functions defined by IAEA, which are to (1) control reactivity, (2) remove heat from the reactor core, and (3) confine radioactive material. Much of the current discussions revolve around how to address distinctions between "design-basis events" and "beyond-design-basis events." This sometimes translates into discussions of how to consider PRAs since "design-basis events" are often thought of as those addressed by traditional deterministic analyses and "beyond-design-basis events" are associated with PRAs. As discussed elsewhere in this report, the concept of "design-enhancement conditions" has been introduced to address some concerns regarding the transition between design and beyond-design-basis events. Various activities (e.g., NUREG-1860, NGNP, IAEA DS-367) have highlighted the relationships between licensing-basis event selection and SSC safety classification and the need to consider and resolve these topics as part of a more holistic regulatory approach.

- The ability to develop and implement alternatives is complicated by having plants in various stages of their life cycle (e.g., operation, construction, design). Operating plants have available an option to adopt the classification scheme defined in 10 CFR 50.69. However, the degree to which the traditional system has been ingrained into codes and standards, procurement processes, and other areas of plant design and operation is an obstacle to the implementation of 10 CFR 50.69. New plant designs undergoing review have either adopted the existing system or introduced the RTNSS concept to address passive safety features. The resolution of some implementation issues with RTNSS during the reviews of the AP1000 and ESBWR designs should make it easier for small modular light-water reactors to adopt the concept for this class of plants (see SECY-11-0024, "Use of Risk Insights to Enhance the Safety Focus of Small Modular Reactor Reviews" (NRC, 2011a)). Generation IV reactor designs generally involve significant differences from the currently operating fleet in terms of reactor coolant, fuel design, and safety features. Many of these designs have also been developed considering international safety standards and could therefore more easily adopt approaches discussed in NUREG-1860, the NGNP white papers, or IAEA DS-367.

- It may be an opportune time to evaluate possible changes to the safety classification approach in concert with the development of a design and licensing strategy for small modular light-water reactors and Generation IV reactor designs. Although NRC staff activities are underway for assessing a longer-term Risk-Informed approach (as described in SECY-11-0024), the various industry players are acting independently and considering slightly different licensing approaches within the general framework that has been used for other new reactor designs. The small modular light-water reactors appear to prefer the stability, predictability, and existing infrastructure of the current processes developed for new passive plant designs (i.e., safety-related and RTNSS classifications). It may be possible to use the flexibility within the existing definitions (e.g., design-basis events and design-basis accidents) to improve the consideration of risk insights and construct a more logical set of licensing-basis events and related equipment special treatment requirements within the current classification construct (or a variation supported by a limited number of exemptions). The preferred approach of the gas-cooled reactor community is fairly well established and is reflected in the NGNP white papers and ANS 53.1. The sodium-cooled fast reactor community is currently developing a standard (ANS 54.1) that will propose a licensing and safety classification scheme for its reactor designs. International efforts are likewise underway, and while they have some commonalities with U.S. proposals, the IAEA and WENRA approaches are likely to use designations and criteria that are different from the various U.S. proposals.

G.8   References

(ANS, 1983)   American National Standards Institute/American Nuclear Society, ANSI/ ANS 51.1 – 1983, "Nuclear Safety Criteria for the Design of Stationary Pressurized Water Reactors," ANS, LaGrange Park, IL.

(ANS, 2011)   American National Standards Institute/American Nuclear Society, ANSI/ ANS53.1 – 2011, "Nuclear Safety Design Process for Modular Helium-Cooled Reactor Plants," ANS, LaGrange Park, IL.

(IAEA, 1996)   International Atomic Energy Agency, "Defence in Depth in Nuclear Safety," INSAG-10, A Report by the International Nuclear Safety Advisory Group, 1996.

(IAEA, 2007)   International Atomic Energy Agency, "IAEA Safety Glossary, Terminology Used In Nuclear Safety and Radiation Protection," 2007 Edition.

(IAEA, 2010)   International Atomic Energy Agency, "Draft – Safety of Nuclear Power Plants: Design," Draft Safety Guide DS-414, September 2010.

(IAEA, 2011)   International Atomic Energy Agency, "Draft – Safety Classification of Structures, Systems, and Components in Nuclear Power Plants," Draft Safety Guide DS-367, April 2011.

(NEI, 2005)   Nuclear Energy Institute, "10 CFR 50.69 SSC Categorization Guideline," NEI 0004, July 2005, Agencywide Documents Access and Management System (ADAMS) Accession No. ML052910035.

(NEI, 2000)   Nuclear Energy Institute, "Guidance and Examples for Identifying 10 CFR 50.2 Design Bases," NEI 97-04, Revised Appendix B, November 2000, ADAMS Accession No. ML003771698.

(NGNP, 2010)   Idaho National Laboratory, "Next Generation Nuclear Plant Structures, Systems, and Components Safety Classification White Paper," INL/EXT-10-19509, September 2010.

(NRC, 1981)   U.S. Nuclear Regulatory Commission, "Standard Definitions for Commonly-used Safety Classification Terms," memorandum dated November 20, 1981, ADAMS Accession No. ML111230450.

(NRC, 1984)   U.S. Nuclear Regulatory Commission, "NRC Use of the Terms 'Important to Safety' and 'Safety Related,'" Generic Letter 84-01, January 5, 1984, ADAMS Accession No. ML031150515.

(NRC, 1985)   U.S. Nuclear Regulatory Commission, "Quality Assurance Guidance for ATWS Equipment That Is Not Safety Related," Generic Letter 85-06, April 16, 1985, ADAMS Accession No. ML031140390.

(NRC, 1985a)   U.S. Nuclear Regulatory Commission, "Policy Statement on Severe Reactor Accidents Regarding Future Designs and Existing Plants," August 1985, ADAMS Accession No. ML003711521.

(NRC, 1988)   U.S. Nuclear Regulatory Commission, "Loss of All Alternating Current Power," Final Rulemaking, published in the *Federal Register* on June 21, 1988 (53 FR 23215).

(NRC, 1988a)   U.S. Nuclear Regulatory Commission, "Station Blackout," Regulatory Guide 1.155, August 1988, ADAMS Accession No. ML003740034.

(NRC, 1995)   U.S. Nuclear Regulatory Commission, "Policy and Technical Issues Associated with the Regulatory Treatment of Non-Safety Systems (RTNSS) in Passive Plant Designs (SECY 94-084)," SECY-95-132, May 22, 1995, ADAMS Accession No. ML003708005.

(NRC, 1998)   U.S. Nuclear Regulatory Commission, "Options for Risk-Informed Revisions to 10 CFR Part 50 – "Domestic Licensing of Production and Utilization Facilities," SECY-98-300, December 23, 1998, ADAMS Accession No. ML992870048.

(NRC, 2000)   U.S. Nuclear Regulatory Commission, "Guidance and Examples for Identifying 10 CFR 50.2 Design Bases," Regulatory Guide 1.186, December 2000, ADAMS Accession No. ML003754825.

(NRC, 2006)   U.S. Nuclear Regulatory Commission, "Guidelines for Categorizing Structures, Systems, and Components in Nuclear Power Plants According to Their Safety Significance," Regulatory Guide 1.201, May 2006, ADAMS Accession No. ML061090627.

(NRC, 2007)   U.S. Nuclear Regulatory Commission, "Seismic Design Classification," Regulatory Guide 1.29, Revision 4, March 2007, ADAMS Accession No. ML070310052.

(NRC, 2007a)   U.S. Nuclear Regulatory Commission, "Quality Group Classifications and Standards for Water-, Steam- and Radioactive-Waste-Containing Components of Nuclear Power Plants," Regulatory Guide 1.26, Revision 4, March 2007, ADAMS Accession No. ML070290283.

(NRC, 2007b) U.S. Nuclear Regulatory Commission, "Feasibility Study for a Risk-Informed and Performance-Based Regulatory Structure for Future Plant Licensing," NUREG-1860, December 2007, ADAMS Accession No. ML080440170.

(NRC, 2009) U.S. Nuclear Regulatory Commission, "Consideration of Aircraft Impacts for New Nuclear Power Reactors," Final Rule, published in the *Federal Register* on June 12, 2009 (74 FR 28111).

(NRC, 2011) U.S. Nuclear Regulatory Commission, "Feasibility of Including Risk Information in Categorizing Structures, Systems, and Components as Safety-Related or Non-Safety-Related," SECY-11-0156, November 2, 2011, ADAMS Accession No. ML112690353.

(NRC, 2011a) U.S. Nuclear Regulatory Commission, "Use of Risk Insights to Enhance the Safety Focus of Small Modular Reactor Reviews," SECY-11-0024, February 18, 2011, ADAMS Accession No. ML110110691.

(WENRA, 2008) Western European Nuclear Regulators' Association, "WENRA Reactor Safety Reference Levels," Reactor Harmonization Working Group, January 2008.

APPENDIX H

NUCLEAR POWER REACTORS – CHAPTER 4 ALTERNATIVES

H.1   Introduction

The main report identifies three options for implementing a risk management framework and adopting a risk-informed and performance-based defense-in-depth approach for the U.S. Nuclear Regulatory Commission's (NRC's) various regulatory programs. Option B involves adopting the proposed Risk Management Regulatory Framework through Commission policy statements, rule changes, and revisions to guidance documents. There are many alternatives that could be pursued under Option B, ranging from selected adjustments of guidance documents to the development of new requirements and regulatory programs (i.e., graduated changes between guidance changes and developing a significantly different regulatory framework). Several possible alternatives for operating and new power reactors are provided in this appendix as a means to support discussions and show the range of choices available under Option B.

A question that inevitably arises when considering a regulatory framework and categorizing events or equipment is whether requirements are needed for adequate protection of the health and safety of the public. The NRC establishes regulatory requirements for protecting public health and safety and common defense and security using a two-tier structure. This two-tier structure is consistent with the statutory requirements in the Atomic Energy Act of 1954, as amended. (See Union of Concerned Scientists v. NRC, 824 F.2d 108, 120 (D.C. Cir. 1987)). The top tier consists of those requirements needed to ensure adequate protection of public health and safety and to be in accord with the common defense and security. The adequate protection standard is also described in terms of ensuring that licensed activities pose no undue risk to public health and safety or to common defense and security. In the above cited case, the Court of Appeals for the D.C. Circuit recognized that adequate protection (no undue risk) does not equate to "zero risk." The Court stated:

> Similarly, under the adequate-protection standard of section 182(a), the NRC need ensure only an acceptable or adequate level of protection to public health and safety; the NRC need not demand that nuclear power plants present no risk of harm.

Measures taken to prevent, contain, and mitigate events or concerns in this tier historically correspond to the traditional design-basis accidents described in Appendix F and several programmatic requirements for dealing with beyond-design-basis events (e.g., emergency planning and loss of large areas because of fires or explosions). NRC requirements to address concerns of adequate protection are developed and imposed without consideration of cost.

The second tier of the NRC's safety structure was described by the Court as follows:

> If it so desires, however, the Commission may impose safety measures on licensees or applicants over and above those required by section 182(a)'s adequate-protection standard. As we have noted, section 161 of the Act empowers the Commission to issue rules, regulations, or orders to "protect health or to minimize danger to life or property." 42 U.S.C. Sec. 2201(b), (i). This section cannot be read simply to permit the Commission to provide adequate

protection; another section of the Act requires the Commission to do that much. We therefore must view section 161 as a grant of authority to the Commission to provide a measure of safety above and beyond what is "adequate." The exercise of this authority is entirely discretionary. If the Commission wishes to do so, it may order power plants already satisfying the standard of adequate protection to take additional safety precautions. When the Commission determines whether and to what extent to exercise this power, it may consider economic costs or any other factor. The Commission, after all, need not exercise the authority granted by section 161 at all; given this fact, the Commission certainly may use cost-benefit analysis to decide whether exercising the authority conferred by section 161 makes economic or policy sense.

In the development of the alternatives presented within this appendix, the Risk Management Task Force (RMTF) considered how some requirements would be associated with the first tier mission of the NRC to ensure an adequate protection of public health and safety and common defense and security, and how other requirements would address the second tier by going beyond adequate protection in an attempt to minimize danger to life or property. As described by the Court of Appeals and within longstanding NRC regulations and guidance documents, the requirements established to minimize risk (beyond measures needed for adequate protection) will not attempt to eliminate all risk, but will instead pursue reasonable reductions. An evaluation of the costs and benefits of proposals falling within the second tier will be used as part of the determination of what is a reasonable requirement to minimize risks to public health and safety and common defense and security.

The RMTF has incorporated the two-tier safety structure discussed above, along with the risk management goal described in Chapter 2, into the alternatives discussed in this appendix. The risk management goal is to provide risk-informed and performance-based defense-in-depth protections to:

1. Ensure appropriate barriers, controls, and personnel to prevent, contain, and mitigate exposure to radioactive materials, according to the hazard present, the relevant scenarios, and the associated uncertainties.

2. Ensure that the risks resulting from the failure of some or all of the established barriers and controls, including human errors, are maintained acceptably low.

Taken together, the above criteria address both tiers of the safety structure. However, it is not the case that the first criterion aligns with the first tier (adequate protection) and the second criterion aligns with the second tier (additional protections). Instead, a specific design feature or operating control can be seen as addressing one or more of the factors within the risk management goal, and the importance of the protection provided will determine whether it serves to ensure adequate protection or provides additional protections to further minimize dangers to life or property.

As discussed in Appendix F, "Nuclear Power Reactors – Licensing-basis Events," the selection and analysis of relevant scenarios play an important role in the design and operation of nuclear power reactors. For the purpose of this appendix, the interrelated processes used to define the proposed Risk Management Regulatory Framework are discussed in terms of the following phases of developing and maintaining design features and operating controls:

- selection of events and conditions to be considered in establishing plant design features and operating controls

- selection and implementation of plant design features and operating controls to meet the defined acceptance criteria for selected events

- preservation of the risk-informed, performance-based defense-in-depth outcomes provided by the selected plant design features and operating controls

Each of the above activities involves the regulatory decisionmaking process discussed in Chapter 3. Event selection is an engineering exercise and includes defining the types of internal transients and external challenges to be analyzed in support of establishing design features and operating controls. An important part of event selection is defining categories, such as design-basis and design-enhancement; related standards and conventions, such as applying the single-failure criterion and determining equipment safety classifications; and defining expectations for the appropriate combination of mechanistic and risk assessment analyses techniques. The output of the event selection phase includes relevant scenarios and related conditions, forces, and other parameters used in determining the appropriate design features and operating controls. The selected design features, operating controls, or other actions provide appropriate risk-informed and performance-based defense-in-depth protections against the scenarios and associated uncertainties. The implemented actions can, in turn, be considered the starting point for future decisions on maintaining or changing designs or operating practices. This third part of the interrelated activities is intended to ensure that the outcome of the previous steps—providing risk-informed defense-in-depth protections—is maintained even though changes to facilities and procedures will occur over the lifetime of each nuclear power reactor.

The event selection, action selection, and outcome preservation terms are used to describe roles and responsibilities of the NRC and licensees and other attributes of the alternatives described in this appendix. The alternatives will also be presented consistent with the event categories described in Appendix F and elsewhere in this report, namely:

- Design-basis Events

  o normal operation

  o anticipated operational occurrences

  o design-basis accidents

  o design-basis external hazards

- Beyond-design-basis Events

  o design-enhancement events

    – internal events

    – external hazards

- o residual risk scenarios

  - – internal events

  - – external hazards

Design-basis events have traditionally been associated with mechanistic analyses, reliance on and protection of safety-related equipment, and use in establishing technical specifications and other licensing-related operational controls. Beyond-design-basis events have more often included the use of best-estimate-type analyses, probabilistic risk assessments (PRAs), and use in establishing additional plant protections to further minimize risks to the public health and safety and common defense and security (e.g., additional protections for station blackout (SBO) conditions and aircraft impacts).

A specific example is the selection of a weather event, such as a hurricane, with related wind speeds and surge levels. Licensees address the selected event by actions, such as designing and building structures with appropriate strengths and installing watertight doors to protect safety-related equipment installed at elevations below the defined surge level. Preserving the outcome is achieved through programs to evaluate plant changes to ensure that the desired protections against a hurricane are maintained over the life of the facility. Event categories can be used to distinguish between "design-basis" forces and surge levels and storm conditions considered less likely or otherwise different from the design-basis scenario. The evaluations of design features and operating controls for the design-basis event currently credit only safety-related systems, structures, and components. The proposed design-enhancement category would capture other storm conditions and could allow licensees to credit other available equipment and procedures.

Two key concepts related to event selection, categorization, and subsequent design features and operating limits are: 1) the threshold to define when an event needs to be considered within a category, and 2) the acceptance criteria to define when a design feature or operating limit provides the desired protection from the defined scenario(s). Differences in how these concepts are incorporated into the event selection, action selection, and outcome preservation phases define the three alternatives described in this appendix for improving the NRC's development of the proposed category of design-enhancement events.

As discussed in Chapter 4 and elsewhere in this report, the RMTF recommends that the NRC recognize "design-enhancement events" as a specific category of beyond-design-basis events. The purpose of the design-enhancement category is to address gaps that exist between the regulatory controls that are appropriate to address the risk management goal (e.g., risk-informed, performance-based defense-in-depth) and current controls involving a combination of design-basis events and ad hoc requirements added in reaction to specific events or other concerns. The goal would be to define a consistent approach for such events in terms of analysis techniques, safety classification, change control, reporting, and other regulatory requirements that have been defined previously on a case-specific basis. Any of these alternatives could also be used to clarify the role of programs, such as severe accident management guidelines (SAMGs) within the revised or added regulations for beyond-design-basis events. The Task Force envisions that the combination of design-basis events, design-enhancement events, and various programs such as emergency preparedness collectively define the risk-informed and performance-based defense-in-depth protections that are the centerpiece of the proposed Risk Management Regulatory Framework.

The RMTF's proposed creation of a design-enhancement category is generally consistent with recommendations by the Fukushima Near-Term Task Force (NRC, 2011) and the Western European Nuclear Regulators' Association (WENRA, 2008). The WENRA design extension category was discussed in Appendix G, "Nuclear Power Reactors – Safety Classification," and is described as follows:

> The design extension analysis shall examine the performance of the plant in specified accidents beyond the design basis, including selected severe accidents, to minimize, as far as reasonably practicable, releases harmful to the public and the environment in cases of events with very low probability of occurrence.
>
> Beyond-design-basis events shall be selected *(based on a combination of deterministic and probabilistic assessments, as well as engineering judgment)* and considered in the safety analysis to determine those sequences for which reasonable, practicable, preventive, or mitigative measures can be identified and implemented.
>
> Realistic assumptions and modified acceptance criteria may be used for the analysis of the beyond-design-basis events. (WENRA)

The concept of design-extension events or conditions has also been incorporated into draft standards prepared by the International Atomic Energy Agency (IAEA) related to the design of nuclear power plants (IAEA, 2010) and the safety classification of structures, systems, and components (IAEA, 2011). For example, the draft standard related to the safe design of nuclear power plants states:

> Requirement 21: Design extension conditions (IAEA)
>
> A set of design extension conditions shall be derived from engineering, deterministic and probabilistic considerations for the purpose of establishing further boundary conditions for the plant to withstand without acceptable limits for radiological consequences being exceeded. These conditions shall be used to identify the additional accident scenarios to be addressed in the design and to plan practicable provisions for the prevention and mitigation of such accidents.

Within the IAEA discussions of design extension conditions are related NRC requirements and activities to address beyond-design-basis events (e.g., severe accident design features for new reactors and severe accident management guidelines for operating reactors).

H.2   Design-Enhancement Events – Possible Alternatives

The following figure represents a general framework for the proposed risk-informed, performance-based defense-in-depth approach:

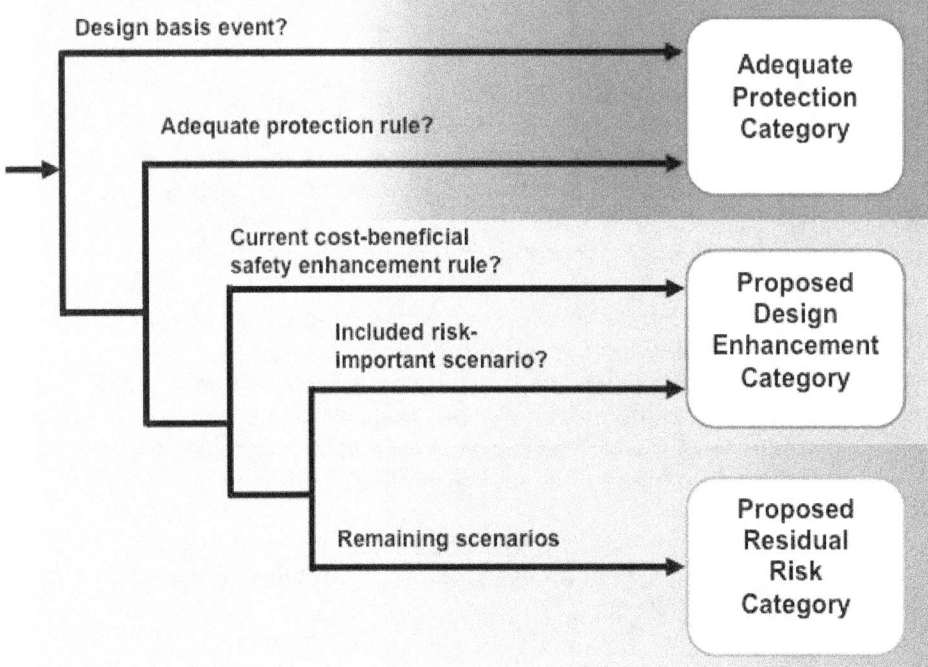

Figure H-1 Regulatory Framework for Nuclear Power Reactors

Integrating the above discussions, the RMTF developed several alternatives for moving forward on the proposed Risk Management Regulatory Framework for nuclear power reactors. The alternatives considered variations in the following three questions:

- Who identifies and decides on the appropriate actions to address design-enhancement events?

- What thresholds are defined for inclusion in the design-enhancement category?

- What acceptance criteria are defined for actions taken to address events and conditions in the design-enhancement category?

Possible variations in handling the above choices result in the three alternatives depicted below:

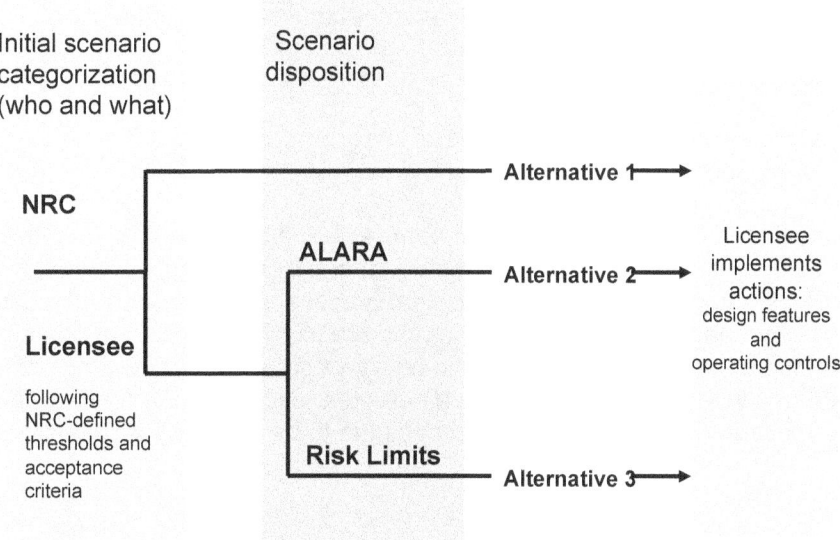

Figure H-2 Possible Alternatives for Nuclear Power Reactors

A discussion of the three selected approaches to design-enhancement is provided below. An additional discussion (Section H.3) is provided following the description of the three alternatives regarding possible changes to the design-basis events that may be facilitated by the introduction of the design-enhancement category for operating and new power reactors.

H.2.1   Alternative 1: NRC Identifies Design-enhancement Events

In the first alternative, the identification of a new category of design-enhancement events takes advantage of existing studies and lessons learned from Fukushima and other historical events. The NRC would identify specific events for inclusion in the design-enhancement category. The NRC would likewise define the acceptance criteria and other requirements associated with each of the specified events within this new category. This basic process (without defining common attributes such as reporting and change control) was used when the NRC imposed requirements for events or conditions, such as SBO and aircraft impact assessments, for new reactors. A possible approach for the event selection phase of the overall process to address design-enhancement events is shown in Figure H-2 as the following steps:

1.   The NRC identifies initiating event(s)

    •   Considers internal transients, external hazards, and
         specific events and conditions

2.   The NRC performs technical analyses

    •   Considers safety benefits, operating experience, and
         regulatory analysis

3.   The NRC defines design-enhancement events

    •   Defines specific events or conditions (e.g., SBO)

4.   The NRC defines acceptance criteria

    •   Considers examples such as fuel integrity, coping time, and codes and standards

The above steps would provide a methodical approach and consistent logic to the identification of design-enhancement events. The identification of and defining acceptance criteria for events on an ad hoc basis is considered to be the current practice and is described in Chapter 4 as Option A. While informed by risk assessments and the NRC's historical practice of treating design-enhancement events within the second tier of its safety structure (i.e., added protection), the determination of whether protection from a particular event is important for adequate protection or as a safety enhancement would continue to be made on a case-by-case basis.

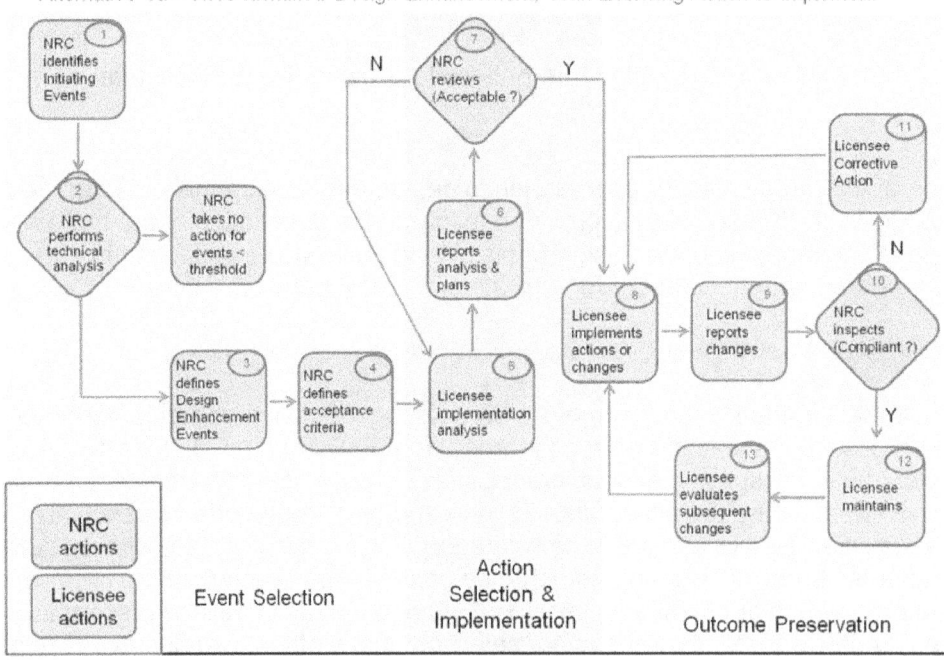

Figure H-3 Alternative 1a (NRC identified, NRC approval)

Note that the acceptance criteria and evaluation techniques could be prescriptive such as they are in the SBO rule or performance-based such as in the aircraft impact rule for new reactors. In accordance with the proposed Risk Management Regulatory Framework promoted in the main body of this report, the NRC's objective is to allow performance-based approaches wherever practical. In addition, the design-enhancement events should be part of the overall combination of events and related actions that ensure that risk-informed and performance-based defense-in-depth protections are provided. Given the significant role of the NRC in defining events within this example, it would not be necessary to require designers or licensees to perform additional studies or identify design or site-specific hazards for possible inclusion as design-enhancement events. The possible need for plant-specific PRAs or other analytical tools would depend on the requirements defined for the specific design-enhancement events (e.g., the analyses to support licensee implementation (step 5) and change control (step 13) in Figure H-3). The lack of site-specific risk assessments and evaluations of appropriate actions is a significant shortcoming of this alternative.

A possible approach to the action selection and implementation phase consists of the licensee performing a facility-specific implementation analysis for the NRC defined event and acceptance criteria, and reporting its results and planned actions to the NRC for review and approval (steps 6 and 7 in Figure H-3). This approach has been used for past and ongoing activities, such as implementing the SBO rule and the adoption of National Fire Protection Association 805, "Performance-based Standard for Fire Protection for Light-Water Reactor Electric Generating Plants" (NRC, 2009). This approach may be required in those cases involving amendments to licenses. However, the NRC review and approval of licensees' analyses and implementation plans can add significantly to the resources and schedules needed to resolve the addition of or changes to events incorporated into the licensing-basis for nuclear power reactors. An approach taken for aircraft impact assessments for new reactors is shown in Figure H-4. This approach emphasizes NRC reviews and inspections during the implementation phase and is usually associated with a more performance-based approach to issue resolution.

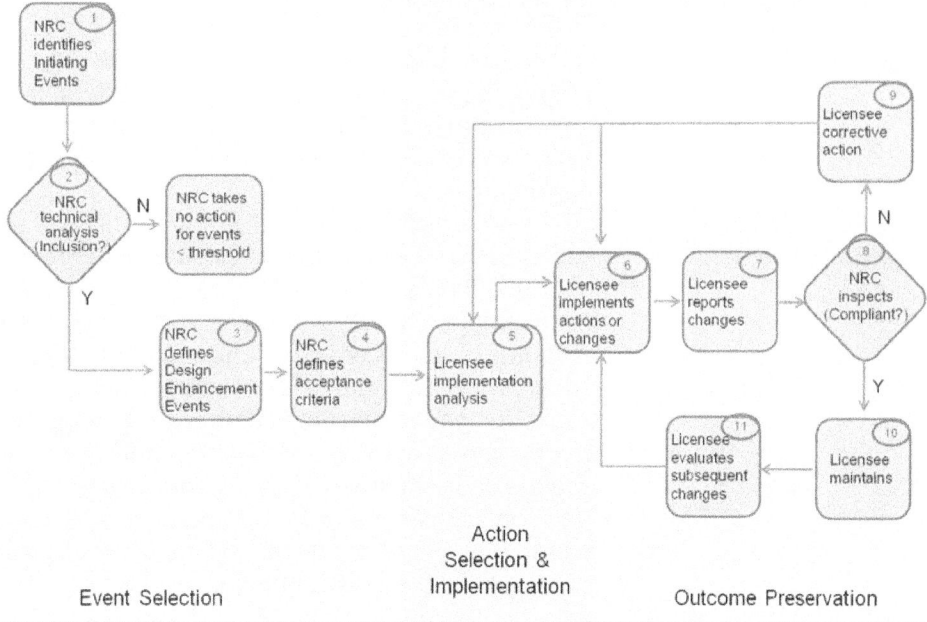

Figure H-4 Alternative 1b (NRC identified, performance-based)

An historical example of event and action selections relates to the availability of electrical power for plant equipment. The design-basis events include several events, such as the loss of offsite power scenario as an anticipated operational occurrence, and assumptions regarding offsite power for the loss of coolant design-basis accident. These events, in turn, led to actions such as the design and installation of safety-related diesel generators, sequencing of loads to be added to the electrical busses, and technical specifications to maintain diesel generator availabilities and capabilities. To address insights from PRAs and operational experience, the NRC imposed additional requirements for plants to withstand for a specified duration and recover from a station blackout event, which assumes the loss of both offsite and onsite emergency power sources. In the aftermath of the attacks of September 11, 2001, the NRC and stakeholders developed additional measures to provide electrical power to selected equipment in the event of a loss of a large area of a plant because of explosions or fires. The Fukushima accident is resulting in additional studies and may possibly identify additional risks, events, or conditions related to responding to a prolonged loss of electrical power. All of these scenarios and related actions form part of the construction of risk-informed and performance-based defense-in-depth for each nuclear power reactor licensed by the NRC.

The third phase shown in the above figures involves licensee and NRC actions to preserve the risk-informed defense-in-depth protections for the relevant scenarios and associated uncertainties. The needed elements to maintain defense-in-depth protections include reporting requirements, inspection activities, configuration control, change control, and corrective action programs. This phase has sometimes been referred to as maintaining the design basis for facilities. The movement to a risk management strategy provides an opportunity to revise the emphasis from "maintaining design basis," which can be perceived as a general resistance to change, to emphasizing performance-based outcomes in terms of preserving risk-informed and performance-based defense-in-depth through ensuring appropriate barriers and controls.

Several possible changes to regulations to implement Alternative 1 are described below:

50.2	Definitions	Add definition for design-enhancement events, risk management, and risk-informed and performance-based defense-in-depth. Add or clarify other terms to clearly distinguish event categories and related regulatory treatments.
50.x	Design-Enhancement Events	Add specific regulation(s) or revise existing regulations to introduce the design-enhancement event category. In this example, the rule would likely define common attributes, such as change control and reporting (e.g., placement within final safety analysis report (FSAR)) and would designate specific events for the category. The appropriate treatment of equipment and operating controls would also be defined. Existing requirements, such as SBO, anticipated transient without scram (ATWS), and aircraft impact assessment (AIA) could be incorporated or referenced in the new regulation.

This alternative might be the most straightforward and perhaps quickest to implement of the three described. It would not, however, easily address site-specific hazards and the "generic requirements" could require actions at some plants to address very unlikely scenarios for their specific locations. The NRC could periodically assess new information and pursue regulatory changes through existing programs, such as the operating experience program, generic safety issue process, generic communication program, and rulemaking.

The RMTF has presented Alternative 1 as a limited improvement over the NRC's existing approach. Historically, this approach has been able to address emerging regulatory issues, but it has had problems in areas of timeliness and consistency (e.g., the patchwork mentioned in the Fukushima Near-Term Task Force (NTTF) report) (NRC, 2011)). Although not the favored approach of the RMTF, the system described as Alternative 1 could be used as a transition to the more risk-informed, performance-based alternatives. As an example, the NRC could identify specific events or conditions identified following the Fukushima accident, and these could be pursued to ensure a timely response to concerns such as possible flooding scenarios. In the longer term, plant-specific risk assessments would be performed and additional actions would be taken as called for in approaches such as those described as Alternative 2 or 3.

H.2.2 Alternative 2: NRC Identifies Thresholds for Event Sequences, Acceptance Criteria Are Based on ALARA Principles

As discussed in other sections of this report, a key to the implementation of risk management systems is the definition of thresholds and acceptance criteria to support the deliberative process. Under this alternative, the NRC would define the threshold for events falling within the design-enhancement category using the risk-informed and performance-based defense-in-depth principle to address possible shortfalls (or excesses) in design-basis events and ensure that the risks resulting from the failure of established barriers and controls are maintained acceptably low. The threshold would, as much as possible, build upon existing practices, such as the requirements and guidance for regulatory analyses, backfits, severe accident mitigation alternatives (SAMAs), and risk-informed licensing (e.g., Regulatory Guide 1.174, "An Approach for Using Probabilistic Risk Assessment in Risk-informed Decisions on Plant-specific Changes to the Licensing-basis," (NRC, 2011a)).

There are many considerations in the identification of appropriate thresholds, including the analysis techniques (e.g., scope and level of PRAs), the handling of uncertainties, and the definition of relevant scenarios. One element of the thresholds could, for example, be defined in terms of relevant scenarios (defined by a single or group of PRA sequences) with an estimated core damage frequency (CDF) greater than 10^{-5} per year or a large early release frequency (LERF) greater than 10^{-6} per year, which is generally consistent with the NRC's guidelines for performing regulatory analyses (NRC, 2004)[1]. At the point where Level 3 PRAs are available, the NRC's Quantitative Health Objectives (QHO) or other societal measures could be directly considered as part of the event categorization criteria. Other criteria or relevant scenarios could be included, as needed, to complement the frequency-related parameters (e.g., security events, aircraft impact for new reactors). It would also be necessary to define approaches for addressing initiating events (e.g., seismic) for which either the conditional failure probability or the uncertainties increase dramatically as frequencies are reduced to levels approaching the thresholds for inclusion or the acceptance critieria defined for the design-enhancement category (e.g., 10^{-5} per year) (Johnson and Apostolakis, 2012).

Alternative 2 would require licensees to perform periodic reviews and analyses to identify relevant scenarios and determine appropriate actions to address design-enhancement events. A requirement similar to 10 CFR 50.71(h) to have and upgrade PRAs would be necessary

---

1   The values used in this report (e.g., criteria such as CDF greater than 10^{-5} per year) are used to describe the alternatives and are not specific RMTF recommendations. Although the values used are generally consistent with current NRC guidance documents, the development of a more detailed proposal would need to carefully consider defining the thresholds and acceptance criteria and ensure the criteria support the overall risk management goal.

for operating reactors, and a new or revised regulation would require the periodic analysis to identify relevant scenarios, considering plant changes, operating experience, and new information. As such, this alternative changes the focus from simply clarifying and expanding the design-enhancement category and would require licensees to adopt a risk management approach for plant design and operation (i.e., the identification and resolution of events falling within the design-enhancement category).

The acceptance criteria for the periodic analysis performed by licensees for design-enhancement events for this alternative would be defined primarily in terms of the "as low as is reasonably achievable" or ALARA principle, similar to that used in the radiation protection arena. The process would employ the decisionmaking process described in Chapter 2. It could use much of the existing guidance (updated, as needed) mentioned previously for regulatory analyses, backfits, SAMAs, and risk-informed licensing actions. For example, existing guidance includes several criteria, one of which is a factor of dollars per person-rem (roentgen equivalent man) avoided through the installation of additional design features to address severe accident scenarios. The development of this alternative would include defining, an appropriate, periodicity for the analyses to identify and address the new design-enhancement events.

Adoption of this alternative would require an update of regulatory analysis guidelines to ensure an effective cost-benefit analysis is performed by licensees when considering ways to address design-enhancement events. As with the discussion on thresholds, it may be appropriate to include considerations for different types of design-enhancement events. For example, the AIAs required for new reactors would be an example of design enhancement, but they have acceptance criteria defined in terms of maintaining barrier functions and not specifically frequency of core damage or radiological release. Considerations within the ALARA criteria also depend on assessment techniques with current Level 1 PRA-based criteria using surrogate measures, such as CDF, and future Level 3 PRAs expected to support criteria based on estimated health effects or other societal risk measures. Other factors might be considered to supplement those identified for various scenarios (e.g., CDF or health objectives) by including measures based on the importance of design features or operating controls in preventing or mitigating multiple scenarios (e.g., PRA importance measures) or in addressing core safety functions (i.e., reactivity control, heat removal, confinement of radioactive materials). These types of consideration are integral to current programs addressing special treatment requirements for different risk-informed equipment categories defined in 10 CFR 50.69, "Risk-informed Categorization and Treatment of Structures, Systems and Components for Nuclear Power Reactors," and the regulatory treatment of non-safety systems (RTNSS) for new reactors (NRC, 1995).

The current requirements in 10 CFR 50.71(h) require new reactor licensees to upgrade their PRA every 4 years, with the upgraded PRA covering initiating events and modes of operation contained in NRC-endorsed consensus standards on PRA in effect 1 year before each required upgrade. The RMTF considers the 4-year interval to be a reasonable starting point for discussions about the proposed risk-management rule. In addition, the current requirements include provisions to improve risk assessments using the NRC endorsement of consensus standards as the vehicle to upgrade the PRA to cover additional modes and initiating events. A similar approach could be taken for the PRA requirements within this alternative and could support the ultimate movement to Level 3 PRAs, which could better support the site-specific ALARA evaluations for design-enhancement conditions.

This alternative is illustrated in Figure H-5. It is readily apparent that this alternative changes the emphasis from the NRC identifying specific scenarios to the NRC defining criteria for design and site-specific risk management to be performed by licensees or applicants.

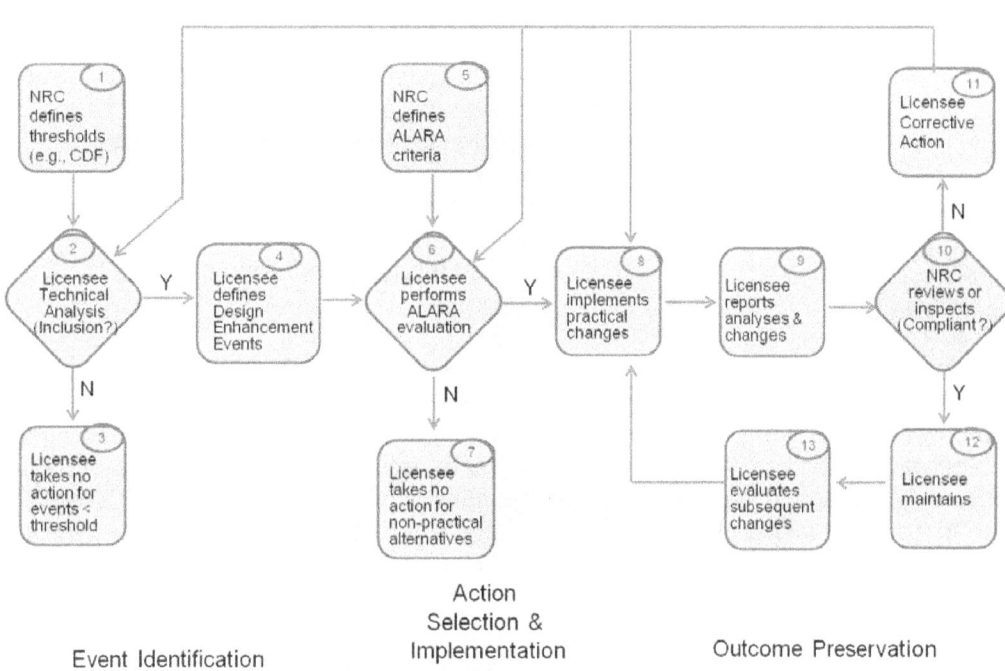

Figure H-5 Alternative 2, Risk Management (ALARA)

Several possible changes to regulations to implement Alternative 2 are described below:

50.2   Definitions

Add definition for design-enhancement events, risk management, and risk-informed and performance-based defense-in-depth. Add or clarify other terms to clearly distinguish event categories and related regulatory treatments.

50.x   Design-enhancement Events

Add specific regulation or revise existing regulations to introduce the design-enhancement event category and define the thresholds and acceptance criteria for the scenarios. The rule would likely define common attributes, such as change control, documentation, and reporting. It would also define the appropriate treatment of equipment and operating controls. Existing requirements, such as SBO, ATWS, and AIA, could be superseded by, incorporated (with either existing or revised requirements), or referenced in the new regulation.

50.y Risk Management	A requirement would be included for licensees to periodically assess and address potential events meeting thresholds derived from the risk-informed and performance-based defensein depth definition. The technical analyses would involve risk assessments (e.g., PRAs) and other techniques, as necessary, to address relevant scenarios. This approach could address matters such as GSI-199, "Implications of Updated Probabilistic Seismic Hazard Estimates in Central and Eastern United States on Existing Plants," and the periodic assessments recommended in the Fukushima NTTF report.
50.71(h) Probabilistic risk assessment (PRA)	This requirement exists for new reactors and would be added for operating reactors. This rule could be revised to address the risk management requirements mentioned above.

In constructing the requirements under this alternative, it would be made clear that specific, NRC regulations (existing or future) define the necessary requirements for adequate protection of the public health and safety. The design-enhancement and risk management regulations discussed here would be confined to the second tier of requirements that provide protection beyond that level. It would be possible under this approach to maintain the NRC's longstanding preference to not have a specific definition of adequate protection in terms of event frequencies and consequences. The thresholds (e.g., 10 CFR Part 20, "Standards for Protection Against Radiation," requirements for high frequency–low consequence events and CDF greater than 10^{-5} per year for low frequency–high consequence events) are a logical extension of current practice. Licensees would be required to assess scenarios within the design-enhancement category to reduce risk to ALARA levels. The possible design features or other measures would be evaluated against their efficacy in reducing the identified risk in a cost-effective manner. The whole approach would be based on the decisionmaking process described in Chapter 2. The determination of "reasonable" or cost-effective would use existing or future guidance used by the NRC in its own determinations (e.g., analyses for SAMAs, rulemakings, or backfits).[2] This should help provide consistency between evaluations performed by licensees and the NRC staff.

The proposed additional risk management requirement would involve licensees periodically (e.g., every 4 years or upon new information becoming available) assessing postulated initiating events, evaluating riskreduction measures, and, where cost-effective, implementing those measures to address relevant scenarios exceeding the defined threshold. A summary of the assessment and related actions would be provided to the NRC, possibly in an existing reporting requirement, such as updating of the FSARs. It should be noted that the periodic assessments and consideration of cost-effective measures to reduce risk would not replace routine (i.e., day-to-day) risk management functions, which would continue to be supported by technical specifications, maintenance rule, and other programs related to configuration management and plant operations.

---

2   A question to address during the development of the design-enhancement category and ALARA approach would be the threshold for including relevant scenarios for consideration. For example, 10 CFR 50.109, "Backfitting," requires that the NRC impose a backfit only when it determines that the new requirements would result in a "substantial increase in the overall protection of public health and safety." NRC guidance documents (NRC, 2004) equate a substantial increase to a reduction in CDF of greater than 10^{-5} per year.

Although arguably consistent with existing requirements and practices related to thresholds and acceptance criteria, Alternative 2 would likely present a communications challenge given that it requires specific discussion of the distinction between adequate protection and added protection. Whereas some would see this alternative as an imposition on licensees to do additional assessments and take additional actions, others would likely see it as allowing licensees to forgo risk reductions based on financial considerations. A recent case that highlights the agency's longstanding difficulty in explaining the subtleties of these concepts is the discussions surrounding some recommendations from the Fukushima NTTF.

Regarding new plant designs and licensing, the adoption of the ALARA principle would reinforce the stated objective in the Advanced Reactor Policy Statement (NRC, 2008) for new designs to provide enhanced margins of safety and use simplified, inherent, passive, or other innovative means to accomplish their safety functions. SECY-10-0121, "Modifying the Risk-informed Regulatory Guidance for New Reactors" (NRC, 2010), discusses several issues related to a possible discrepancy between the Advanced Reactor Policy Statement goals for enhanced safety for new plants and the actual regulation and oversight of new reactors based on the same rules and guidance applied to operating reactors. The proposed ALARA approach for the design-enhancement conditions could be a practical means to require new plants to define specific measures to reduce risks since such measures could be developed in a more cost-effective way before completing plant designs or construction. This approach would be similar to that taken for AIAs and recognizes that integrating protections into plant designs can be cost effective even if major plant changes to an operating reactor would be impractical. This alternative has the advantage that the desired outcome is achieved without having different regulatory requirements for new and existing reactors, and yet the risk reductions achieved by newer plant designs would be maintained by the proposed risk management rule. Some new reactor designs have even included specific measures to address the RTNSS as a way to reduce risks identified in PRAs while controlling costs by crediting existing systems and programs.

Additional improvements might be pursued along with the specific changes needed to implement this alternative to ensure consistency between various regulations for event analyses, change control, and reporting. For example, various external events (e.g., weather, industrial facilities, seismic, and flood) that include event frequencies as a key parameter use different thresholds and acceptance criteria for those frequencies. Other regulations governing activities such as changes, tests, and experiments; maintenance; and fire protection also use risk-insights to govern plant operations or interactions with the NRC. While not necessarily essential for this alternative, the development of the design-enhancement and risk management requirements would be an opportunity to review and revise various requirements to ensure a consistent treatment of the relevant scenarios. The development of the risk management requirements and associated requirements for the treatment of equipment and operating controls could also improve the alignment between the licensing and regulations for power reactors and the reactor oversight process (ROP), which already relies heavily on risk assessment techniques. For example, plants currently install safety enhancements to improve plants' risk profiles and thereby avoid adverse performance indicators or reduce the significance of possible violations when evaluated using the ROP significance determination process. The NRC encourages such enhancements, but it is not always clear what, if any, regulatory controls are applicable to them. This alternative would likely capture these enhancements within the proposed design-enhancement and risk-management regulations. A similar discrepancy has been identified between the NRC's environmental regulations, which require an assessment of severe accident mitigation alternatives, and the "safety regulations," which do not require the implementation of cost-justified enhancements. In addition to the opportunities to improve consistency, there would likely be some necessary changes to peripheral regulations to maintain an effective regulatory system. For example, the requirements within the proposed risk management rule would need to be developed to work with the change control provisions of other NRC regulations (e.g., 10 CFR 50.59, "Changes, Tests, and Experiments") to avoid conflicting findings regarding a proposed plant modification.

H.2.3 Alternative 3: NRC Identifies Thresholds for Initiating Event Frequencies, Acceptance Criteria Are Fixed

Alternative 3 is similar to Alternative 2 except it replaces the ALARA principle for the design-enhancement analyses with more definitive acceptance criteria in terms of event frequencies and consequences. The process shown in Figure H-3 would therefore represent this alternative, except the thresholds and acceptance criteria would be revised. Although this alternative, when compared to Recommendation 2, does not as clearly address Recommendation PR-R-2 by considering costs in establishing design-enhancement events, a proposal similar to Alternative 3 has been made in various forums and is presented here for the sake of discussion. Note that in this alternative, the design-basis events continue to be addressed by existing regulations and traditional approaches and the risk management requirements focus on the design-enhancement category.

As with Alternative 2, there are numerous ways to define the thresholds and acceptance criteria, depending on the chosen risk assessment techniques and other considerations. In the example included here to describe this alternative, the threshold for design-enhancement events is based on postulated initiating event (PIE) frequency, and the acceptance criteria are defined in terms of parameters, such as the QHO, or, as in Alternative 2, surrogate measures, such as CDF (and possibly adding LERF as another measure). Additional considerations in defining the thresholds could bring in measures based on the importance of design features or operating controls in preventing or mitigating multiple scenarios (e.g., PRA importance measures) or in addressing core safety functions (i.e., reactivity control, heat removal, confinement of radioactive materials). For the purpose of explaining this alternative, the example uses a PIE frequency of 10^{-5} per year as the threshold and a CDF of 10^{-6} per year as the acceptance criteria. In other words, licensees in this example would be required to:

1. Perform the periodic reassessment discussed in Alternative 2.

2. Identify scenarios (e.g., single or multiple PRA sequences) in which the PIE frequency exceeds 10^{-5} per year and the related CDF exceeds 10^{-6} per year (or equivalent if using LERF or QHO).

3. Identify and implement measures to reduce the frequency of the initiating event or resultant core damage for the relevant scenarios from Step 2 such that threshold or acceptance criteria are not exceeded.

To differentiate between those requirements needed for adequate protection (Tier 1) and those pursued for additional safety (Tier 2), this alternative would emphasize differences in analysis techniques and special treatment requirements for credited equipment and procedures. As with Alternative 2, adequate protection would be addressed primarily through existing specific rules, traditional design-basis accidents, safety-related equipment, and well established procedures (e.g., emergency operating procedures).[3] Design-enhancement events would be addressed similarly to Alternative 2 and would involve best-estimate risk assessments, reliable but not necessarily safety-related equipment, and response guidelines for severe accidents and extreme damage conditions. While not specifically including a cost-benefit assessment, licensees would be able to develop the most cost-effective approaches to address the relevant scenarios.

As with Alternative 2, pursuit of this option could also include assessments and revisions to other regulations to improve the consistency between various thresholds and decision criteria.

This example would also present communication challenges given that it involves discussions of the differences in treatment of design-basis and design-enhancement events. In addition, the use of risk-informed acceptance criteria that include a specific CDF (e.g., 10^{-6} per year) or other risk measure might be interpreted as a new and clear definition of adequate protection since this alternative does not include a cost-benefit analysis for the features and controls added to address design-enhancement events. Such cost-benefit analyses are normally performed for the added protections associated with the second tier of the NRC's safety structure, as defined by the Court of Appeals. As previously mentioned, the NRC has historically avoided a specific definition of adequate protection to ensure it maintains a needed degree of latitude in the regulation of nuclear power reactors.

H.3   More on Design-Basis and Design-Enhancement Events

The focus of the first three alternatives was on establishing a category for certain beyond-design-basis events. Requirements and processes would be established to identify the relevant scenarios, to define and implement measures to meet the acceptance criteria set for those scenarios, and to preserve the outcome of the effort by maintaining appropriate barriers and controls to satisfy the risk-informed defense-in-depth goal. Existing regulatory programs that relate to the design-enhancement concept include the RTNSS category of equipment in the new reactor arena and the ROP for operating reactors, which includes the significance determination process and performance indicators that consider beyond-design-basis events. There is also an overlap between some aspects of the alternatives discussed in this appendix and regulatory initiatives such as 10 CFR 50.69, and proposed changes to 10 CFR 50.46, "Acceptance Criteria for Emergency Core Cooling Systems for Light-Water Nuclear Power Reactors."

The previous alternatives could, in theory, be pursued without addressing the current handling of design-basis events. The alternatives could, however, support or be developed in conjunction with changes contemplated for regulations, such as 10 CFR 50.46. One of the difficulties in developing changes to that regulation is that the initiative attempts to separate the spectrum of loss of coolant accidents, all of which are currently considered design-basis events, into those remaining as design-basis events and those to become beyond-design-basis events. The development of the design-enhancement category could facilitate this change by providing a regulatory structure for the breaks beyond the break-size kept in the traditional design-basis event category. In the absence of a common category with defined acceptance criteria and rules for change control and reporting, the draft regulations for the 10 CFR 50.46 initiative included specific new regulations in these areas for the breaks being removed from the design-basis event category. This, in turn, is seen by some as defeating the underlying

---

3   Note that the RMTF proposed approach for Alternative 3 would maintain design enhancements as measures taken beyond that needed for adequate protection. This approach differs from the NTTF recommendation for the design-extension category to be imposed on the basis of adequate protection.

goal of removing unnecessary regulatory burden associated with extremely unlikely events and has stymied the initiative. Performance-based approaches like Alternatives 2 and 3 could be developed to provide the necessary combination of regulatory structure and licensee flexibility to enable progress on 10 CFR 50.46 and similar initiatives.

The development of and subsequent experience with 10 CFR 50.69 highlight the difficulty of trying to introduce risk-informed methodologies to address previously unregulated scenarios, leaving traditional design-basis events intact, and acknowledging that relief may be warranted from requirements for some equipment credited in responding to design-basis events. As discussed in Appendix G, "Nuclear Power Reactors – Safety Classification of Systems, Structures, and Components," a major focus of 10 CFR 50.69 is the classification of equipment. The risk-informed safety class (RISC) categories developed for 10 CFR 50.69 are presented in Figure G-1, and repeated below:

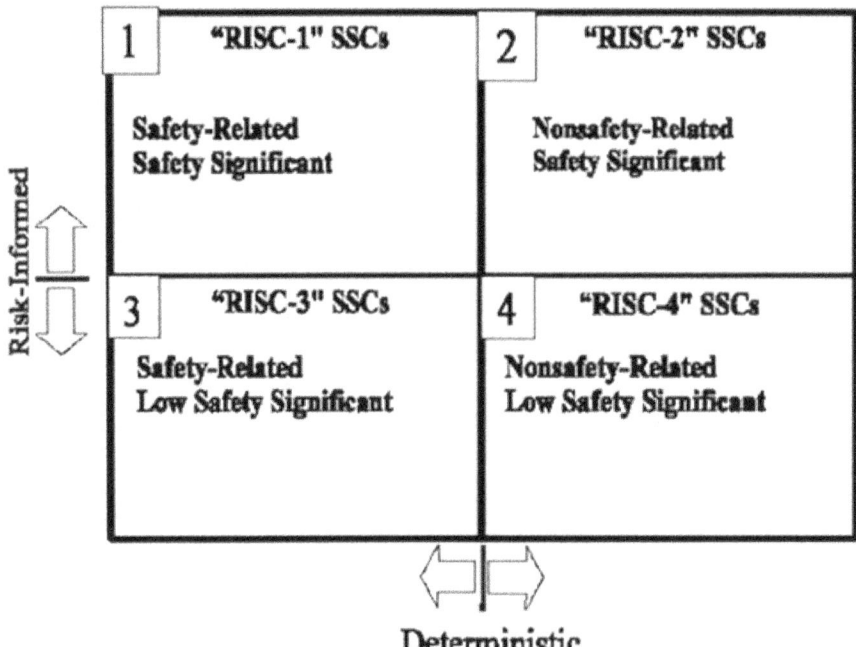

Figure H-6   10 CFR 50.69 Risk-Informed Safety Class Categories

There has been limited interest from the industry in adopting 10 CFR 50.69, largely because, although the available guidance details the categorization of equipment into the above classes, there does not appear to be agreement among parties as to the allowable relaxation from the special treatment requirements for the RISC-3 equipment.

There is a correlation between the actions that would be taken to address design-enhancement events under the proposed risk management requirements and the actions taken to address the RISC-2 equipment under 10 CFR 50.69 and the RTNSS equipment for some new reactors. In each case, the resultant requirements improve the assessment of and the actions taken to respond to relevant scenarios that are not addressed well by the design-basis events but that nevertheless contribute to the risk profile of a nuclear power reactor. The previously discussed three alternatives do not address the possible changes to specific areas of nuclear power plant design and operation that result from the existing approach to design-basis events. Possible changes in this area have been recognized in various agency initiatives, including the creation of the RISC-3 category in the above figure. The pursuit of Alternative 2 or 3 without some changes to the set of design-basis events and related regulatory requirements could be perceived as a one-sided application of risk insights, which is the perception that has limited the acceptance of 10 CFR 50.69.

A simplified summary of the current approach to design-basis events is provided within the following figure and table (Figure H-7 and Table H-1, repeated from Appendix F):

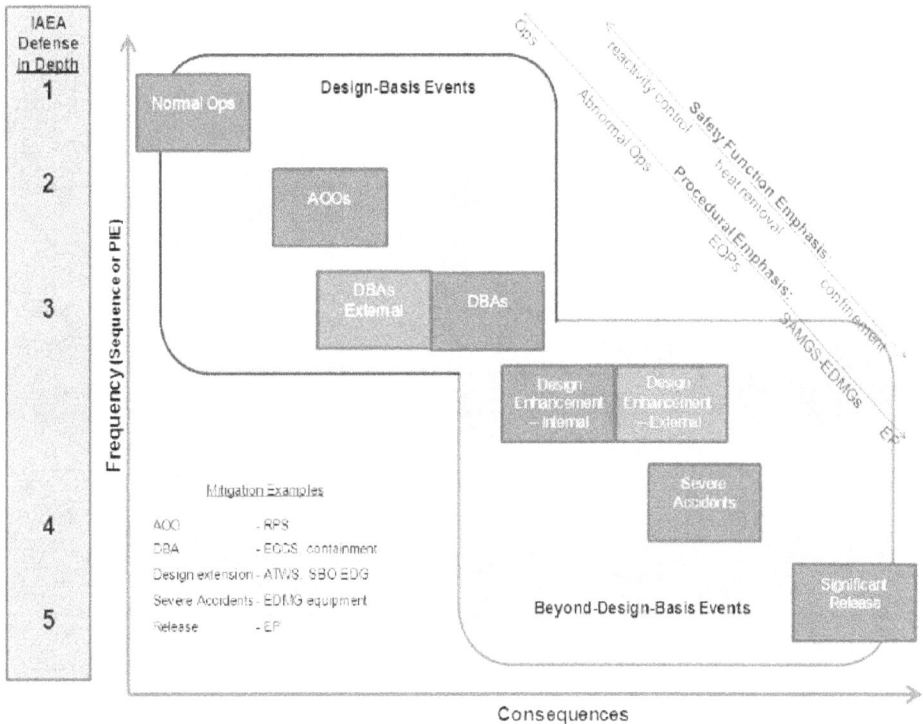

Figure H-7 Event Categories and IAEA Defense-in-depth

Table H-1 Licensing Approach to Events (current approach)

	PIEs	Key Assumptions	Acceptance Criteria
Normal Ops	n/a	n/a Maintain initial conditions, standby equipment	10 CFR Part 20, ALARA
AOOs	Functional list (e.g., ANS 51.1/52.1 Cond II) (PIE frequency < ~10^{-2}/yr)	Credit SR SSCs single failure	Maintain safety limits such as DNB, RCS pressure
DBAs	Functional list (e.g., ANS 51.1/52.1 Cond III/IV) (PIE frequency < ~10^{-4}/yr)	Credit SR SSCs single failure	Maintain coolable core, containment
Design-basis External Events	maximum credible scenario (e.g., probability of exceedance < ~10^{-7} for tornado conditions)		Engineering analysis, safety margins	Codes and standards related to barrier design and equipment qualification
Design-Enhancement (in concept but term not currently used)	Specific events ATWS SBO AIA	As defined in each rule – generally best estimate	As defined in each rule (alternate shutdown, coping time, AIA functions)
Residual Risks (in concept but term not currently used)	n/a	n/a	n/a *(related items include severe accident features (new rxs), SAMDA, ROP, and risk metrics)*
Beyond-Design Basis External Events	n/a	n/a	n/a

The current construct includes (1) thresholds for including events (historically addressed by standards such as American Nuclear Society Standard ANS 51.1(ANS, 1983)), (2) key assumptions and analysis techniques, such as the single failure criterion, and (3) acceptance criteria, such as safety limits (e.g., departure from nucleate boiling (DNB), reactor coolant system (RCS) pressure, peak cladding temperature, and offsite dose). Although the specific events and acceptance criteria are often not specifically addressed within the NRC's regulations, the alignment of events, analysis methodologies, and acceptance criteria with the related plant equipment, operational programs, and other measures has evolved over many years and is ingrained in guidance documents, codes and standards, and institutional memory within the NRC and other organizations. It should be noted that previous evaluations and initiatives have not identified a fundamental flaw with the makeup or handling of design-basis events, and the RMTF agrees that, overall, the regulations and guidance have ensured that nuclear power plants pose no undue risk to public safety. And so, although possible improvements have been developed for the consideration of events within the NRC's licensing processes, a major obstacle is the difficulty in mapping out a strategy for changing from the current state, which is generally considered acceptable, to any revised construct of events and related protected measures.

The addition of a design-enhancement category and related requirements for identifying events and developing measures to keep risks as low as is reasonably achievable or below an established level may change the landscape currently defined by design-basis events and several selected conditions (e.g., SBO, ATWS, and aircraft impact for new reactors). This change may enable changes to the current handling of design-basis events by providing a defined regulatory program for items that may not warrant the requirements associated with design-basis events but should have some measures to manage the risks introduced by nuclear power reactors. A review of the existing design-basis events and the overall combination of events within all categories could be performed to determine whether some elements of the current design-basis events could be better addressed within the design-enhancement category. Any relocations from the design-basis categories to the design-enhancement category would likely support a more performance based way of ensuring risk-informed and performance-based defense in depth. In some ways, this effort would resemble previous NRC initiatives, such as conversion to the improved standard technical specifications, which involved relocating various requirements based on the criteria defined in 10 CFR 50.36, "Technical Specifications," from plants' technical specifications to licensee-controlled documents. This limited approach would not include significant changes to the general structure of the design-basis events and would not require immediate or dramatic actions by licensees to reanalyze events, change licensing-basis documents, or revise plant equipment or procedures. It is possible that such an approach could enable licensees to proceed in a piecemeal fashion, revising the treatment of specific events as needed to address particular plant modifications, equipment procurement issues, or other specific problems.

A premise for this limited approach is that adequate protection for a nuclear power reactor is provided by a licensee's compliance with the requirements derived from the current set of design-basis events and regulatory requirements for specific programs, such as emergency planning. These requirements define the minimum set of protective measures that the NRC deems necessary to allow the operation of a nuclear power reactor. The situation is complicated somewhat by the mix of regulatory requirements used to ensure adequate protection and the fact that some are prescriptive in nature and some are performance-based. An example of prescriptive requirements is plant technical specifications, with requirements for plant shutdowns within specified time periods for certain conditions involving inoperable equipment.

Performance-based approaches include 10 CFR 50.54(hh)(2), which defines adequate protection requirements to develop and implement strategies to address core safety functions under circumstances associated with the loss of large areas of a plant because of explosions or fire but provides licensees flexibility in selecting and controlling equipment used to support those strategies.

The decisionmaking process will need to consider the relationship of the specific event with other design-basis and beyond-design-basis events and related regulatory requirements, an assessment of options, the various technical analyses, implementation issues, stakeholder views, and other factors as part of the deliberation and resultant decision on the handling of specific events or conditions. Although not proposing to make recommendations on specific design-basis and design-enhancement events, the RMTF offers the following observations regarding the event types, thresholds, assumptions, and acceptance criteria discussed in Appendix F.

H.3.1   Normal Operations

Current requirements regarding normal plant operations include standard radiation protection and effluent controls and ensuring that initial conditions (e.g., power level and shutdown margin) are consistent with assumptions in design-basis accident analyses and that safety equipment (e.g., diesel generators and emergency core cooling) are available and capable of performing safety functions. There are several ongoing initiatives, such as efforts to risk-inform technical specifications that are leading to changes in requirements for how licensees maintain plant readiness for possible events. An example is the extension of the allowable times that equipment can be out of service if a licensee implements a risk-informed safety function determination program. In addition to continuing such initiatives, the Task Force envisions some changes to requirements for normal operation as a result of adding design-enhancement events and the possible movement of some current design-basis events (discussed below) to the design-enhancement category.

H.3.2   Anticipated Operational Occurrences

The current list of anticipated operational occurrences for nuclear power reactors has evolved over the course of licensing various reactors, the development of regulatory guides, and the availability of consensus codes and standards. These transients are usually grouped by initiating events, such as loss of coolant flow, reactivity addition, and loss of heat removal. Anticipated operational occurrences are generally considered to include initiating events with an approximate frequency greater than 10^{-2} per year. The initial conditions for the transients are established as limits on normal operation and include parameters such as power levels, temperatures and pressures, and power distributions within the reactor core. The performance of the reactor protection system is modeled for bounding types of events within the category to provide confidence that safety limits associated with certain barriers (e.g., fuel cladding, reactor coolant system) are not exceeded. Conservative assumptions on plant parameters, coincident failures, the inclusion of the single failure criterion, and the reliance on only safety-related equipment are incorporated into the analysis.

With the addition of the design-enhancement category, the NRC and stakeholders could revisit the handling of some or all of the current AOOs. Given that preventing minor fuel damage is not necessarily part of the requirements for safety-related equipment and is as much an asset protection feature as it is a public health concern, a review of current AOOs could result in some scenarios being moved to the design-enhancement category. AOOs are also evaluated using analysis assumptions that may go beyond that needed for adequate protection (e.g., coincident failures) and the design-enhancement category could support movement of the less likely scenarios out of the design-basis events. However, the current safety limits and analyses are well established and changes would need to be evaluated to determine if any significant changes to the handling of AOOs were warranted. In the absence of new evaluations on the handling of AOOs, insights might be available from PRA studies, evaluations of 10 CFR 50.69, or other existing technical analyses that could determine whether some AOOs result in unwarranted protective measures and would be candidates for deletion or movement to design-enhancement events.

H.3.3   Design-Basis Accidents

Design-basis accidents generally address possible but unlikely initiating events, such as loss of coolant accidents. The analytical approach includes assumptions such as single failures, coincident failures, and reliance on only safety-related equipment. Acceptance criteria for design-basis accidents are established to limit fuel damage, maintain ability to cool the core, and maintain the integrity of containment structures. Ongoing activities related to design-basis accidents include proposed changes to 10 CFR 50.46 (differentiating between very unlikely large breaks and breaks below a transition break size), and GSI-191 (evaluating long-term cooling requirements and issues related to blockage of containment sumps). As previously mentioned, the creation of the design-enhancement category could facilitate these activities by providing an alternative to the regulatory controls associated with design-basis accidents. For example, IAEA Draft Safety Guide DS-367, "Safety Classification of Structures, Systems, and Components in Nuclear Power Plants" (IAEA, 2011), which is described in Appendix G, distinguishes between equipment used to reach a controlled state following a design-basis accident and equipment used to subsequently achieve and maintain a safe shutdown state. Such a distinction could support assessing specific phases within the current design-basis accidents. Risk assessments might inform decisions on other design-basis accidents, such as secondary-side high-energy line breaks and control rod ejection accidents, which could, in turn, support changes to limits and equipment associated with those events.

H.3.4   Design-Basis External Events

External events included in the design basis were intended to address credible scenarios related to external hazards, such as from neighboring industrial facilities, transportation routes, weather, floods, earthquakes, and security events. The analytical approach generally consists of defining a conservative challenge to the plant based on site-specific considerations. Conservative values for weather-related forces (e.g., wind speeds), flood levels, and seismic-induced motions are used as inputs to determine the needed protections for safety-related equipment. For example, a room containing emergency core cooling equipment would be evaluated to ensure that it would withstand design-basis winds (including debris), design-basis floods, and could withstand a design-basis earthquake. Confidence is provided by showing that required safety-related equipment is protected against all of the identified external threats, up to the design-basis values. Plants are also evaluated to show that operational programs related to fire protection and security are able to provide the required levels of protection from fire or malevolent acts.

As mentioned in Chapter 4, one issue with plant-level design-basis events relates to the differences in thresholds for defining those events. An example would be establishing requirements for structures at a site to ensure they could withstand a design-basis tornado determined using a probability of exceeding the design value set at 10^{-7} while the same structures might be based on a historical review of precipitation and flooding levels that are not easily expressed in probabilistic terms. The selection of design-basis values for external hazards is admittedly made difficult by the limited availability of historical information and predictive models. The Task Force has recommended that the NRC should evaluate its handling of external hazards and, to the degree possible, improve the consistency of regulatory treatment between the various threats (including security). The addition of the design-enhancement category would provide regulatory controls for any initiating events that, upon further review, would not warrant treatment as design-basis events. Should the evaluations associated with the Fukushima accident or other assessments determine that design-basis external hazards were not adequately defined; the NRC would pursue the needed regulatory reforms to require analyses and plant changes. Although independent from the efforts of this Task Force, redefining design-basis external hazards could consider the desire to improve consistency and the possible introduction of risk management requirements for licensees. Improved consideration of possible cliff-edge effects (i.e., catastrophic results from exceeding design values) is envisioned as a result of adopting the risk-informed defense-in-depth goal and will be discussed within the section on beyond-design-basis external events.

H.3.5 Design-Enhancement Events

Several approaches to identify, analyze, and protect against design-enhancement events were provided in this appendix as Alternatives 1–3.

H.3.6 Residual Risks

There are few specific regulatory requirements related to residual risks, which for the purpose of this report will be considered as any scenario not meeting the threshold of a design-enhancement or design-basis event (e.g., a scenario with estimated CDF below 10^{-6} per year in the previously discussed Alternatives 2 and 3). The limited consideration of these events currently include consideration of severe accident features in new plant designs and the contribution of these events to the overall risk profile for comparison to the NRC safety goals or guidance in Regulatory Guide 1.174. The alternatives considered by the Task Force would not significantly change the treatment or role of the events within the residual risk category. The risk-informed, performance-based defense-in-depth approach would include a careful assessment of possible cliff-edge effects and other insights from the risk assessment to ensure that uncertainties and other factors were adequately considered for those scenarios in the residual risk category.

H.3.7 Beyond-Design-Basis External Events

Beyond-design-basis external events have not generally been addressed within the NRC's regulatory programs. The evaluations performed under the Individual Plant Evaluation – External Events activities did identify some site-specific vulnerabilities related to external hazards, and licensees took some corrective actions. The degree to which such actions were incorporated into licensing-basis documents (e.g., FSARs) varies.

The adoption of the design-enhancement category could be especially useful in addressing site-specific external hazards that are not currently addressed by (or that would be relocated from) the design-basis events category. The periodic reanalysis included in Alternatives 2 and 3 would require licensees to address new information on hazards, such as revised seismic hazard information that currently requires handling through the NRC's generic safety issues and generic communications programs. An important aspect of the envisioned risk assessments is the identification and resolution of possible cliff-edge effects in which an external hazard exceeding a certain level has a very high probability of causing core damage. The flooding and resultant core damage events at the Fukushima reactor facility are examples of such vulnerabilities. There would continue to be events, however, that would be considered too improbable to be addressed within the design-enhancement category and would therefore go unaddressed or be considered only so far as they contribute to the overall risk profile of the nuclear power reactor. In addition, whatever approach is selected will need to recognize and resolve the limitations of dealing with seismic risks or other external hazards at extremely low event frequencies (Johnson and Apostolakis, 2012).

Special attention should be given to events or conditions that, although considered beyond the design basis for nuclear power reactors, resulted in actions deemed necessary for adequate protection of the public health and safety. Emergency planning is an example of an operational program that might not be necessary to address design-basis events but was added as a defense-in-depth measure and has been deemed necessary for adequate protection for operating nuclear power reactors. The requirements for licensees to develop a strategy for the loss of large areas of a plant because of explosions or fires were likewise issued on the basis that they were required for adequate protection. However, given the performance-based nature of that requirement, the resultant extreme damage mitigation guidelines and related equipment are treated in a manner usually associated with design-enhancement events.

H.4   References

(ANS, 1983)   American National Standards Institute/American Nuclear Society, ANSI/ANS 51.1 – 1983, "Nuclear Safety Criteria for the Design of Stationary Pressurized Water Reactors," ANS, LaGrange, Park, IL.

(IAEA, 2010)   International Atomic Energy Agency, "Draft – Safety of Nuclear Power Plants: Design," Draft Safety Guide DS-414, September 2010.

(IAEA, 2011)   International Atomic Energy Agency, "Draft – Safety Classification of Structures, Systems, and Components in Nuclear Power Plants," Draft Safety Guide DS-367, April 2011.

(Johnson and Apostolakis, 2012)   B.C. Johnson and G.E. Apostolakis, "Seismic Risk Evaluation within the Technology Neutral Framework," *Nuclear Engineering and Design*, 242 (2012) 341– 352.

(NRC, 1995)   U.S. Nuclear Regulatory Commission, "Policy and Technical Issues Associated with the Regulatory Treatment of Non-safety Systems (RTNSS) in Passive Plant Designs (SECY-94-084)," SECY-95-132, May 22, 1995, Agencywide Documents Access and Management System (ADAMS) Accession No. ML003708005.

(NRC, 2004)   U.S. Nuclear Regulatory Commission, "Regulatory Analysis Guidelines for the U.S. Nuclear Regulatory Commission," NUREG/BR-0058, Revision 4, August 2004, ADAMS Accession No. ML042820192.

(NRC, 2008)   U.S. Nuclear Regulatory Commission, "Policy Statement on the Regulation of Advanced Reactors," Final Policy Statement, October 7, 2008, ADAMS Accession No. ML082750370.

(NRC, 2009)   U.S. Nuclear Regulatory Commission, "Risk-Informed, Performance-Based Fire Protection for Existing Light-Water Nuclear Power Plants," Regulatory Guide 1.205, Revision 1, December 2009, ADAMS Accession No ML092730314.

(NRC, 2010)   U.S. Nuclear Regulatory Commission, "Modifying the Risk-informed Regulatory Guidance for New Reactors," SECY-10-0121 September 14, 2010, ADAMS Accession No ML102230076.

(NRC, 2011)   U.S. Nuclear Regulatory Commission, "Recommendations for Enhancing Reactor Safety in the 21st Century: The Near-Term Task Force Review of Insights from the Fukushima Dai-Ichi Accident," July 2011, ADAMS Accession No. ML112510271.

(NRC, 2011a)   U.S. Nuclear Regulatory Commission, "An Approach for Using Probabilistic Risk Assessment in Risk-informed Decisions on Plant-specific Changes to the Licensing-basis," Regulatory Guide 1.174, Revision 2, May 2011, ADAMS Accession No. ML100910006.

(WENRA, 2008)   Western European Nuclear Regulators' Association, "WENRA Reactor Safety Reference Levels," Reactor Harmonization Working Group, January 2008.

APPENDIX I

STAKEHOLDER INTERACTIONS

I.1   Background

The value of soliciting insights from the U.S. Nuclear Regulatory Commission (NRC) staff and external stakeholders regarding the Risk Management Task Force (RMTF) activities was recognized from its formation. The RMTF charter included several questions that were considered in the development of surveys for internal and external stakeholders. The internal surveys were prepared and distributed to a cross-section of NRC staff and managers in several key program offices. A summary of the responses to the internal surveys is provided in Section I.2. The request for public input to the process took the form of a notice in the *Federal Register* (published on November 22, 2011 (76 FR 72220)). A summary of the responses to the *Federal Register* notice is provided in Section I.3. In developing its findings and recommendations, the RMTF also benefited greatly from numerous informal discussions with NRC staff and managers, and external stakeholders. RMTF members accompanied Commissioner George Apostolakis on visits to all of the NRC regional offices. These interactions with the regional staffs were especially useful because they provided the RMTF with an opportunity to hear firsthand the experiences and issues associated with the use of risk-informed and performance-based approaches in the NRC's oversight process.

I.2   Internal Stakeholders

An internal survey was developed and distributed to a cross-section of NRC staff and managers in several key program offices that deal with the regulation and oversight of byproduct, source, and special nuclear materials. A summary of the survey responses is provided below:

1.   Do you believe there is a common understanding and usage of risk-informed, performance-based approaches, and defense in depth within the NRC, industry, and other stakeholders? Which concepts are especially unclear?

Reactor respondents noted that the five elements of a "risk-informed approach" are defined in Regulatory Guide (RG) 1.174, "An Approach for Using Probabilistic Risk Assessment in Risk-informed Decisions on Plant-specific Changes to the Licensing-basis," and the four guidelines for establishing a "performance-based approach" are in SRM/SECY-98-144, "White Paper on Risk-informed and Performance-based Regulation (Revised)."

While there was significant disagreement as to which terms are the most or the least understood, the majority of respondents—in both the reactor and non-reactor program areas— believed that none of the terms are well understood. Several of those who believed there was a general conceptual understanding still stated that they did not believe this carried over to the implementation process. Two respondents suggested using the maintenance rule as an example of a successful implementation framework for how the concepts can work together.

Many respondents noted the common misunderstanding between risk-informed and risk-based. One respondent claimed NRC staff and members of the public are generally comfortable using risk insights to tighten standards, while the industry is interested in using risk assessments to weaken standards, rather than applying risk in both directions. Another noted that the implementation of defense-in-depth and safety margins are within divisions with staff trained in deterministic analyses and not probabilistic risk approaches (PRAs), so they do not have the training to appropriately consider risk.

Some of the conceptual difficulties with defense-in-depth included the definition of a barrier: How effective does it need to be to properly apply defense-in-depth? The relationship between defense-in-depth and safety margins is also not well-defined. Additionally, one respondent noted the danger of using defense-in-depth to apply any requirement that the NRC wants without a strong basis, undermining the risk-informed approach. Respondents from NRC materials areas noted that the concept of defense-in-depth, in particular, was not one that was understood or applied within that program.

Finally, some respondents found a tendency to generalize the term performance-based to any regulations that are not prescriptive. In using performance-based measures, it is important that the measurement criteria are directly related to safety and security. There is concern that certain performance-based requirements are difficult to enforce.

2.   ***Are the agency's current practices, in terms of policies, procedures, and training, adequate for accomplishing the goal of a more holistic, risk-informed, performance-based regulatory structure?***

The majority of reactor respondents believed that the current practices were insufficient. Most agreed that the current policies were effective for their areas of applicability, but there was concern that this did not go much beyond the use of quantitative risk estimates, as defined in RG 1.174, to address uncertainties, model quality, the interplay of risk and defense-in-depth, or a more general risk management framework. There was not as clear a consensus from respondents in the agency's non-reactor programs. In particular, these respondents noted that that one of the challenges the agency faces is recognizing that one approach to risk does not fit the wide diversity of the NRC's programs (e.g., reactors, materials, waste, and fuel cycle).

Several thought that the policies were sufficient but that procedures were not as well defined. One respondent noted that National Fire Protection Association 805 seemed to run into problems because it attempted to be more risk-based than risk-informed.

Most agreed that additional training for the staff would be useful. While current training courses have been established, they have been sparsely attended. Additionally, the focus has been primarily on PRA rather than on risk management more broadly.

One respondent focused on the holistic aspect of the RMTF mission. The respondent discussed the need for flexibility to account for the differences in licensing decisions. The response continued by noting that some of the differences in the application of these concepts could also come from genuine differences in core beliefs about the industry's viability and the NRC's level of responsibility in improving safety rather than just maintaining it.

3.   ***How effective have past and ongoing risk-informed initiatives been? What are the relevant lessons learned from these initiatives? For example:***

   a.   ***Regulatory Guide 1.174***

   b.   ***Risk-informed technical specifications***

c.   *Risk-informed In-Service inspection*

d.   *Risk-informed In-Service testing*

e.   *Title 10 of the* **Code of Federal Regulations (10 CFR)** *50.69, "Risk-informed Categorization and Treatment of Structures, Systems and Components for Nuclear Power Reactors"*

f.   *NUREG-1556 licensing guidance*

g.   *NUREG-1520, Fuel Cycle Standard Review Plan*

h.   *Materials inspection frequencies*

i.   *Enforcement policy*

j.   *Rulemaking activities*

k.   *Other*

The general consensus seems to be that the first four were generally successful, but 10 CFR 50.69 was not. Additional successful initiatives offered by the respondents included Generic Issues, station blackout rule, individual plant examinations and individual plant examinations of external events, combustible gas control requirements, 10 CFR 50.65(a)(4), reactor oversight process (ROP), and risk-informed performance-based (RIPB) pressurized thermal shock rule.

There were several attempts at describing why 10 CFR 50.69 failed, but there seemed to be general agreement that the resulting rule did not provide any incentive for the industry to adopt it. Furthermore, it may have been the large relaxation of requirements suggested by risk insights that led the staff to compensate by making the final approved structure too burdensome for voluntary implementation. One respondent also noted that the process was poorly defined, requiring a continual changing of requirements to satisfy concerns of the Commission, Advisory Committee on Reactor Safeguards, and the industry.

Several responders noted that the implementation of risk-informed approaches can be predicted based on whether or not they will lessen the burdens for the industry. If the industry does not see a cost benefit in a voluntary approach, it will not adopt it.

Even some of the successful approaches required a long time through a poorly defined process before they were implemented.

The NUREG-1556 series was cited as a particularly helpful series in improving the efficiency of the materials licensing process and greatly improved understanding of that process, but some noted that the documents were outdated and in need of revision. Risk-informing of materials inspection frequencies were considered to be a means to save resources and not really an effort at applying risk to the inspection program. More frequent inspections of higher risk activities such as industrial radiography and well-logging are needed. One respondent noted that risk-informing has been most effective and lasting when applied to rulemaking and guidance development.

Finally, two of the responders discussed the idea that just because an approach is used does not necessarily mean that it has been successful. One suggested that research needs to be conducted to look at safety outcomes of the risk-informed approaches that have been attempted.

4.   *How effective have past and ongoing performance-based initiatives been? What are the relevant lessons learned from these initiatives?*

There were few detailed responses to this question, and the answers were split. Those who thought performance-based initiatives have been successful pointed to the requirements for looking at uncertainty, and the NRC's flexibility in enforcement to allow better focus on maintaining safety. Also mentioned were the particular examples of maintenance and surveillance, which can focus resources on important concerns while also providing incentives to the industry through the potential for a reduced burden. Some noted that performance-based inspections in the materials area have been successful and focus on the right things, but expressed a concern over the frequency of inspections, especially in light of the 10 year license renewal cycle.

The negative responses focused on the difficulty and time burden in defining appropriate criteria, which can make the respective initiatives inefficient and can have the potential that the industry can satisfy poorly defined criteria without addressing real safety concerns.

5.   *How effective have past and ongoing risk-informed and performance-based initiatives been? What are the relevant lessons learned from these initiatives?*

There was general agreement among the respondents that the maintenance rule and the ROP have been successful, but several suggested that the ROP is currently overly risk-based and should move to a more risk-informed approach. One respondent credited the ROP's success, in part, to the flexibility in its implementation. Many found it too early to judge the fire-protection initiatives. A couple found the early work promising, but others thought the early process has been too slow and overly contentious. Part 36 (pool irradiators) and the two-person rule were highlighted as rule changes that appropriately and effectively addressed safety. One respondent noted that revisions to Part 35 were risk-informed, but that the training and experience requirements for authorized users were difficult to understand and difficult to implement.

One respondent conducted a survey in 2002 of the nuclear industry that found that the maintenance rule and risk-informed technical specifications had been the best received, and the respondent thought that these results would still apply today. (see Appendix E)

6.   *How have other major initiatives or program changes incorporated risk-informed, performance-based approaches? What are the relevant lessons learned from these activities in terms of how they could have incorporated RIPB approaches or the limitations of RIPB approaches*

One respondent discussed the response to September 11, concluding that in the end, it was primarily based on the subjective opinion of NRC security experts on the assumption that PRA could not be applied to security. However, this person felt the B.5.b[1] implementation

---

1   B.5.b refers to an item within security-related Orders issued by the NRC following the events of September 11, 2001. This item was subsequently codified in Section 50.54(hh) of 10 CFR Part 50.

represented severe mitigation actions that would be useful for a range of beyond-design-basis events and was an impressive early example of risk-informed decisionmaking. A more detailed discussion of frequencies and consequences associated with security would be classified. Non-reactor respondents on the issue of security requirements were split, with some considering that risk was appropriately considered and others noting that the requirements have gone too far.

The GSI-191[2] was regarded by most respondents as a failing to properly use risk insights. They noted that there was no motivation for the industry to use it, and the uncertainties were so large that the regulations would need to be added to the full PRA structure for them to be defensible.

Most responses also thought the SAMAs were of limited use because only agerelated cost beneficial improvements could be requested.

Several general lessons learned were included by individual respondents. A selection of these is provided below:

- Ensure that defense-in-depth is not weakened when implementing a risk-informed performance-based approach.

- Because of the wide variety of regulated uses, there have been a variety of risk-informed, performance-based approaches used. What is needed is a more top-down effort that aligns RIPB through a broad agency wide policy statement.

- General license program is one example of regulation which is not RIPB. The minimal requirements of the program combined with little or no inspection oversight creates a regulatory program that does not work well.

- Use care when risk assessments point to excessive margin because the margin may also be necessary to account for uncertainty elsewhere in the system.

- Build on risk-informed programs on previous work avoiding the revision or elimination of the current implementation.

- Think of what the NRC response would be if the regulatory chronology were reversed: If the current risk-informed suggested policies were in place for decades and somebody suggested increasing to the design basis, would it still make sense to add that conservatism?

7.   *If you had to choose just one area to focus on for risk-informing your programs, what would it be and why? If none, why not?*

The respondents had a variety of answers, depending on their program areas. There were multiple calls for looking at beyond-design-basis and severe accidents and for making greater use of more advanced PRA techniques. One response focused on the potential benefits of agerelated passive system degradation. Non-reactor respondents identified several areas, including the licensing process generally, the waste classification system in Parrt 61, and materials security requirements.

---

2   GSI-191 is Generic Safety Issue 191, "Assessment of Debris Accumulation on PWR [pressurized water reactor]. Sumps." The issue has been the subject of numerous generic communications, meetings, and reports. A report , NUREG-0933, "Resolution of Generic Safety Issues," is periodically updated to provide a status and closure plans for generic safety issues.

A more general response suggested that the focus should be on better understanding, as well as use, and communication of risk results. Two responses suggested that currently there was too much attention being paid to risk insights and more focus should be on defense-in-depth.

8.   If you had to choose just one area to focus on using a performance-based approach, what would it be and why? If none, why not?

The few responses to this question focused on inspection procedures and qualification requirements for several applications. Positive responses considered the potential for fewer NRC resources in the future, and the ability of performance-based approaches to better focus those resources on public health and safety.

Two responses mentioned the concern that a performance-based approach that is not risk-informed must be carefully planned to avoid risk side effects. In using performance-based approaches for procedures with limited data, one response emphasized the importance of ensuring the data are sufficient to measure the relevant criteria; proxy data may be unsuitable for performance measurement. The general license program was again identified as an area for improvement given that many general licensees do not understand their responsibilities, the risks associated with the devices that they have and, as a result, material ends up being improperly disposed.

9.   If you had to choose just one area to focus on using a risk-informed and performance-based approach, what would it be and why? If none, why not?

Several responses looked at inspection, safety-grade components versus performance-based reliability, and the beyond-design-basis categories of events. One response also suggested the most important area to apply a risk-informed approach is security, especially any future rulemaking processes.

One response suggested rewriting the 1998 "white paper" on risk-informed efforts (versus risk-based) and performance-based approaches, with a focus on emphasizing the maintenance rule, addressing tactical aspects of implementing the policy statement on PRA, and establishing the framework as a new policy statement.

Two respondents noted that the fuel cycle oversight program needs to be more risk-informed and performance-based, especially given that the ISA requirements in Subpart H to Part 70 already reflect an RIPB approach.. Another noted that the regulations overall need to be examined based on their contribution to safety, starting with those items that have the greatest overall contribution to risk.

The respondent working with research and test reactors thought the application of risk-informed, performance-based approaches could be useful for licensing and license renewal of research and test reactors (RTRs) if it was found that the initial effort was justified.

The respondent working in emergency preparedness mentioned that an attempt to apply a risk-informed approach was already underway, and this person looks to further include risk insights as the program moves forward.

10. *How would you go about transitioning from the current approach to the risk-informed, performance-based approach in the area you have chosen?*

For the proposed projects, responses included ensuring high-level buy-in from the Commission through the managerial ranks, public meetings to discuss the new approach, and allowing fewer exceptions to the current rules (notices of enforcement discretion and emergency or exigent technical specification changes). There were conflicting answers about the need for industry collaboration. One response mentioned that RIPB approaches to safety-related equipment should only be done on a voluntary basis, while another said that, generally, voluntary approaches would not work. Also, some suggested low-level changes to start, while one suggested rewriting the entire framework, starting with the safety goal. Another response mentioned that the most realistic place to begin is with the implementation of the Fukushima Near-Term Task Force recommendations. One respondent recommended complementary "forward-looking" and "backward-looking" reviews of regulations. The former would look at unaccepatable event or outcomes and then identify those portions of the regulations that assurre those events or outcomes do not take place. The latter would involve developing a list of key regulatory requirements and the consequences that these requirements are designed to prevent. The information from these evaluations would provide a basis for determining the adequacy of regulatory requirements from a RIPB perspective.

11. *Taking all things into consideration (e.g., additional modeling development, data collection, skill sets, legal considerations, and organization factors), what needs to be in place for No. 10 to occur, and how long do you think it would take?*

Several responses focused on the need for Commission-level buy-in, and the need for convincing the public and the industry of the need for a change. Other suggestions included the integration of additional risk staff into deterministic divisions, more training on the interpretation of PRA and risk in general, and allowing fewer industry exceptions to risk-informed rules. The time estimates were in the 5-year to 10-year range. Some suggested the approach is evolutionary and should continue for the foreseeable future.

12. *How should the key concepts in the deterministic regulatory framework (e.g., design-basis accidents, dose limits, safety-related and defense in depth) be integrated with the key features of a risk-informed, performance-based approach?*

There were several categories of responses. The first pointed out that RG 1.174 already discusses how to integrate probabilistic and deterministic approaches. One person noted that in developing alternative rule making, the guideline to comply with existing regulations may not make sense. The response suggested that the concept, however, should be preserved by looking at the motivation for relevant existing regulations and using risk insights to meet the intention of the regulation.

The second category discussed the use of defense-in-depth as a way to account for uncertainties in the model. One response suggested using risk-informed approaches to outline the scope, which would allow the use of defense-in-depth and other deterministic approaches to be used in a more rational and focused manner.

Materials respondents noted that deterministic requirements are attractive to many licensees and NRC staff because of their simplicity and clarity. One stated that dose limits, for example, are by their very nature RIPB: they provide a limit which licensees have flexibility on the means to achieve that limit.

One response suggested that risk-informed approaches should lead to the optimization of risk reduction given fixed resources rather than allowing only the loosening of specific rules. Any rule relaxation would lead to some increase in risk and would be difficult to justify, this respondent said.

13.   Is it clear in what situations and circumstances the NRC should be receptive to, or on its own introduce, risk-informed approaches, performance-based approaches, or risk-informed and performance-based approaches?

The respondents pointed out that the NRC currently can do very little without industry buy-in, and it would be difficult to pursue a new approach without the backing of stakeholders. One suggested it might be possible for the NRC to introduce approaches on its own if they could be shown to lead to increased public safety and security, but this would be difficult to put into practice. Several respondents suggested the NRC should always be open to new approaches by the industry (one noted that there may be some exceptions), and when they prove to be cost beneficial and technically feasible, these approaches should be pursued.

14.   Which external stakeholders do you consider critical for the Task Force to consult in its work? Suggestions on how best to engage them?

All but two of the reactor responses suggested talking to industry leaders early in the process. One suggested talking to specific industry experts, but not through the Nuclear Energy Institute (NEI); the other thought that for emergency preparedness, interagency partners and relevant contractors (offsite response organizations) should be contacted first. One commenter suggested that stakeholders need to know they have something to gain or lose in the project to become fully engaged. Non-reactor respondents considered NEI and NRC licensees as critical groups to be engaged.

Most of the non-reactor respondents identified the need to involve the Agreement States, given their role in regulating 85% of the materials licensees in the US as well as all of the LLW disposal sites. Specifically, the Organization of Agreement States and the Conference of Radiation Control Program Directors were cited as key stakeholders.

Several comments mentioned the senior reactor analyst and general PRA and probabilistic safety analysis community. One response suggested engaging prescreened academicians to engage in a workshop to identify implementation ideas. Most suggested the public at different stages of development, including critics. Some also suggested like-minded agencies (International Atomic Energy Agency, U.S. National Aeronautics and Space Administration, U.S. Department of Energy, U.S. Environmental Protection Agency, U.S. Department of Homeland Security), national labs, and professional standards groups (e.g., American Society of Mechanical Engineers, American Nuclear Society).

15. *What do you feel are the two or three key obstacles to making further progress on moving toward more risk-informed or performance-based regulatory approaches? Are the current capabilities and limitations of probabilistic risk assessment, as used in your area, one of the major obstacles? (Also see Question 11.)*

Many responses focused on internal and external communications. Some suggested a more broad-based risk education of staff, especially leadership and management. This training would move beyond the use of PRA to the use of risk information to make decisions in all NRC regulatory programs. Additionally, one response suggested comparing the limitations of probabilistic approaches to those of deterministic analyses. The response suggests that when the focus is only on the limitations of probabilistic approaches, there is an assumption that the deterministic approaches have fewer limitations. Some respondents identified the lack of risk information and appropriate tools and the cost of obtaining such information as obstacles. Others identified inertia in moving away from regulatory approaches that are known and predictable, while another identified the Agreement States as a challenge.

Many also discussed the limitations of PRAs, both at the NRC (addressing all internal and external initiators, all modes of operation, human factors, etc.) and within the industry. Also, one suggested the safety goals should better reflect public risk tolerance (e.g., Fukushima may not have violated the safety goals but would not be tolerated by the public).

I.3 External Comments

The RMTF published a request for public comment regarding an adoption of risk management concepts in the *Federal Register* (published on November 22, 2011 (76 FR 72220)). The notice asked for stakeholder input on the following questions:

1. Do you believe there is a common understanding and usage of the terms risk-informed, performance-based, and defense-in-depth within the NRC, industry, and other stakeholders? Which terms are especially unclear?

2. What are the relevant lessons learned from the previous successful and unsuccessful risk-informed and performance-based initiatives?

3. What are the relevant lessons learned from the previous successful and unsuccessful deterministic regulatory actions?

4. What are the key characteristics for a holistic risk management regulatory structure for reactors, materials, waste, fuel cycle, and security?

5. Should the traditional deterministic approaches be integrated into a risk management regulatory structure? If so, how?

6. What are the challenges in accomplishing the goal of a holistic risk management regulatory structure? How could these challenges be overcome?

7. What is a reasonable time period for a transition to a risk management regulatory structure?

8.   From your perspective, what particular areas or issues might benefit the most by transitioning to a risk management regulatory approach?

The responses to the *Federal Register* notice are available in the NRC's Agencywide Documents Access and Management System (ADAMS) (Case/Reference No. NRC20110269). A summary of the responses is provided below:

Table I-1 Summary of Comments from External Stakeholders

Organization (ADAMS Accession No.)	Relevant Report Section(s)	Comments
Energy Solutions (ML12023A038)	Decommissioning, waste disposal, transportation	NRC should hold a series of workshops for all program areas

Defense-in-depth (DID) well defined; not Risk-Informed (RI) or Performance-based (PB)

MARSSIM [Multi-agency radiation survey and site investigation manual] is an example of a good RIPB regulation

Bad deterministic example – Regulatory Guide (RG) 1.86 uses standards based on 1974 technology limits

Avoid a one-size-fits-all approach

Difficulties include available resources, criteria for success, consistency in interpretation and application, data availability and sufficiency, and qualitative vs. quantitative standards

This should be implemented piecemeal after workshops with stakeholders

10 CFR 61 would benefit from risk management regulatory approach (some aspects don't make sense at certain facilities) |

Council on Radionuclides and Radiopharma-ceuticals (CORAR) (ML12018A230)	Materials	Suggest workshops and training in risk management before implementation. Some regulations that claim to be risk-informed put a burden on all licensees when only some are of concern Risk information could help incentivize self-identification of deficiencies The framework should only apply to those considerations that affect all licensees Risk management framework could help to exempt licensees from regulations that are not appropriate for them
Organization of Agreement States (OAS) (ML12013A123)	Materials	Performance-based is well understood in terms of inspections, but not sure about licensing. Risk-informed and DID do not seem to make sense for materials as outlined in the FRN Can't answer Qs 3-8 because risk management concepts are not used as outlined in FRN Could use a performance-based approach to reduce frequencies of inspections for good licensees
Society for Nuclear Medicine (SNM) (ML12079A135)	Materials	Current definitions and examples of RI, PB, and DID are for reactors, need examples for nuclear medicine Dose-based release criteria have been very effective in nuclear medicine community Holistic risk management approach would require rewrite of 10 CFR 35.75 and would take 5 years SNM believes a RIPB approach would be extremely useful for nuclear medicine.

NEI Attachment 2 (ML12027A141)	Materials and Fuel Cycle	Each sector developed its own meaning of RI, PB, and DID that is acceptable to NRC. Believe NRC understands different risks and should build upon current flexible framework Decision to stop Cesium-137/Chloride rule was based on RIPB insights; several others Many requests for additional information don't seem to consider risk Need to have engagement and workshops like for Safety Culture Policy Statement Could use risk information to determine licensing fees (Level 1 facility vs. power reactor) Fuel cycle facilities could potentially benefit from RIPB approach depending on implementation
Next Generation Nuclear Plant (NGNP) (ML12010A098)	Power Reactors	Need to better define DID; Idaho National Laboratory (INL) has suggestions INL has reported on how risk concepts affect NGNP in several white papers referenced Need a blend of deterministic and probabilistic Defining licensing-basis events (LBEs) are a good place to start in looking at risk-informed approach to NGNP The white papers could provide a basis to develop the entire process.

Pressurized Water Reactor (PWR) Owners Group (ML12032A029)	Power Reactors	NRC needs to better define DID There is a current overreliance on numbers coming from PRA Analyses are too risk-based (especially Significance Determination Process (SDP)), rather than risk-informed Sometimes conservatisms in model are not considered Maintenance rule and Mitigating System Performance Indication (MSPI) work well Generally, NRC too focused on refining methods and not on context of decisions Important to have early buy-in, pilot cases, and a common vision of regulatory shift Risk-informed approach needs to percolate through staff level and would be helpful in GSI-191 and Fukushima lessons learned
PWROG – WCAP Attachment (ML12032A029)	Power Reactors	An attached Risk-Informed safety analysis developed in 2003
NEI Attachment 1 (ML12027A141)	Power Reactors	RI is not well defined, too much of a focus on numbers. DID is sometimes used to avoid RI applications Maintenance rule has been successful PB approach ATWS and station blackout rule were based on PRA insights and an example of a positive use of risk information NFPA 805 and MSPI have been challenging to implement; General structure of MSPI has been good, but too much focus on making it a "true" PRA The SDP can lead plants to spend resources refining assumptions in PRA rather than fix the problems Decades to fully implement risk management regulatory structure. Can use risk insights immediately GSI-191 would be a good place to use an RI approach

National Fire Protection Association (NFPA) (ML12010A026)	Power Reactors	Agree with the need for RIPB. NFPA 805 is a good example of PB. NFPA 806 will become an important part of oversight. NFPA 804 is prescriptive but should be part of NRC's oversight and regulatory process
Farshid Shahrokhi (ML12013A125)	Overall	Should establish high-level goals for all sectors; these would then naturally lead to individual sector criteria Need a collaborative environment with open-ended ways to meet safety criteria Precedent has played a role in past decisions, but this does not necessarily lead to better safety regulations
NEI Main (ML12027A141)	Overall	Current risk-informed approaches should be encouraged and sustained Don't layer PRA on top of deterministic regulations Currently, focus too much on numerical thresholds
Prasad Kadambi (ML12005A212)	Power Reactors	Worried about focusing too much on numbers coming out of a PRA Likes reactor oversight process (ROP) and seismic margins analysis Believes we should implement risk management in a looser, case-by-case decision analysis context NRC needs to be more explicit in performance-based approaches Put nuclear consequences in context of total consequences (Fukushima: reactors vs. earthquake, tsunami)

STP (ML12079A134)	Power Reactors	The RIPB and DID definitions are fuzzy; need criteria, details for implementation and should apply beyond reactors to materials, security, etc. Need to add oversight and documented expectations for RIPB approaches. The Risk Managed Technical Specifications is an example of successful risk-informed initiative Deterministic regulations lead to resource misallocation. Risk should be in regulatory policy statements (work hour rule, cybersecurity) and risk methods to determine significance criteria for regulatory response Successful implementation requires Commission buy-in and pilot tests; full implementation $5 million to $10 million/ year and 50 FTEs This approach would lead to improved safety, better operational flexibility and decisionmaking; improved regulatory decisionmaking; better societal quality of life; better worker safety and quality of life
Organization of Test, Research and Training Reactors (TRTR) (ML12013A122)	Nonpower Reactors	The ratio of regulation to risk from NPRs is higher than for the other regulatory licensees NPRs should not be required to do PRAs Biggest issue to implementation is public risk perception Switching to a risk management framework may benefit RTRs by lessening their regulatory burden
Penn State (ML12010A027)	Nonpower Reactors	Prefer less regulation and more consultation NRC seems to be less focused on risk to public than on political considerations and other difficult-to-quantify risks Would like to move to a more performance-based approach

Kennecott Uranium Company (ML12013A154)	Uranium Recovery	Supports RIPB because of low risk in handling radioactive materials RI, PB, DID are well understood Rad risks should be ranked against other risks at reactors, materials, waste, fuel cycle, and security sites Main difficulty is perceived vs. real risk, especially with the public Benefits of RIPB include review ground water restoration and remediation, soil remediation, standards for Radon222, etc. Many favorable supporting documents
National Mining Association (NMA) (ML12019A111)	Uranium Recovery	RIPB approach promotes efficient use of limited resources RPB approach should apply to licensing actions, policy development, and inspection/enforcement actions The Regulatory Issue Statement is a good example of a RIPB success Currently too much of a focus on numerical (baseline) recovery when in certain sites that doesn't make sense
Wyoming Mining Association (ML12013A124)	Uranium Recovery	Similar to Kennecott comments…

APPENDIX J

FINDINGS AND RECOMMENDATIONS

The various sections of the report provide the key RMTF findings and recommendations related to proposed Risk Management Regulatory Framework and its application to specific NRC program areas. A summary of the RMTF findings and recommendations is provided in Table J-1.

Table J-1 Risk Management Task Force Findings and Recommendations	
Chapter 2 – A Proposed Risk Management Regulatory Framework	
Finding 2.1	Whether used explicitly, as for power reactors, or implicitly, as for materials programs, the concept of defense-in-depth has served the NRC and the regulated industries well and continues to be valuable today. However, it is not used consistently, and there is no guidance on how much defense-in-depth is sufficient.
Finding 2.2	Risk assessments provide valuable and realistic insights into potential exposure scenarios. In combination with other technical analyses, risk assessments can inform decisions about appropriate defense-in-depth measures.
Recommendation 2.1	The goal to adopt risk-informed and performance-based approaches, where practical, should continue and should be incorporated into the revised regulatory framework.
Recommendation 2.2	The general regulatory approach of the NRC should be defined in terms of "managing the risks" posed to workers and the public from the various uses of byproduct, source, and special nuclear materials.
Recommendation 2.3	In defining requirements for the protection of workers and the public, the NRC should recognize and address uncertainties associated with the hazards and the events, including human errors, which could challenge or degrade barriers and controls. A balanced approach that considers traditional and risk assessment techniques should be used to identify barriers and controls so that appropriate requirements are defined to prevent, contain, and mitigate exposures to radioactive materials.
Recommendation 2.4	The NRC should formally adopt the proposed Risk Management Regulatory Framework through a Commission Policy Statement.

Chapter 4 – Implementation Options	
Section 4.2.1 Nuclear Power Reactors	
Finding PR-F-1	The concept of design-basis events and accidents continues to be a sound licensing approach, but the set of design-basis events and accidents has not been updated to reflect insights from power reactor operating history and more modern methods, such as PRA.
Recommendation PR-R-1	The set of design-basis events and accidents should be reviewed and revised, as appropriate, to integrate insights from power reactor operating history and more modern methods, such as PRA.
Finding PR-F-2:	Requirements for beyond-design-basis accident scenarios (e.g., station blackout) were established at different times and in different ways. Differences in implementation approaches have reduced the efficiency and consistency of the NRC's regulatory and oversight activities.
Finding PR-F-3:	The extent to which licensee activities undertaken as part of voluntary industry initiatives can be credited has been a source of contention in the Reactor Oversight Process and has reduced the efficiency of that process.
Recommendation PR-R-2	The NRC should establish through rulemaking a *design-enhancement category* of regulatory treatment for beyond-design-basis accidents. This category should use risk as a safety measure, be performance-based (including the provision for periodic updates), include consideration of costs, and be implemented on a site-specific basis.
Finding PR-F-4	The processes for establishing the external hazard design bases do not use consistent event frequency and magnitude methods.
Finding PR-F-5	New information that would provide the basis for external hazard frequency updates is not systematically collected, evaluated, and communicated.
Finding PR-F-6	PRA methods for assessing external hazard risks are available, but expertise in performing such studies is very limited. Uncertainty analyses and the recognition of the limitations of available scientific knowledge are a key element of these methods.

Recommendation PR-R-3	The NRC should reassess methods used to estimate the frequency and magnitude of external hazards and implement a consistent process that includes both deterministic and PRA methods. Consideration of the risks from beyond-design-basis external hazards should be included in the design-enhancement category described in Recommendation PR-R-2.
Recommendation PR-R-4	The NRC should establish a program to systematically collect, evaluate, and communicate external hazard information.
Finding PR-F-7	The availability and broad-scale use of quantitative risk assessment methods (PRA) for power reactors provide an opportunity for a more quantitative characterization of defense-in-depth.
Recommendation PR-R-5	The NRC should apply the risk-informed and performance-based defense-in-depth concept to power reactors in a more quantitative manner.
Finding PR-F-8	Vulnerability assessments performed to assess security have important similarities in scope (e.g., facility equipment and radioactive hazards considered) and methods to risk assessments.
Finding PR-F-9	Differences in regulatory language and approaches between power reactor security and safety regulation may have reduced the efficiency and effectiveness of the NRC's work.
Finding PR-F-10	In the past decade, considerable research has been performed on estimating security risks, much of it sponsored by the U.S. Department of Homeland Security.
Recommendation PR-R-6	The NRC should develop and implement guidance for use in its security regulatory activities that uses a common language with safety activities and harmonizes methods with risk assessment and the proposed risk-informed and performance-based defense-in-depth framework.
Section 4.2.1.1 - Operating Power Reactors	
Recommendation OR-R-1	For operating reactors, the establishment of the design extension category can be followed by a review of design basis events/accidents and related revisions to AOOs and DBAs to integrate insights from operating history and more modern methods. The NRC need not impose such a requirement but should be amenable to related industry initiatives should they pursue revisions to design basis events based on the introduction of the design extension category.

Recommendation OR-R-2	For operating reactors, the RMTF recommends that the NRC should establish through rulemaking a design-enhancement category of regulatory treatment for beyond-design-basis accidents.
Recommendation OR-R-3	The NRC should reassess the methods used to estimate the frequency and magnitude of external hazards and implement a consistent process that includes both deterministic and PRA methods. For operating reactors, the RMTF recommends that the design-enhancement category rulemaking include consideration of external hazards.
Recommendation OR-R-4	For operating reactors, the RMTF recommends that the NRC develop and implement guidance for the collection and dissemination of external hazard information.
Recommendation OR-R-5	The NRC should apply the risk-informed and performance-based defense-in-depth concept to power reactors in a more quantitative manner. For operating reactors, the RMTF recommends that this recommendation be implemented in the form of guidance to the NRC staff and in future requirements established for operating reactor licensees.
Recommendation OR-R-6	For operating reactors, the RMTF recommends that guidance be developed and implemented to better harmonize terminology and methods used for reactor safety and security.
	Section 4.2.1.2 - New Reactors
Recommendation NR-R-1	For new reactors, the RMTF recommends that the NRC be amenable to and promote, where practical, the adoption of more risk-informed and performance-based approaches for the selection of more relevant scenarios for design-basis events. Changes pursued for operating reactors (OR-R-1) should also consider applicability to new reactors.
Recommendation NR-R-2	Apply Recommendation PR-R-2 (design-enhancement category) to new reactors.
Recommendation NR-R-3	Apply Recommendation PR-R-3 (include external events in design-enhancement category) to new reactors.
Recommendation NR-R-4	Apply Recommendation PR-R-4 (periodically evaluate new information regarding external hazards) to new reactors.

| Recommendation NR-R-5 | Apply Recommendation PR-R-5 (issue guidance to adopt risk-informed and performance-based defense-in-depth) to new reactors. |
| Recommendation NR-R-6 | Apply Recommendation PR-R-6 (develop guidance and consistent approach between safety and security) to new reactors. |

Section 4.2.1.3 - Generation IV Reactors

Recommendation GIV-R-1	For Generation IV reactors, the RMTF recommends that the concept of design-basis accidents be maintained, but the NRC should be amenable to and promote, where practical, the adoption of more risk-informed approaches for the selection of relevant scenarios (e.g., alternatives to the single failure criterion) for design-basis accidents.
Recommendation GIV-R-2	Apply Recommendation PR-R-2 (design-enhancement category) to Generation IV reactors.
Recommendation GIV-R-3	Apply Recommendation PR-R-3 (include external events in design-enhancement category) to Generation IV reactors.
Recommendation GIV-R-4	Apply Recommendation PR-R-4 (periodically evaluate new information regarding external hazards) to Generation IV reactors.
Recommendation GIV-R-5	Apply Recommendation PR-R-5 (issue guidance to adopt risk-informed and performance-based defense-in-depth) to Generation IV reactors.
Recommendation GIV-R-6	Apply Recommendation PR-R-6 (develop guidance and consistent approach between safety and security) to Generation IV reactors.

Section 4.2.2 - Nonpower Reactors

| Finding NPR-F-1 | The concept of defense-in-depth for NPRs remains relevant. Knowledge gaps and uncertainties continue to be effectively addressed by the defense-in-depth approach. |
| Finding NPR-F-2 | The analysis of design basis and the maximum hypothetical accidents based on conservative design limits, acceptance criteria, safety margins, and assumptions in conjunction with the application of a defense-in-depth philosophy continues to be a sound but highly conservative licensing approach to ensuring adequate safety of NPRs. |

Recommendation NPR-R-1	The proposed defense-in-depth framework should be applied to the NPR licensing process to ensure that the current amount of defense-in-depth is appropriate given the relatively small radioactive hazard. This application should include safety and security licensing matters.
Finding NPR-F-3	The application of modern risk assessment methods at NPRs could provide valuable insights into accident scenarios not previously identified by the earlier deterministic safety assessment and could be valuable in focusing the application of licensing and oversight resources on areas of risk importance. Risk assessment insights, in conjunction with a formal risk management decisionmaking process, could significantly contribute to the development of a more efficient and effective NPR regulatory framework. NPR PRA models developed by others could be used as a starting point for facility-specific PRA models at NRC-licensed NPRs. Even with this background, however, funding such assessments could be problematic for NPRs.
Recommendation NPR-R-2	The NRC should evaluate the utility of performing a pilot risk assessment, including consideration of external hazards, using modern risk assessment methods at an NPR. This evaluation would assess the value of the risk insights gained from the risk assessment on the basis of possible safety enhancements and possible contributions to a more efficient and effective risk-informed and performance-based regulatory framework for NPRs.
Finding NPR-F-4	The traditional NPR licensing approach shares the same limitations as the power reactor approach for methods used to estimate frequencies and magnitudes of external events.
Recommendation NPR-R-3	NPRs should be considered, to the extent practical, in the implementation of power reactor Recommendation PR-R-3 and Recommendation PR-R-4.
Finding NPR-F-5	NPR security requirements are not risk-informed beyond the use of a graded approach based on NPR power levels.
Finding NPR-F-6	The development of risk-informed security regulations or guidance would enhance efficiency by focusing resources on identified areas important to facility security.
Recommendation NPR-R-4	If the NRC decides to develop and implement a risk-informed and performance-based defense-in-depth regulatory framework to ensure the safety of NPRs, then the agency should also develop guidance for use in its NPR security regulatory activities that uses a common language with safety activities and harmonizes methods with the risk-informed and performance-based defense-in-depth framework.

Section 4.3 - Materials Uses	
Finding-M-F-1	The materials program has successfully developed and incorporated risk insights and performance considerations into its rulemaking, policy development, and routine licensing and inspection activities.
Finding-M-F-2	Deterministic approaches such as dose limits and possession limits are useful concepts across the wide range of materials uses that should be retained, but broader use of deterministic approaches can lead to ineffective use of limited licensee, NRC, and Agreement State resources.
Finding-M-F-3	Buy-in of the 37 Agreement States is essential to the success of risk management process implementation given their role in regulating more than 85 percent of the materials licensees in the United States.
Recommendation-M-R-1	The NRC materials program should continue to apply risk insights and performance-based considerations, as appropriate, in rulemaking, guidance and policy development, and implementation in accordance with the proposed risk management framework.
Recommendation-M-R-2	The development and rollout of the recommended Risk Management Policy Statement should be closely coordinated with the leadership of the Agreement States and a joint NRCAgreement State Working Group should be established to guide risk management implementation in the materials area.
Finding M-F-4	Differences in regulatory language and approaches between safety and security in the materials area may have reduced the efficiency and effectiveness of the NRC and Agreement State regulatory programs.
Recommendation M-R-3	The NRC should apply common risk approaches to safety and security based on the proposed risk management and defense-in-depth regulatory framework.
Finding M-F-5	The terminology of defense-in-depth is not used consistently across the agency's materials regulatory programs.
Recommendation M-R-4	As part of the implementation of the proposed risk management regulatory framework, the RMTF recommends that the materials program should more explicitly consider the defense-in-depth philosophy in rulemaking, guidance, and program implementation, and modify appropriate parts of staff training to make these concepts a central part of such training.

Finding M-F-6	The materials program could benefit from a more structured application of the risk management process in resource allocation. This process would allow program managers to more systematically apply resources to those areas where the safety or security risk warrants it.
Recommendation M-R-5	Headquarters and Region managers should undertake a more formal review of operational data and risk considerations to help determine if and where budgeted resources need to be adjusted. This option takes into account the fact that budgets are developed well in advance of budget implementation and that operational realities may change in the intervening time.
Section 4.4 – Waste	
Section 4.4.1 - Low Level Waste	
Finding LLW-F-1	The regulatory framework for LLW disposal is, for the most part, implicitly followed a risk-informed and performance-based approach. However, changes in the LLW environment and the maturing of the performance assessment method over the past 30 years have underscored the need to provide a stronger risk basis to the program. The staff is currently implementing Commission direction to make it more so and will be sending various products to the Commission over the next year or two. However, as noted in earlier sections of this report, the terms risk-informed and performance-based are not well understood and have not been consistently applied across NRC regulatory programs.
Finding LLW-F-2	Certain aspects of the LLW regulatory framework readily lend themselves better to the risk management approach, such as waste classification or concentration averaging. Applying the proposed risk management approach to comprehensive LLW licensing decisions, however, may be more challenging because it involves estimating facility performance due to events potentially far into the future.
Recommendation LLW-R-1	The NRC should adopt the concept of risk management to the LLW program, including any revisions proposed to 10 CFR Part 61 (including performance assessment requirements) and related guidance documents.

Finding LLW-F-3	The interlocking and reinforcing systems approach in 10 CFR Part 61 (site suitability, waste form and classification, intruder barrier, institutional controls, etc.) represents an implicit consideration of defense-in-depth features, based on the risk posed by various classes of waste.
Finding LLW-F-4	The NRC has not developed an explicit characterization of defense-in-depth considerations for the LLW program.
Recommendation LLW-R-2	The NRC should develop an explicit characterization of how defense-in-depth within the proposed risk management framework applies to the LLW program and build this into current and future staff guidance documents and into training and development activities for the staff.
Finding LLW-F-5	Consideration of environmental risks as well as safety risks is a central part of the LLW regulatory program.
Recommendation LLW-R-3	The NRC should include environmental reviews within the scope of its risk management framework.
Section 4.4.2.2 - High Level Waste	
Finding-HLW-F-1	The development of regulations for geologic disposal of HLW at Yucca Mountain was based significantly on risk information developed from performance assessments and closely followed the proposed Risk Management Regulatory Framework.
Finding-HLW-F-2	The NRC's regulatory philosophy of defense-in-depth is reflected in the multiple-barrier requirement for post-closure in 10 CFR Part 63. Compliance with the multiple barrier requirements is demonstrated through the performance assessment.
Finding-HLW-F-3	As performance assessment capabilities and experience increased at the NRC during the past 30 years, so did the use of risk insights to help guide the HLW program. Risk insights and performance assessment capabilities have been used to improve the efficiency and effectiveness of guidance documents; inform pre-licensing interactions with DOE; and help identify and direct data needs and experimental activities.
Recommendation-HLW-R-1	Any future revisions to the regulatory framework for geologic disposal of HLW should be done in accordance with the proposed risk management framework to ensure that risk information continues to be appropriately considered in the development of requirements and appropriately reflect any future HLW disposal paradigm.

Finding-HLW-F-4	The risk-informed, performance-based approach for regulating geologic disposal (YMRP and regulations) was intended to provide flexibility to the applicant in developing the best approach in meeting the regulatory requirements, and allow the use the risk information to focus regulatory review on safety-significant issues.
Section 4.5 – Uranium Recovery	
Finding-UR-F-1	Uranium recovery facilities are of low radiological risk to workers and members of the public under normal operational conditions and most accident scenarios.
Finding-UR-F-2	Although the NRC staff has made inroads to risk-informed, performance-based licensing of uranium recovery facilities, the regulatory framework is largely a deterministic one. NRC regulations in 10 CFR Part 40 and Appendix A to 10 CFR Part 40 reflect the requirements of UMTRCA and EPA's rules in 40 CFR Part 192, which pre-date the Commission's move to a risk-informed, performance-based framework in the mid-1990s. Similarly, program guidance, especially for conventional mills, could benefit from a greater risk basis.
Finding-UR-F-3	The exact nature of the ISR rule under development by the EPA is not clear at this time; therefore, presents an uncertainty to adoption of the proposed risk management regulatory framework.
Recommendation-UR-R-1	Notwithstanding the current uncertainty associated with the EPA rulemaking, the NRC should adopt the proposed risk management regulatory framework to the uranium recovery program to provide greater efficiency, effectiveness, and predictability in policy development and regulatory decisionmaking.
Recommendation-UR-R-2	The NRC should work closely with the Agreement States and the regulated community to guide implementation of risk management in the uranium recovery program.
Finding-UR-F-4	Consideration of environmental risks is a central part of the uranium recovery regulatory program.
Recommendation-UR-R-3	The NRC should include environmental reviews within the scope of its risk management framework.

Section 4.6 – Fuel Cycle	
Finding F-F-1	The current fuel cycle regulatory approach incorporates several elements of the proposed risk management regulatory framework, such as the use of ISAs to identify safety significant items, and the implementation of a revised fuel cycle oversight program as directed by the Commission.
Finding F-F-2	The concept of defense-in-depth, as embedded in fuel cycle regulatory requirements and practices, is consistent with Commission guidance. Its implementation changes as the processes change at the fuel cycle facilities.
Recommendation F-R-1	The fuel cycle regulatory program should continue to evaluate the risk and the associated defense-in-depth protection by using insights gained from ISAs. ISAs should continue to evolve to support regulatory decisionmaking.
Section 4.7 – Interim Spent Fuel Storage	
Finding S-F-1	The regulatory approach for SNF storage is largely based on meeting applicable industry consensus standards and conservative guidance to ensure adequate safety margins in the facility and cask designs and operations. More recently, insights from a limited number of risk studies have been gradually factored into this regulatory approach. Furthermore, though qualitative, a systematic approach that parallels answering the risk triplet was used in the latest revision of the Standard Review Plan.
Recommendation S-R-1	While elements of the proposed risk management approach have been used in the SNF storage regulatory approach to evaluate the acceptable level of risk and the sufficiency of defense-in-depth (physical barriers, controls or margins) more consistently, the NRC should develop the necessary risk information, the corresponding decision metrics, and numerical guidelines. This is important in guiding further changes to the existing SNF storage regulatory approach and the evaluation of strategies for extended SNF storage activities.
Finding S-F-2	The concept of defense-in-depth is not explicitly or consistently applied in the SNF storage regulatory program.

Recommendation S-R-2	As part of the implementation of the proposed risk management regulatory framework, the NRC should more consistently consider the concept of defense-in-depth explicitly and evaluate its proper use in the SNF storage regulatory program. The NRC should also improve appropriate parts of staff training to make this concept a central part of such training.
Section 4.8 – Transportation	
Finding T-F-1	While the U.S. transportation regulatory approach is governed by the IAEA transportation regulations, the current NRC transportation regulatory approach uses several elements of the proposed risk management framework.
Finding T-F-2	Risk assessments have been conducted on the safety of transportation of spent fuel. However, there is a lack of risk information on the transportation of other radioactive materials.
Recommendation T-R-1	Considering the strong international regulatory basis for transportation and the need to conform U.S. standards to those of the IAEA and other member states, application of the proposed risk management regulatory framework should focus on implementation guidance.
Recommendation T-R-2	The risk management process should be used to influence the future outcome of IAEA deliberations on proposed changes in international transportation regulations.
Recommendation T-R-3	The NRC should explore the value of using risk insights to justify regulations different from the IAEA's for domestic use only, such as regulations dealing with domestic storage and transportation of high burnup fuel. Risk information could be used to develop a more flexible approach toward implementing and making gradual changes to current transportation regulations.

NRC FORM 335
(12-2010)
NRCMD 3.7

U.S. NUCLEAR REGULATORY COMMISSION

BIBLIOGRAPHIC DATA SHEET

(See instructions on the reverse)

1. REPORT NUMBER
(Assigned by NRC, Add Vol., Supp., Rev., and Addendum Numbers, if any.)

NUREG-2150

2. TITLE AND SUBTITLE

A Proposed Risk Management Regulatory Framework

3. DATE REPORT PUBLISHED

MONTH	YEAR
April	2012

4. FIN OR GRANT NUMBER

5. AUTHOR(S)

Commissioner George Apostolakis, Mark Cunningham, Christiana Lui, George Pangburn, and William Reckley

6. TYPE OF REPORT

NUREG

7. PERIOD COVERED (Inclusive Dates)

8. PERFORMING ORGANIZATION - NAME AND ADDRESS (If NRC, provide Division, Office or Region, U. S. Nuclear Regulatory Commission, and mailing address; if contractor, provide name and mailing address.)

Risk Management Task Force
Office of Commissioner George Apostolakis
U.S. Nuclear Regulatory Commission
Washington, D.C. 20555-0001

9. SPONSORING ORGANIZATION - NAME AND ADDRESS (If NRC, type "Same as above", if contractor, provide NRC Division, Office or Region, U. S. Nuclear Regulatory Commission, and mailing address.)

same as above

10. SUPPLEMENTARY NOTES

11. ABSTRACT (200 words or less)

At the request of Chairman Gregory B. Jaczko, a task force headed by Commissioner George Apostolakis prepared this report. The task force's charter was to develop a strategic vision and options for adopting a more comprehensive, holistic, risk-informed, performance based regulatory approach for reactors, materials, waste, fuel cycle, and transportation that would continue to ensure the safe and secure use of nuclear material. The proposed risk management regulatory framework builds upon well established practices, such as the NRC's defense-in-depth philosophy and its policies to incorporate risk-informed and performance-based approaches to the agency's regulation and oversight of byproduct, source, and special nuclear materials. Risk management is being adopted by many different organizations, including Federal agencies, and would seem to be the logical next step in the evolution of the NRC's regulatory programs. The report describes a proposed risk management regulatory approach that could be used to improve consistency among the NRC's various programs and discusses implementing such a framework for specific program areas.

12. KEY WORDS/DESCRIPTORS (List words or phrases that will assist researchers in locating the report.)

risk management
regulatory framework
risk assessment
defense in depth
risk informed
performance based
design enhancement

13. AVAILABILITY STATEMENT

unlimited

14. SECURITY CLASSIFICATION

(This Page)

unclassified

(This Report)

unclassified

15. NUMBER OF PAGES

16. PRICE

NRC FORM 335 (12-2010)

NUREG-2150
April 2012